1985

GENES
Structure and Expression

Horizons in Biochemistry and Biophysics Series

Editorial Board

Horizons in Biochemistry
and Biophysics

Volume 7

GENES
Structure and Expression

Volume Editor
A. M. Kroon
Laboratory of Physiological Chemistry,
University of Groningen,
Groningen, The Netherlands

Series Editors
E. Quagliariello
Department of Biochemistry,
University of Bari

and

F. Palmieri
Department of Pharmaco-Biology,
Laboratory of Biochemistry,
University of Bari

A Wiley–Interscience Publication

JOHN WILEY & SONS
Chichester · New York · Brisbane · Toronto · Singapore

Library of Congress Catalog Card Number: 75-647868

British Library Cataloguing in Publication Data:

Genes.—(Horizons in biochemistry and biophysics; v.7)
  1. Genetics
  I. Kroon, A. M.
  574.87'322   QH426
  ISBN 0 471 90264 0

Printed in Great Britain

Contents

List of contributors

BLUMENTHAL, R.   *Laboratory of Theoretical Biology, NCI, National Institutes of Health, Bethesda MD 20205, U.S.A.*

BONEN, L.   *Section for Molecular Biology, Laboratory of Biochemistry, University of Amsterdam, Kruislaan 318, 1098 SM Amsterdam, The Netherlands*

BORST, P.   *Antoni van Leeuwenhoekhuis, The Netherlands Cancer Institute, Plesmanlaan 121, 1066 CX Amsterdam, The Netherlands*

BRYAN, P. N.   *Institut für Molekularbiologie II der Universität Zürich, Hönggerberg, CH-8093 Zürich, Switzerland*

BUTLER, P. J. G.   *MRC Laboratory of Molecular Biology, Hills Road, Cambridge, CB2 2QH, U.K.*

DE BOER, H. A.   *Genentech, Inc., 460 Point San Bruno Boulevard, South San Francisco, California 94080, U.S.A.*

DESTREE, O. H. J.   *Anatomisch-Embryologisch Laboratorium, Universiteit van Amsterdam, Mauritskade 61, 1092 AD Amsterdam, The Netherlands*

DE VRIES, H.   *Laboratory of Physiological Chemistry, State University at Groningen, Bloemsingel 10, 9712 KZ Groningen, The Netherlands*

GARRETT, R. A.   *Department of Chemistry, Division of Biostructural Chemistry, University of Aarhus, Langelandsgade 140, 8000 Aarhus C, Denmark*

GRIVELL, L. A.   *Section for Molecular Biology, Laboratory of Biochemistry, University of Amsterdam, Kruislaan 318, 1098 SM Amsterdam, The Netherlands*

HANSON, R. W.   *Department of Biochemistry, Case Western Reserve University, School of Medicine, 2119 Abington Road, Cleveland, Ohio 44106, U.S.A.*

HENNIG, B.   *Institut für Physiologische Chemie, Georg-August Universität, Humboldtallee 7, D3400 Göttingen, Deutschland*

KEMPF, C.   *Laboratory of Biochemistry and Metabolism, NIADDK, National Institutes of Health, Bethesda, MD 20205, U.S.A.*

KLAUSNER, R. D.   *Laboratory of Biochemistry and Metabolism, NIADDK, National Institutes of Health, Bethesda, MD 20205, U.S.A.*

KLUG, A.   *MRC Laboratory of Molecular Biology, Hills Road, Cambridge, CB2 2QH, U.K.*

KROON, A. M.   *Laboratory of Physiological Chemistry, State University at Groningen, Bloemsingel 10, 9712 KZ Groningen, The Netherlands*

NEUPERT, W.   *Institut für Physiologische Chemie, Georg-August Universität, Humboldtallee 7, D-3400 Göttingen, Deutschland*

NOLLER, H. F.   *Division of Natural Sciences, University of California, Thimann Laboratories, Santa Cruz, CA 95064, U.S.A.*

SACCONE, C.   *Istituto di Chimica Biologica, Università di Bari, Via Amendola 165/A, 70126 Bari, Italy*

SHEPARD, H. M.   *Genentech, Inc., 460 Point San Bruno Boulevard, South San Francisco, California 94080, U.S.A.*

VAN KNIPPENBERG, P. H. *Laboratory of Biochemistry, State University, Wassenaarseweg 64, 2333 AL Leiden, The Netherlands*

VAN RENSWOUDE, J.   *Laboratory of Biochemistry and Metabolism, NIADDK, National Institutes of Health, Bethesda, MD 20205, U.S.A.*

VAN'T SANT, P.   *Laboratory of Physiological Chemistry, State University at Groningen, Bloemsingel 10, 9712 KZ Groningen, The Netherlands*

VOORMA, H. O.   *Laboratory of Molecular Cellbiology, State University Utrecht, Transitorium III, Padualaan 8, 3584 CH Utrecht, The Netherlands*

WYNSHAW-BORIS, A. J. *Department of Biochemistry, Case Western Reserve University, School of Medicine, 2119 Abington Road, Cleveland, Ohio 44106, U.S.A.*

Preface

Volume 7 of Horizons in Biochemistry and Biophysics is completely devoted to gene structure and expression. Many books on this subject have been published in the past. Most of these are highly specialized and not in the first place directed to the interested colleague or student not yet involved in this field of science. Since Horizons in Biochemistry and Biophysics primarily aim to reach this group of scientists, the editors thought it useful to bring together some information concerning topics, which at first glance may appear rather distant, but which are all pertinent to the fascinating biological process of the flow of genetic information from the carrier of this information, the DNA, to the final functionally active gene products, the proteins. In this volume emphasis is put on the eukaryotic cell. An exception is made for the two chapters dealing with the ribosomal RNAs and proteins. This choice was made on the one hand because as yet much more relevant information is available for bacterial ribosomes than for eukaryotic cytoplasmic ribosomes and on the other hand since the observations most likely allow, at least to a certan degree, extrapolation to the latter type of ribosomes.

The chapters 1 and 2 deal with the structure and function of nucleosomes and chromatin, with the role of histones and with the possible mechanisms of activation for replication and transcription and of the selectivity of the information retrieval. The structure and function of ribosomal RNA and ribosomal proteins are described in the chapters 3 and 4. From sequence analysis quite detailed information on the various ribosomal building blocks is available and the way to the integration of this knowledge as to understand the complete architecture and functioning of the ribosomes in protein synthesis, seems open now. At the level of initiation of protein synthesis good progress has been made with respect to the sequential reactions of the various factors and the regulatory steps (chapter 5).

The major issues in euykaryotic gene expression concern the regulatory mechanisms. Of course also bacteria need their regulation processes, but the demands of eukaryotic cells, especially of multicellular organisms with different cell types and tissues, are much stronger. Firstly because of the considerable compartmentation of these cells and the presence of two (the nucleocytoplasmic and mitochondrial), in plants even three (also the plastidic) genetic systems. Secondly because of the high degree of differentiation of the various cell types in higher organisms. Seven chapters are dealing with various aspects of this regulatory processes. They do not give the final words on either of the topics treated, but they point to the problems, develop the possible approaches to tackle the various problems and summarize the observations and facts gathered so far. So at the end it is hoped that a clear picture has emerged, showing the needs for further research at the one side and a better understanding of the complexity of the overall process of gene expression at the other side.

The transport of proteins from the sites of genetic expression to their sites of functional expression is discussed in chapter 6. The many different membranes, each surrounding specific compartments of the cell, put strong barriers to many molecules, especially macromolecules like proteins. None the less there has to be a vivid protein traffic wthin the cell to enable the proteins to function properly. Besides via signal peptides, other conformational properties and thermodynamic conditions may play a role in this transport circuit.

As an example of differential expression of genes, hormonal regulation is the subject of chapter 7. It is well recognized these days that information about the signals on RNA and DNA responsible for differential expression must reside in specific base sequences. To analyse these sequences one may use the recombinant DNA technique. For this reason chapter 8 dealing with the strategies for optimizing foreign gene expression in *E. coli* is included.

The chapters 9 to 12 are focusing on various aspects of the biogenesis of mitochondria. The interplay between the different genetic systems in one cell is discussed for mitochondria. The type of problems seen in this context and the needs for concerted action of these two systems are essentially similar, although different in details for plastids.

The description of the mosaic genes of mitochondria has a more general range as well. Also many nuclear genes are mosaic and need proper processing of the primary transcripts. The way this processing is accomplished may be different in detail again but not in essence. The assembly of the mitochondrial proteins is rather well documented and adds further information to the phenomena described in chapter 6. Finally the non-universality of the genetic code is an unexpected finding, but interesting especially from an evolutionary point of view.

A. M. Kroon
F. Palmieri
E. Quagliariello

Genes: *Structure and Expression*
Edited by A. M. Kroon
© 1983 John Wiley & Sons Ltd.

THE STRUCTURE OF NUCLEOSOMES AND CHROMATIN

A. Klug and P.J.G. Butler

MRC Laboratory of Molecular Biology,
Hills Road, Cambridge, CB2 2QH, U.K.

INTRODUCTION

Chromatin is the nucleoprotein complex found in the nucleus of all eukaryotic cells. It undergoes major changes in morphology during the cell cycle, varying in form from the highly compact chromosomes of metaphase, which are readily visible even with the light microscope, to the much more diffuse structure present in the nucleus during many of the other phases of the cycle. However, even this "diffuse" structure must involve considerable compaction of the cellular DNA since the genome length of about a metre, for a typical organism, must be packed into a nucleus less than 10 μm in diameter.

The earliest work on the structure of chromatin was concerned with defining its composition and elucidating the components. As found in the nucleus, chromatin contains approximately twice the mass of protein as of DNA. The protein consists of roughly equal amounts of two categories of protein, a number of very basic (i.e. positively charged) proteins collectively called "histones" and the other, so-called "non-histone", proteins which have a much wider range of properties and range from acidic to mildly basic in composition.

The high positive charge and abundance of the histones made them obvious candidates for interacting directly with the DNA, neutralising much of the negative charge and enabling the folding to occur which was essential for compaction. It was also once thought that they might well exercise a controlling role in the functioning of the DNA. The very nature of the histones made them extremely unfavourable for the separation techniques then available and there appeared to be tens of different components, allowing ideas of a moderate specificity for interaction with the DNA. However, steady improvements in fractionation and careful characterisation reduced the apparent number of individual histones so that, in a review in 1969, Stellwagen and Cole (1969) concluded that there were less than ten molecular species in any organism and hence that "the multiplicity of histones seems inadequate for them to be active repressors functioning individually, but it also seems too great for histones to have a very passive role". They were thinking in terms of a direction of histones by other proteins and

1

A. Klug and P.J.G. Butler

TABLE 1

The Histones

Histone	Molecular weight (M_r)	Degree of conservation
H3	15,400	Highly conserved
H4	11,340	Highly conserved
H2A	14,000	Moderate variation between tissues and species
H2B	13,770	Moderate variation between tissues and species
H1	21,500	Varies markedly between tissues and species
H1° ∿	21,500	Variable, mostly present in non-replicating cells
H5	21,500	Very variable, only present in transcriptionally inactive cells of some species

suggested that such direction could be "to specific genetic sites, or for specific or general derepression".

Within the next two years, the basic characterisation of the histones had been completed and there were known to be five types of histone, with a complete primary sequence for at least one member of each type. Moreover, in a review in 1971, De Lange and Smith could point out the extremely high conservation of sequence in both H3 and H4 and our general picture of the properties of the histones has not changed since then (Table 1). However, despite this successful characterisation of the histones, their function still was unknown and De Lange and Smith (1971) concluded that "the precise functions of histones, either collectively or individually, have not been established. It seems likely that histones function in the regulation of transcription and replication and as important structural components which may additionally serve to protect genetic material from various degradative processes".

The non-histone proteins have also been studied and several of the major ones isolated and sequenced (see Huntley and Dixon, 1972, for review). However, unlike the hisones, there are multiple non-histone proteins and individually they are present in much smaller amounts than any histone. Thus while the five histones are present in comparable molar amounts, even the major non-histone proteins are present at concentrations only about an order of magnitude lower and many of them are present in very small amounts indeed. Moreover, the non-histone proteins tend to be less tightly bound to the DNA than the histones and typically are washed off by 0.35 M salt, while the histones do not begin to come off significantly until above 0.45 M salt. This mode of identification is, however, potentially unreliable - one protein identified as a histone in trout (H6) was later found by sequence analysis to be closely related to some of the major non-histone proteins and not to the authentic histones (Huntley and Dixon, 1972; Walker, Hastings and Johns, 1977; Goodwin, Walker and Johns, 1978).

Contemporaneously with this characterisation of the histones, the first steps were being taken to investigate the structural organisation within chromatin, using X-ray scattering from fibres (Wilkins, Zubay and Wilson, 1959; Pardon, Wilkins and Richards, 1967; Pardon and Wilkins, 1972). This showed a simple series of low angle reflections or bands at about 11, 5.5, 3.7, 2.7 and 2.2 nm and this pattern could be observed from intact nuclei as well as fibres of extracted chromatin. This meant that there was some repetitive sub-structure in chromatin, on the scale of about 11 nm. In the absence of any biochemical evidence for the likely structure, the X-ray pattern was interpreted as coming from a superhelical coiling of the DNA, with a diameter of about 10 nm and a helical pitch of 11 nm. The histones were thought to lie generally along the DNA, possibly in the major groove, and to induce the superhelical coiling by a torsion effect due to inexact matching of the spacings of the positively charged residues on the proteins and the negative phosphate groups of the DNA.

The transition from these early ideas to our current more specific structural picture came as a result of the application of nuclease digestion to probe the organisation of chromatin (Clark and Felsenfeld, 1971; Hewish and Burgoyne, 1973; Burgoyne, Hewish and Mobbs, 1974). Thus, Hewish and Burgoyne (1973) reported that an endogenous nuclease in the nuclei isolated from rat or mouse liver would digest chromatin to produce multimers of 200 base pair lengths of DNA. This result was rapidly confirmed and also shown to be produced by other nucleases, in particular micrococcal nuclease (Noll, 1974), showing that it was not a property of the specific enzyme but rather of the substrate. In other words, it could be concluded that there was a basic repeating pattern in the organisation of the majority of chromatin. While the length of this repeat has since been shown to vary somewhat not only between types of nucleus but also within an individual nucleus, the overall picture suggested by Kornberg (1974), with a repeating unit of approximately 200 base pairs of DNA, two each of the four histones H2A, H2B, H3 and H4 and a single molecule of histone H1, is still the basis of our current ideas.

GROSS ORGANISATION

One problem with the folding and compaction of DNA lies in its stiffness. This can be measured by the "persistence length", which gives the distance apart of two regions on the rod-like DNA double helix for it to have bent sufficiently that the directions of the rod in these two places are not correlated with each other. The more flexible the rod, the shorter will be the persistence length. For double helical DNA the persistence length is largely

independent of ionic strength above 1 mM and equals about 50 nm
(Hagerman, 1981).   This corresponds to approximately 150 base
pairs of DNA and correlates well with the shortest lengths of DNA
which can readily circularise, by having their ends joined to give
covalently closed circles (Shore, Langowski and Baldwin, 1981).   A
circle with a circumference of 50 nm would have a radius of about
8 nm and this therefore is the smallest radius of curvature around
which free DNA will easily bend.

In practice, electron micrographs of chromatin show a variety of
appearances from "beads-on-a-string", with the "beads" about
10 nm diameter (Olins and Olins, 1974), through a rather uniform
fibre of 10 nm diameter (Ris and Kubai, 1970) to a less uniform
30 nm fibre (Finch and Klug, 1976) (these different forms are
discussed below).   The length of the DNA will not allow it to be
simply stretched straight along the 10 nm fibres, so it must be
folded with a maximum radius of curvature of about 5 nm.   Such
compaction can only come about by a close interaction with the
histones, with a binding energy sufficiently high to overcome the
unfavourable bending of the DNA which is necessary.   Moreover,
in order to minimise this bending the DNA needs to lie on the
outside of the fibre.   While this idea was generally unexpected
when it was first suggested (Kornberg, 1974), as it is in marked
contrast to the situation in viruses to which chromatin had been
likened, with hindsight it has the obvious advantage of allowing
relatively unencumbered access to the DNA for the many enzymes
and regulatory proteins which must interact with it if it is to
function, even while the DNA is in its folded state.

The size of the DNA in a chromosome suggests that it is impossible
for it to be folded in a unique and specific fashion, but rather it
must be packaged into a repeating structure, as indicated by the
X-ray diffraction pattern, so that the same motif can be used many
times.   This is the obvious cause of the nuclease digestion pattern
(Hewish and Burgoyne, 1973;   Burgoyne, Hewish and Mobbs, 1974;
Noll, 1974;   Kornberg, 1974) and, as already described, the
repeating unit is the 200 base pairs of DNA with nine histone
molecules.   This subunit is called the "nucleosome" (Oudet,
Gross-Bellard and Chambon, 1975) and appears to be the
fundamental unit for the structure of all chromatin.   The picture
is, however, complicated by the fact that not all nucleosomes are
identical, even within a given nucleus.   One obvious variable is
the exact length of the DNA in the nucleosome.   While the general
nature of the nucleosomal repeating structure was being
established, it rapidly became clear that the repeat length varied
from about 198 base pairs, as found in many cells, up to 210 base
pairs in some metabolically less active nuclei (e.g. the nucleated
erythrocytes of birds or amphibia) and down to 163 base pairs in
more active nuclei (e.g. yeast cells or the neuronal cells from
brain) (Morris, 1976;   Thomas and Thompson, 1977;   Lohr et al.,
1977;   Weintraub, 1978).   More recently a range of repeat length
has even been found within individual nuclei (Prunell and
Kornberg, 1982), although the general repeat is still characteristic
of the particular cell type.

Whatever the cell type, digestion with micrococcal nuclease reduces the DNA length from the full-length monomer, through a brief plateau around 165 base pairs, to an end product of 147 base pairs (Fig. 1). Only substantially more extensive digestion causes further degradation of the DNA of these "nucleosome cores", which are therefore relatively stable. The cores contain this discrete length of DNA and two molecules each of the four "core histones", H2A, H2B, H3 and H4 (Sollner-Webb and Felsenfeld, 1975; Axel, 1975; Bakayev et al., 1975; Shaw et al., 1976). H1 appears to be lost concomitantly with digestion below 165 base pairs (Noll and Kornberg, 1977), and the particle containing this length of DNA, the eight core histone molecules and one molecule of H1 has been described and called a "chromatosome" (Simpson, 1978). The extra DNA in the nucleosome repeat, which joins the nucleosome cores or, perhaps more precisely, the chromatosomes, is frequently referred

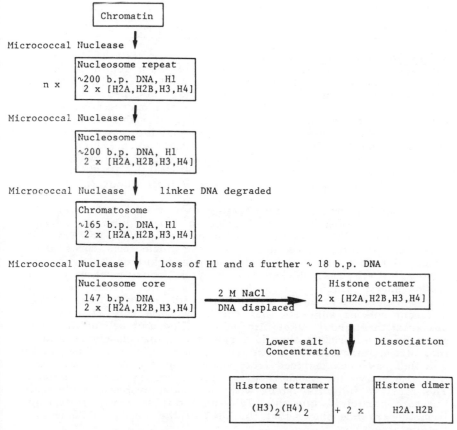

Fig. 1   Schematic diagram of breakdown of chromatin to subunits by nuclease digestion and dissociation of subunits with salt.

to as "linker DNA".   However, in native chromatin at normal ionic strengths it certainly is not stretched out, but is structured into the 10 nm fibres.   It appears to be variation in the length of this linker DNA which allows the differing repeat lengths.

The closely packed nucleosomes form the 10 nm fibre, which can then be further folded to give the 30 nm fibres, as discussed below, and at some stages of the cell cycle to even higher levels of structure up to the chromosomes.   Little is known about these highest levels of structure, but it is apparent that, even by the 30 nm fibre, although the structures present are ordered, they are not essentially completely regular, as can be deduced from the fact that the bands in the X-ray patterns are relatively broad.   Thus while the nucleosome cores, and perhaps also the chromatosomes, appear to have an unvarying structure, the varying lengths of the linker DNA mean that the nucleosome repeat is not constant and the regularity falls off further on going to higher levels of structure, even while reasonable order is maintained.

STRUCTURE OF THE NUCLEOSOME CORE

X-Ray and Neutron Diffraction Studies of Crystals.   The core particles in the earliest available crystals were found to have partially proteolysed histones, but their physico-chemical properties were still very similar to those with intact histones.   From a combination of X-ray diffraction and electron microscopy, it was found that the core particle had the shape of a flat circular disk, somewhat wedge-shaped and strongly divided into two layers (Finch et al., 1977).   The overall dimensions were about 11 nm x 11 nm x 5.7 nm.   A model was proposed in which there was a central core of the histone octamer with the DNA wound into about 1.8 turns of a shallow superhelix of pitch about 2.7 nm.

More recently, crystals have been grown which contain intact nucleosome cores and which diffract to a resolution of about 0.6 nm, and a full analysis is being carried out.   The crystals obtained are indistinguishable when the nucleosome cores are prepared from a wide variety of sources, including rat liver, beef kidney, calf thymus, chicken erythrocytes, mouse myeloma cells and scallop sperm (Finch et al., 1981).   The uniformity of the crystal unit cell parameters and X-ray intensity distribution from all these different species suggests that the nucleosome core structure is universal throughout eukaryotic cells.   The unit cell of the earlier proteolysed material is slightly larger than, but closely related to, that of the intact cores.   The arrangement of the particles in the unit cell has been deduced from analysis of the X-ray data, and this also shows the presence of a dyad symmetry in the particles. The data are fully consistent with the model proposed earlier for the core particle, while the high angle diffuse X-ray scattering from the crystals shows that the DNA of the core particle is in a B-type structure.

Nucleosome core particles are very suitable for study by neutron scattering using the contrast variation method (described in the next paragraph), as they contain approximately equal proportions of protein and DNA.   Several groups have investigated the scattering from solutions of core particles (Richards et al., 1977;   Hjelm et al., 1977) and the model described above is consistent with the solution scattering data (Pardon et al., 1978;   Suau et al., 1977). However, in solution scattering all of the three-dimensional information is compounded into one-dimensional data and this has the great disadvantage that solution scattering does not show the orientational relations in the particle of the different features which give rise to the diffraction pattern.   This problem does not arise in crystal diffraction and a low angle neutron diffraction study of crystals of nucleosome cores has recently been reported (Finch et al., 1980;   Bentley, Finch and Lewit-Bentley, 1981).

At low resolution (to about 2 nm) the protein, DNA and solvent components of the crystals scatter neutrons (or X-rays) as uniform domains of different contrast.   Thus by changing the D_2O content of the solvent, its scattering power for neutrons can be varied relative to those of the protein and DNA.   Hence at 39% D_2O the solvent scattering matches that of protein and only the DNA is effectively scattering, while at 65% D_2O only the protein effectively scatters.   Fourier maps have been calculated, to a resolution of about 2.5 nm, for the three principal projections of the nucleosome core, at contrasts of 39% and 65% D_2O, to show the distributions of the DNA and protein components separately (Finch et al., 1980; Bentley, Finch and Lewit-Bentley, 1981).   The maps at 39% D_2O

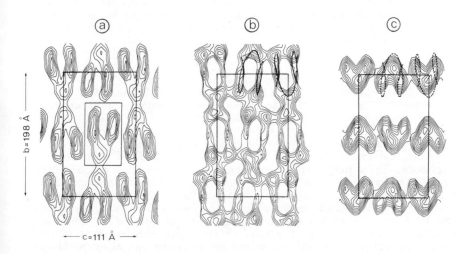

Fig. 2   Fourier projection maps of nucleosome cores.   (a) map from X-ray data, with both DNA and protein visible;   (b) the DNA component (neutron scattering in 39% D_2O), with the path of the superhelix superimposed;   (c) the protein core component (neutron scattering in 65% D_2O).   From Finch et al., 1980.

Projections of the Histone Octamer (from 3D image reconstruction).

Projections of protein distribution in nucleosome core crystals (from neutron
contrast variation).

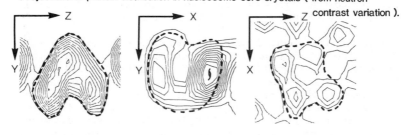

Fig. 3   Comparison of projections of the protein core in the
         directions of the crystal axes obtained from neutron
         scattering (65% D_2O; from Bentley et al., 1981) and those
         of the three-dimensional image reconstruction of the
         histone octamer (Fig. 5 below; from Klug et al., 1980).
         In the crystal there is overlap along the screw axis in the
         c direction.

are consistent with projections of about 1.8 turns of DNA in the
unit cell arrangement (Fig. 2), while at 65% D_2O the maps showed
projections (Fig. 3) similar to those of the image reconstruction of
the histone octamer from electron micrographs.

Histone Association in Solution - The Histone Octamer.   Around
physiological ionic strength, DNA-free histones form a number of
oligomers, in a polymorphic aggregation (reviewed by Isenberg,
1979), and while some of these interactions probably mimic those in
the nucleosome, the picture is not simple.   If, however, the
histones are prepared directly from chromatin by dissociation from
the DNA in 2 M salt a specific aggregate is formed.   This is
thought to be of similar structure to the aggregate present in intact
chromatin since chemical cross-linking between histones produces
very similar patterns in both cases (Thomas and Kornberg, 1975a).
Moreover, the cross-linked aggregate can be reassociated with free
DNA to regenerate many of the properties of chromatin (Stein,
Bina-Stein and Simpson, 1977).

The earliest cross-linking studies, in which the aggregate free in
solution in 2 M salt was shown to be similar to the state in native
chromatin, showed that during a time course, the monomeric

histones became cross-linked in an octameric aggregate and so it was proposed that the histones were associated as an octamer (Thomas and Kornberg, 1975a).   Moreover, it was shown by use of a cleavable cross-linking agent, that the octamer contained two molecules of each of the core histones, H2A, H2B, H3 and H4, and the pattern of dimeric contacts between them was also determined (Fig. 4) (Thomas and Kornberg, 1975b).   The existence of this octamer free in solution was however questioned (Weintraub, Palter and Van Lente, 1975), and it was suggested to be an artefact caused by the cross-linking together of two "heterotypic tetramers", each containing a single copy of each core histone.

Such a "heterotypic tetramer" has been invoked in a number of models for the action of chromatin (Weintraub, Palter and Van Lente, 1975; Weintraub, Worcel and Alberts, 1976).   Its existence appeared to be supported by other studies, either as the only

	H2A	H2B	H3	H4
H2A		dms zl ma	t-Pt	dms
H2B	dms zl ma			dms zl ma t-Pt
H3	t-Pt		dms S-S ma	dms zl
H4	dms	dms zl ma t-Pt	dms zl	

dms   dimethyl suberimidate (12 Å)
ma   methyl acetimide ($\sim$ 4 Å)
zl   "zero-length" linker
S-S   disulphide by oxidation
t-Pt   <u>trans</u> isomer of $Cl_2(NH_3)_2Pt(II)$

Fig. 4   Pattern of the histone-histone cross-linking with short cross-linking agents.   (References to the individual experiments are given by Klug et al., 1980, and also Lippard and Hoeschele, 1979.)

aggregate occurring in 2 M salt (Campbell and Cotter, 1976) or in
equilibrium with the octamer (Chung, Hill and Doty, 1978), but
more careful characterization of the size of the aggregate in solution
showed not only that it indeed had the molecular mass of the
octamer, but also that cross-linked and non-cross-linked material
were indistinguishable (Thomas and Butler, 1977).   This
characterization as an octamer has been further confirmed by both
more sedimentation analysis (Eickbush and Moudrianakis, 1978) and
a determination of the octamer structure (Klug et al., 1980) (see
below).   Moreover, the pattern of dissociation of the octamer is

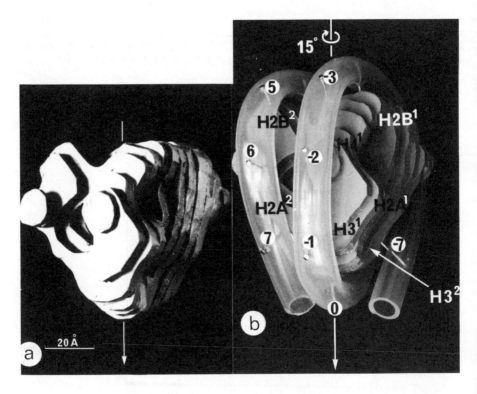

Fig. 5   (a) Model of the histone octamer obtained by 3-D image
           reconstruction from electron micrographs (Klug et al.,
           1980).   The dyad axis is marked.   (b) The ridges on
           the periphery of the model form a left-handed helical ramp
           onto which 1.8 - 2 turns of a superhelix of DNA could be
           wound.   (Note that for clarity, the diameter of the tube
           is smaller than the true scale for DNA.)   Distances along
           the DNA are indicated by the numbers -7 to +7, taking the
           dyad axis as origin, to mark the 14 repeats of the double
           helix contained in the 147 base pairs of the nucleosome
           core.   The assignment of individual histones in the model
           is described in the text.

not to heterotypic tetramers (Weintraub, Palter and Van Lente, 1975), but rather to the $(H3)_2(H4)_2$ tetramer and dimers of H2A.H2B (Thomas and Kornberg, 1975a,b; Thomas and Butler, 1978; Eickbush and Moudrianakis, 1978). There is thus no evidence so far for the heterotypic tetramer and if this particular state of association occurs it could do so only transiently or under rather special circumstances.

Image Reconstruction of the Histone Octamer. The occurrence and overall characterization of the histone octamer, which forms the protein core of the nucleosome, has just been described. During attempts to crystallize it, ordered 3-dimensional aggregates - hollow tubes - were obtained which were investigated by electron microscopy. The tubes were found to be composed of stacks of rings, each of which contained ten octamers, and images of the tubes were analysed and the image reconstruction method used to produce a low resolution 3-dimensional map of the octamer (Klug et al., 1980). The overall state of preservation of the material could be checked as, to the resolution of the micrographs, the octamer appears to possess a 2-fold axis of symmetry (as does the nucleosome core particle itself), and data from the best preserved tubes were averaged and a 3-dimensional map of the histone octamer tube computed, to a resolution of about 2 nm. A model of an individual octamer is shown in Figure 5a.

Like the nucleosome core, the histone octamer is a wedge-shaped particle of bipartite character. Its periphery shows a series of ridges which form an almost continuous helical ramp of external diameter 7 nm and pitch about 2.7 nm, exactly suitable to act as a former onto which 1.8 turns of superhelix of DNA could be wound to give the appropriate dimensions for the nucleosome core (Fig. 5b).

Spatial Arrangement of the Core Histones. A different approach to study the structure of the nucleosome core has been to estimate the histone-histone proximities by protein cross-linking (Fig. 4) (e.g. Thomas and Kornberg, 1975b) and also to map the order of the histones along the nucleosomal DNA by chemical cross-linking of histones to DNA (Fig. 6) (Mirzabekov et al., 1978). On the assumption that the histones are too small to be multidomain proteins, this data has been used (Klug et al., 1980) to assign regions of density in the octamer map to particular histones (Fig. 5b and 6). Despite the lack of resolution of individual histones in this map, consideration of all these kinds of data allows the spatial arrangement of histones in the nucleosome core to be deduced. According to this proposal, the helical ramp in the octamer map is composed of a particular sequence of the eight histones, in order H2A-H2B-H4-H3-H3-H4-H2B-H2A, with the dyad in the middle. This structure is consistent with many other observations on the histone aggregates and their association with DNA, some of which are discussed below.

It is worth emphasising some particular points in the logic used to interpret the cross-linking data. Since the cross-links to the two

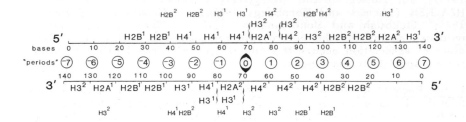

Fig. 6   Assignment of DNA-histone cross-links to individual histone molecules. Based on the models of the octamer and nucleosome core particles (Fig. 5), the superscripts 1 and 2 have been assigned to the pattern of cross-linking observed (Mirzabekov et al., 1978; Shick et al., 1980) to distinguish the two copies of each histone present. Large letters denote the major cross-links and smaller letters above and below denote weaker cross-links. The overall dyad at the centre has the effect of interchanging the superscripts on the histones.

copies of each histone present in the nucleosome core cannot be distinguished, the points of contact of histones along a strand of the DNA, determined by the histone-DNA cross-linking studies, are insufficient alone to fix the spatial arrangements of the histones. An added complication is that, because the two superhelical turns of DNA are close together, the pattern of histone-DNA cross-links need not directly reflect the linear order of histones along the length of the DNA. However, the additional information from the 3-dimensional density map restricts the number of possibilities and enables choices to be made. Thus it is concluded that the same H2A molecule (H2A^2 in Fig. 6) makes contact with both the middle of the DNA (75 bases from the 5'-end of one strand, between "sites" 0 and -1) and one end of it (125 bases from the 5'-end of the other strand, between sites +5 and +6). The possibility of this in three dimensions can be seen in Figure 5b. Furthermore the less frequent cross-linking interactions, observed as weak spots in gels during the more detailed investigations (Shick et al., 1980), readily fit with this model without need to invoke elongated shapes or long overlaps for the histones. Thus the H2B cross-links reported as spanning from 25 to 50 bases (a distance of 8 nm) can be explained without a single H2B molecule having to extend all that way: the weak cross-links reported around 40 to 50 bases from the 5'-end are assigned to molecule H2B^2 (at sites -2 to -3 in Fig. 5b), while its major contacts are at 105-115 bases, with DNA on the superhelical turn below (sites +4 and +5).

Roles of Individual Histones in the Nucleosome Core.   The spatial arrangement of histones just proposed for the octamer leads naturally to specific roles for the individual histones in folding the DNA on the nucleosome (Fig. 7, a-c) (Klug et al., 1980).   The $(H3)_2(H4)_2$ tetramer has the shape of a spring-washer (more accurately a single turn of a helicoidal surface) and defines the central turn of the DNA superhelix.   H2A and H2B could then add as two heterodimers, H2A.H2B, one onto each face of the H3-H4 tetramer, with each binding an extra half-turn of DNA and thereby completing the two-turn superhelix.   Such a distribution of the histones readily explains the many observations that H3 and H4 together, in the absence of H2A and H2B, can confer nucleosome-like properties on DNA, causing supercoiling and a regular pattern of resistance to micrococcal nuclease digestion, whereas H2A and H2B alone cannot, even though they will bind to DNA (Camerini-Otero, Sollner-Webb and Felsenfeld, 1976; Camerini-Otero and Felsenfeld, 1977a;   Stockley and Thomas, 1979). Finally, for reasons which are discussed below, we believe that H1 binds to the unique region where the DNA enters and leaves the chromatosome, stabilising and, as it were, "sealing off" the structure (Fig. 7d).

In this model, the $(H3)_2(H4)_2$ tetramer provides the primary structural framework for the nucleosome and, in turn, the central element of this tetramer is a homodimer of H3, which ends up spanning across the particle dyad axis and interacting with the central region of the DNA in the nucleosome core.   The local dyad of DNA which coincides with the dyad of the H3 dimer becomes the overall dyad of the DNA superhelix.

Only in two rotational orientations of the DNA duplex about its helix axis can a local dyad of the DNA coincide with the overall two-fold symmetry axis of the particle: one or other groove must be directed away from the protein core at the dyad.   Moreover, DNase I digestion data shows that the exact cleavage points on opposite strands of the duplex are, on average, staggered by two base extension in the 3'-direction (Lutter, 1977;   Sollner-Webb and Felsenfeld, 1977).   To accord with the symmetry of the particle, the positions on the phosphodiester backbone around the central site must be symmetrically disposed around the dyad axis and they will be accessible for nuclease attack only if it is the minor groove which faces outwards towards the solution, as pictured in Figure 8.

One consequence of this geometry and of the B-type conformation of the nucleosomal DNA is that the minor groove faces out into solution not only on the dyad, but also at intervals of about 10 b.p. in both directions along the DNA away from the dyad, and so in principle is accessible for cleavage by DNase I at these sites.   In practice, the sites on the dyad and at +10 and -10 b.p. to each side of the dyad (here taken as the origin) are not efficiently cleaved by DNase I (Fig. 9b) (Lutter, 1978) and therefore these regions must be hindered in some way, probably by making substantial contacts with the histone tetramer (Klug et al., 1980). By the same reasoning, other strong histone contacts are likely at

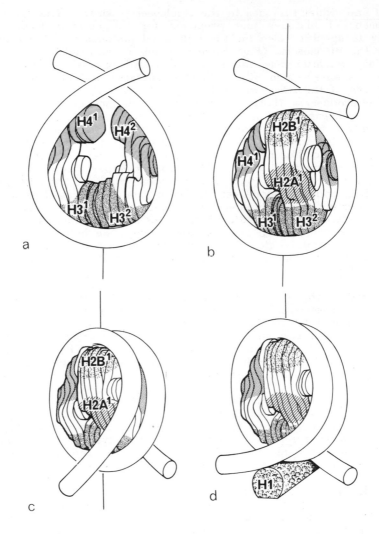

Fig. 7   "Exploded" views of the nucleosome, showing assembly out
of constituent histones, based on Fig. 5.   The patches on
the histone core indicate locations of individual histone
molecules, but the boundaries between them are not known
and are thus left unmarked.   (a) the $(H3)_2(H4)_2$ tetramer
has the shape of a spring washer and can act as a spool
for 70 - 80 b.p. of DNA, forming about 1 superhelical
turn.   (b) an H2A.H2B dimer associates with one face of
the tetramer.   (c) H2A.H2B dimers on opposite faces each
bind 30 - 40 b.p. DNA, or one half superhelical turn, to
give a complete 2-turn particle.   (d) histone H1 interacts
with the unique configuration of DNA at the entry and exit
points to seal off the nucleosome.

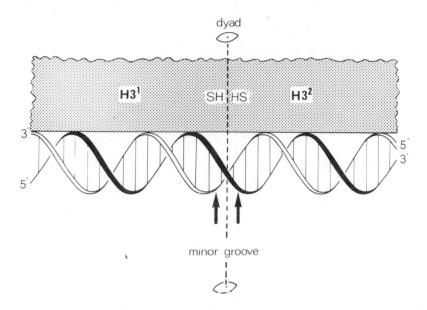

Fig. 8   Orientation of the DNA double helix around the nucleosome
           core dyad.   The minor groove must face away from the
           underlying histone core to account for the staggered
           cleavage sites (arrowed) found with DNase I (Lutter, 1977;
           Sollner-Webb and Felsenfeld, 1977).   An H3.H3 dimer
           straddles the dyad and the invariant cysteine residues
           must lie close to it, to account for the cross-link that can
           be formed between them (Camerini-Otero and Felsenfeld,
           1977b).

the well protected sites on the opposite end of the dyad, i.e. in
the region of the H2B molecules (Fig. 5b), as well as at the ends
of the DNA.

While the H3-H4 interactions and also the H3 dimer contacts are
evident from protein cross-linking on the nucleosome (Fig. 4),
there is no evidence for an isolated $(H3)_2$ dimer in solution without
H4.   One intriguing feature of the H3-H3 dimer contact is the
presence of the invariant sulphydryl group (Cys 110) on H3, which
must lie close to the dyad axis (Fig. 8), since it can be oxidised to
form an S-S bridge between H3's (Camerini-Otero and Felsenfeld,
1977b).   It is tempting to speculate about possible roles for this
potential bridge during various phases of the cell cycle.

The heart of the nucleosome, therefore, consists of a dimer of H3
molecules, interacting with a dyad of the DNA double helix.   This
is strikingly analogous to the situation found for two specific DNA
binding proteins, whose structures have recently been determined

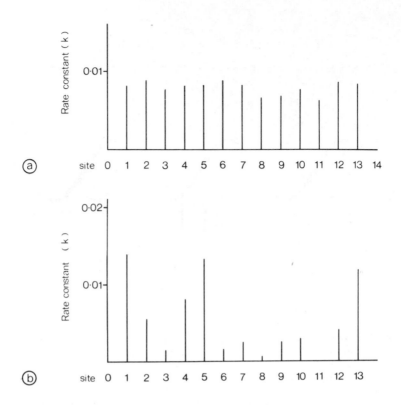

Fig. 9   Comparison of rate constants for DNase I cleavage of
various sites on DNA.   (a) DNA bound on planar surface,
with all sites essentially equivalent (Rhodes and Klug,
1980).   (b) DNA on nucleosome core, where the rates will
be influenced by variations in accessibility and local
conformation (Lutter, 1979).

to high resolution, namely cro (Anderson et al., 1981) and CAP
(McKay and Steitz, 1981).   Here too the molecules are found as
dimers, and the structures are of a form which would permit
interaction with a (specific) dyad of the DNA double helix.

Stabilising Interactions in the Nucleosome Core.   While the core
histones alone can aggregate to form the octamer, which is the
protein kernel of the nucleosome core, this is only stable at high
salt concentrations.   Therefore the protein-DNA interactions must
contribute significantly to the overall stability of the core at low
salt concentrations and, given the opposite and high charges of the
histones and DNA, together with the non-specific interaction, it
might seem logical to expect much of the binding to arise from
simple charge-charge interaction.   This expectation is supported
by the observation that chromatin is dissociated by increasing ionic

strength, with first the H1 (and H5, if any) coming off around 0.5 M, followed by the H2A.H2B dimer around 0.85 M and finally the $(H3)_2(H4)_2$ tetramer by 1.25 M.

The variation of the rates of cleavage by DNase I at different cutting sites along the nucleosomal DNA (Fig. 9b) shows the different degrees of protection by the histones and reflects varying strengths of interaction. This effect is observed on nucleosomes prepared from bulk chromatin and is distinct from any possible binding preferences for particular sequences of DNA, for which there is evidence although the effect appears to be relatively weak (e.g. Chao, Gralla and Martinson, 1979; Igó-Kemenes, Omori and Zachau, 1980). One region of the nucleosome cores where it is relatively simple to probe the interaction is at the ends of the DNA, and it was found that the ends could become unbound and "melt" before the central region (Weischet et al., 1978). While unwinding is almost certain to start from an end, yet the fact that micrococcal nuclease digestion generates the specific DNA length on the cores shows that these "end points" must be sites of relatively strong interaction, to prevent further "nibbling" of the DNA. More detailed measurements of the dissociation of the ends of the DNA have shown that over the terminal 20 b.p. at each end, only about 15% of the DNA phosphate charges are neutralized (McGhee and Felsenfeld, 1980), despite the fact that the charge on the histones is sufficient to half neutralize the DNA.

This result was rather unexpected, but fits with the observation that histones which have had their amino terminal tails removed by proteolysis can still be reassembled with DNA to produce particles with many of the properties of nucleosomes (Whitlock and Stein, 1978). However, it is still possible that most of the histone charges are used not in individual nucleosomes but rather in the chromatin structure. Thus one obvious suggestion is that the tails, and the positively charged groups in them, may be interacting with neighbouring nucleosomes rather than the DNA in their own core. Such inter-nucleosomal interaction could provide some of the energy necessary for the condensation of the nucleosome filament, or 10 nm fibre, into higher order structures. It could also explain the compaction which does occur even in the absence of H1. These effects are discussed below.

HELICAL PERIODICITY OF DNA ON THE NUCLEOSOME AND IN SOLUTION.

The model of the nucleosome based on the X-ray and neutron diffraction results contains two superhelical turns of DNA. This is in apparent contradiction to physico-chemical measurements of the superhelicity of closed circular DNA extracted from the simian virus 40 minichromosome, where apparently only one superhelical turn was found per nucleosome (Germond et al., 1975), and this disagreement has become known as the "linking number paradox" (the linking number is the number of times the two strands of the DNA double helix wind round each other). When it was recognised (Crick, 1976) that the physico-chemical results were in fact not measurements of the actual number of superhelical turns in 3-dimensions, but rather the change in linking number, a resolution of the paradox became possible.   One possible reconciliation of the X-ray and physico-chemical results would be for the screw of the DNA double helix to change in going from solution onto the surface of the nucleosome (Finch et al., 1977): the helical periodicity would have to reduce by about 0.5 base pairs per turn.   (Any change of screw makes a further contribution to the linking number, in addition to that made by the superhelical path of the DNA double helix.)   If the DNA of the nucleosome had 10 b.p. per turn, as measured in the first nuclease digestion studies (Noll, 1974), this explanation would require the periodicity of DNA in solution to be close to 10.5 b.p. per turn, rather than the value of 10.0 found from X-ray diffraction of fibres.

Recently measurements have been made of the periodicity of DNA in solution, using both the change in linking number produced by the insertion of known numbers of base pairs of DNA into large closed circular DNA duplexes (Wang, 1979; Peck and Wang, 1981) and also the period of DNase I cutting of linear DNA immobilised onto a flat surface (Rhodes and Klug, 1980, 1981).   This latter technique will be described in some detail, as it also illustrates the method for the determination of the periodicity of DNA on the nucleosome.   If a straight, stiff piece of DNA lies on a flat surface, access of an enzyme to one side of the DNA helix will be hindered in a uniform way along its length.   Hence, only the more exposed phosphodiester bonds will be accessible to the nuclease and these will recur with the periodicity of the double helix.   Thus the distance between cutting sites on a given strand, which will be shown by the lengths of the single-stranded fragments, gives the helical repeat directly.   Fragment lengths over the size range of interest can be measured to a single nucleotide, even in random sequence DNA (Lutter, 1979).   The differences in fragment lengths were found to be independent of the surface used and the cutting rate was the same at all potential sites (Fig. 9a), strongly indicating that there is no effect of the surface and that the repeat is that of mixed sequence DNA in solution.   (Any local sequence-specific variations in DNA periodicity will average out in such a bulk measurement.)   Both methods of determination give a value of 10.6 ± 0.1 b.p., significantly higher than the 10.0 b.p. observed in fibres of B-form DNA.

While this determination of the periodicity of B-form DNA in solution appeared to resolve the linking number paradox, a further complication arose when more accurate measurements of the periodicity of DNase I cutting sites on the nucleosome gave an average distance between sites of 10.4 b.p. (Prunell et al., 1979). Closer examination of the more complex geometry of the nucleosome, however, shows that the angle of attack of the enzyme might well vary with the position of the DNA site (Fig. 10a) and hence there is no reason why the cutting should exactly reflect the helical periodicity, indeed for a left-handed superhelix one would expect it to be greater.   Precise determination of the cutting sites for both DNase I and, more recently, DNase II with respect to the 5'-ends

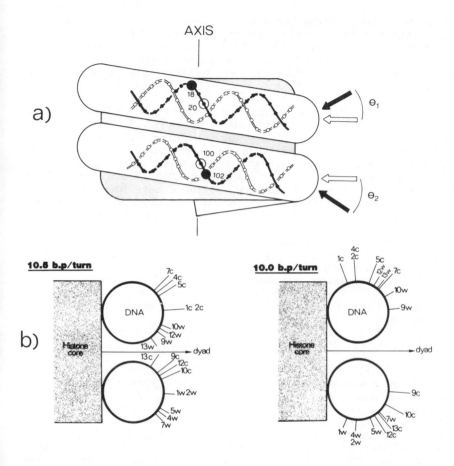

Fig. 10   Schematic picture of possible angles of attack by DNase I
on nucleosomal DNA.   (a) general view of DNA wound on
the core, with white arrows (and open circles) showing
attack on most exposed positions in structure with
indefinite number of superhelical turns, where all sites are
equivalent.   Black arrows (and circles) show most
exposed sites on nucleosome core where the different
repeats of the double helix are in non-equivalent
environments (Prunell et al. 1979).   (b) angular
dispositions of actual cutting sites calculated for two
different screws of the DNA double helix (Klug and
Lutter, 1981).   The two strands of the duplex are
labelled C and W and the sites are numbered from the
5'-ends.   A screw of 10.5 b.p./turn would require the
nuclease to gain access to sites located in the 0.5 nm
crevice between the two superhelical turns of DNA, while a
screw of 10.0 b.p./turn is compatible with the steric
limitations.

shows that although the average separation is 10.4 b.p., there are
significant variations in the distance between different cutting sites
along the DNA (Lutter, 1979, 1981).   In particular the cutting
periodicity of the central 60-70 base pairs of the nucleosomal DNA
is close to 10.6 b.p., whilst that of the 30-40 b.p. at each end is
close to 10.

The locations of the nuclease cutting sites can be analysed to orient
the one-dimensional data, measured in numbers of bases, in spatial
terms in three dimensions (Klug and Lutter, 1981).   Taking the
midpoint of the DNA on the nucleosome dyad as the reference point,
the cutting site locations along a DNA strand were mapped onto
models of the nucleosome core containing DNA of different helical
periodicities.   It was found that, while a periodicity close to
10.5 b.p. per turn leads to a sterically implausible model (Fig.
10b), a periodicity of 10.0 b.p. per turn leads to cutting site
positions which are not only sterically sound, but which also fall
into a pattern which would be expected when the access of the
nuclease is restricted by the histone core on one side and the
adjacent superhelical turn of DNA on the other (Fig. 10b).   Hence
as originally proposed, a value for the helical periodicity close to
10.0 b.p. per turn on the nucleosome, taken together with the
periodicity close to 10.5 for DNA in solution, resolves the so-called
linking number paradox.

LOCATION AND ROLE OF HISTONE H1

Histone H1 is, as already mentioned, the histone most readily
dissociated from chromatin and its removal appears to leave the
underlying nucleosomal structure essentially unaltered, except that
at the salt concentration involved sliding of nucleosomes may occur,
altering the nucleosomal spacing (Steinmetz, Streeck and Zachau,
1978).   Commensurate with this is the finding that H1 readily
migrates from intact chromatin onto H1-depleted chromatin (Caron
and Thomas, 1981) and even between different nucleosome oligomers
during nuclease digestion of chromatin, if care is not taken to avoid
this.   These observations suggest that H1 may be external to the
main nucleosomal structure and this is supported by the absence of
H1 from the nucleosome cores.

Although H1 is much too small to be seen directly in the electron
microscope, its position on the nucleosome can be inferred from its
effect on the appearance of chromatin in a moderately folded state
(discussed further below).   There is a clear difference between
the structures observed below ionic strength 5 mM in the presence
or absence of H1 (Fig. 11).   With H1 an ordered structure is
seen, with the nucleosomes flat on the supporting grid in a more or
less regular zigzag (Thoma, Koller and Klug, 1979), a structure
which is consistent with the measured mass per unit length in
solution under similar ionic strength (Sperling and Tardieu, 1976;
Suau, Bradbury and Baldwin, 1979).   The zigzag form arises from
the DNA entering and leaving the nucleosome at sites close
together, as would be expected for a particle containing about 2
superhelical turns.   In chromatin depleted of H1, the entry and

exit points are much less regular and frequently almost opposite each other. Indeed, at very low ionic strength the nucleosomes without H1 unravel into a rather linear structure in which beads are no longer visible. Such an open structure is consistent with physico-chemical measurements on nucleosome cores at very low ionic strength in solution (Gordon et al., 1978; Martinson, True and Burch, 1979; Dieterich et al., 1980). H1 prevents this happening and one can therefore conclude that at least part of it must be located at, and stabilise, the region where DNA enters and leaves the nucleosome (Fig. 7d).

The biochemical evidence is also consistent with a particle with two full turns of DNA stabilised by an H1 molecule located on one side of it. Thus while the nucleosome core particle is the only relatively stable, and therefore homogeneous, product so far observed during digestion of chromatin by micrococcal nuclease, a

salt concentration

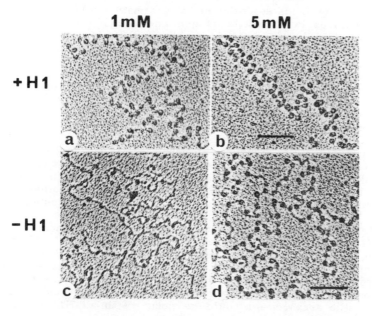

Fig. 11 Electron micrographs of chromatin, with and without H1, at low ionic strengths (Thoma et al., 1979). The structure of the filament containing H1 is more ordered (a and b) with the DNA entering and leaving the nucleosome at sites close together; without H1 there is no order (in the sense of a defined filament direction) and even when nucleosome beads are visible (at higher ionic strength, d) the DNA enters and leaves the nucleosome more or less at random.

transient intermediate (since named the chromatosome (Simpson, 1978)) containing about 165 b.p. is seen and it is during its conversion from about 165 to 147 b.p. that H1 is released (Noll and Kornberg, 1977). When the 147 base pairs of the core particle were found to correspond to 1.8 superhelical turns, it was therefore suggested that the 165 base pair particle contains two full superhelical turns of DNA (Finch et al., 1977). Since the pitch of the superhelix is not much greater than the diameter of the DNA duplex, this brings the two ends of the DNA close together on the nucleosome, so that both can be associated with the same single molecule of H1. If, as suggested (Noll and Kornberg, 1977; Weischet et al., 1978; Thoma, Koller and Klug, 1979), the chromatosome constitutes the basic structural element of chromatin, the length of the linker DNA must vary greatly. Thus while the shortest known chromatin repeat is between 160 and 170 b.p., even in rat liver with a repeat about 200 b.p., and hence an average linker length of 35 b.p., there is a large variation about the mean (Lohr et al., 1977) with the range extending practically to zero (Prunell and Kornberg, 1982).

The presence of H1 is essential for the formation of regular higher order structures in chromatin (Finch and Klug, 1976; Thoma, Koller and Klug, 1979; Renz, Nehls and Hozier, 1977a,b; Strätling, 1979; Butler and Thomas, 1980 - see below). Thus its removal perturbs the salt-induced conformational changes of nucleosome oligomers in solution (Renz, Nehls and Hozier, 1977a,b; Strätling, 1979; Butler and Thomas, 1980) and even precludes the formation of the regular zigzag (Thoma, Koller and Klug, 1979), as just described. In such a zigzag, the H1 regions of adjacent nucleosomes appear to be close together or touching. With increasing ionic strength, more of the H1 regions could interact with one another, to produce a more compacted aggregate. Polymers of H1 have indeed been shown to exist by chemical cross-linking at both low and high ionic strength (Olins and Wright, 1973; Thomas and Khabaza, 1980), but it remains to be shown that they actually play the specific role just suggested.

An additional structural complication arises from the attachment of a single H1 molecule to the two-turn nucleosomal particle. While the two-turn particle has dyad symmetry, this must be broken by the H1 molecule in the chromatosome, which will therefore have a polar structure. It is not known how the H1 molecules in the chromatosomes are arranged within even the 10 nm nucleosome filament: the arrangement could be polar, with all H1 molecules oriented in the same direction, or non-polar, where the H1 molecules could either be randomly oriented or else arranged in some specific pattern, such as alternating in orientation. Any specific pattern of orientation, whether polar or non-polar, would require the incorporation of H1 to be organised so that the orientation of an incoming molecule is determined by its neighbour. Although we think a random orientation is the most probable - it is certainly compatible with the irregularity of higher order structure referred to below - it is not yet possible to determine the arrangement in native chromatin and more data are required to settle this question.

HIGHER ORDER STRUCTURE OF CHROMATIN

Models for the Compaction. The well known "beads-on-a-string" appearance of nucleosomes along the DNA (Olins and Olins, 1974) only occurs when the structure has been stretched out, or at very low ionic strength when the electrostatic repulsion between nucleosomes may produce a similar effect. In the presence of even low concentrations of salt, native chromatin, containing histone H1, occurs as a continuous filament of nucleosomes of diameter about 10 nm. It is the folding of the filament, to give the 30 nm diameter fibres or even more condensed structures, which is usually referred to as "higher order structure" and two main classes of model have been proposed for the first level of folding, from the 10 nm filament to the 30 nm fibre.

The simplest model involves the formation of "solenoids" by a helical winding of the nucleosome filament (Finch and Klug, 1976), with about 6 nucleosomes per turn and the turns in apposition. This was proposed on the basis of both electron micrographs of chromatin and companion X-ray diffraction studies (Sperling and Klug, 1977), which showed that the occurrence of the 10 nm reflection corresponded to the formation of the 30 nm fibres as seen in the electron micrographs. An obvious explanation is that this reflection is generated by the helical pitch of the solenoid which is determined by the diameter of the 10 nm filament. Such a model does not require an exact number of nucleosomes per turn, as many regular helices have non-integral numbers of subunits per turn. Moreover if, as seems probable, the folding is brought about by interaction of increasing regions of H1 with each other, eventually producing a helical polymer along the centre of the solenoid and thus accounting for its geometrical form, slight variations in these contacts could lead to a family of solenoids. Such variations could be similar to the "quasi-equivalent bonding" found in the formation of the protein capsid of many isometric viruses (Caspar and Klug, 1962) and would allow solenoids with varying numbers of nucleosomes per turn. Such polymorphism might be expected not only under different conditions, in particular of ionic strength, but also possibly at different points along a given solenoid, leading to the irregularity which is observed. The variations in linker length (Lohr et al., 1977; Prunell and Kornberg, 1982) might add to this irregularity.

The location of the H1 whose aggregation accompanies, and may even control, the formation of the solenoid near the centre of the solenoid has an important consequence. The flexibility and irregularity of the 30 nm fibre may arise because these essential contacts are those near the centre of the fibre rather than those directly between the faces of the nucleosomes. This mode of construction would allow for easy deformation, so that the fibre could fold back further on itself to give a still more condensed state of chromatin (e.g. Fig. 14 below).

As an alternative to the continuous helical solenoid, a discontinuous type of higher order structure has been proposed (Renz, Nehls and Hozier, 1977a,b; Strätling, 1979), in which groups of nucleosomes form "superbeads" by an H1 dependent aggregation and these superbeads then aggregate further to give the 30 nm fibre.   While such superbeads would provide a facile explanation for the "bumpy" appearance of the 30 nm fibre at intermediate ionic strengths (see below), it is not clear whether they have any defined structure and estimates of the number of nucleosomes in a superbead have varied from 6 to 10 (Renz, Nehls and Hozier, 1977b), 7 or 8 (Hozier, Renz and Nehls, 1977; Strätling, Müller and Zentgraf, 1978), up to as many as 12 (Renz, 1979).   More recently, one of the originators of the idea of superbeads has suggested that they may not have a specific structure (Renz, 1979), in which case their significance is unclear.   The plasticity of the solenoidal structures just discussed means that short lengths of the nucleosome filament could fold up into a variety of bead-like structures, while still possessing an underlying helical character.   The variability in the descriptions of superbeads and the failure to find any structural specificity, strongly suggest that superbeads may well correspond to isolated turns, or groups of a few turns, of a solenoid.   When "superbeads" have been characterized (on sucrose gradients), they are most often observed in preparations where extensive redistribution of H1 has taken place and frequently the fractions containing superbeads have an enhanced H1 content (e.g. Ruiz-Carrillo et al., 1980; Jorcano et al., 1980), indicating that particularly stable turns of solenoid may be those which have bound additional H1 (or H5) molecules and that these may constitute much of the superbead preparation.

Electron Microscope Observations.   As it has not yet been possible to crystallise anything larger than the nucleosome core, the main technique which can be used to try to determine the actual higher order structure of chromatin is electron microscopy.   While isolated nucleosomes are clearly visible in the electron microscope and the zigzag that they form at low ionic strength can be seen, both the 10 nm nucleosome filament and the 30 nm fibre show very little detail, but rather appear to have relatively smooth surfaces and therefore show little contrast along their lengths.   The lack of ordered aggregates of the 30 nm fibres, and also their irregularity along their lengths, means that it has not been possible to employ image analysis techniques on electron micrographs, to enhance the detail sufficiently for the structure to be distinguished directly. It has therefore been necessary to follow the formation of the higher order structure by varying the conditions, and to try to deduce the final structure from the appearance of partially condensed forms.

In the earliest experiments to follow the condensation (Finch and Klug, 1976), divalent cations were found to cause formation of the 30 nm fibres even at very low concentrations, and no intermediate forms could be discerned, although the requirement for H1 was clear.   More recently, monovalent salts have been used to carry out a systematic study of the formation of the 30 nm fibres with increasing ionic strength (Thoma, Koller and Klug, 1979), and it has been possible to obtain a range of structures showing increasing degrees of compaction as the ionic strength for the fixation step was raised. Thus from a filament of nucleosomes around 1 mM (Fig. 11a), the extent of structure increased through the zigzag filament around 5 mM (Fig. 11b) and then a family of more compact intermediate helical structures, each having apparently increasing numbers of nucleosomes per turn, until by 60 mM a compact structure was formed (Fig. 12). Between 40 and 60 mM, the fibre was still compact, though with discontinuities in places, but it nevertheless had a clear three-dimensional appearance with cross-striations about 10-15 nm apart.

Taken in isolation out of the ionic strength series, the partially formed structures around 40-50 mM look similar to the micrographs which have been cited as evidence for superbeads, but when seen in context it is apparent that the clumps of nucleosomes are not discrete structures but only intermediates in a process of continuous aggregation, held together by the fixation necessary for specimen preparation (Thoma, Koller and Klug, 1979; de Murcia and Koller, 1981).

Measurement of the mass per unit length in the scanning electron microscope of the 30 nm fibres gives a value compatible with the solenoidal model (Engel, Sütterlin and Koller, 1980). The 30 nm fibres are stable under physiological conditions of ionic strength and are thought to correspond to the fibres observed in sections of fixed and embedded nuclei (Ris and Kubai, 1970). Moreover, under appropriate conditions the bulk chromatin spilling out of nuclei can have the same appearance as soluble chromatin under the same ionic strength (Labhart and Koller, 1981).

Solution Studies. Preparation of the specimens for electron microscopy involves fixation and this step could be generating some of the structures observed. It is therefore desirable to complement microscopy with physico-chemical studies in solution which, while they cannot determine the structure, are not prone to the same artefacts and can provide valuable confirmation of changes in the structure. The most directly comparable studies are some on the sedimentation behaviour of nucleosome oligomers of defined sizes over a range of monovalent salt concentrations (Butler and Thomas, 1980; Thomas and Butler, 1980), which were carried out under conditions which were designed to be comparable with those for the electron microscopy.

For oligomers up to five nucleosomes, the sedimentation coefficient shows a power-law dependence upon ionic strength up to 25 mM, but is independent above this. Hexamer and larger oligomers show an abrupt change of behaviour, with the power-law dependence over the whole range of ionic strength up to 125 mM (Fig. 13). The change in behaviour between pentamer and hexamer is compatible with the formation of some compact structure which requires at least six nucleosomes. There is, however, no further discontinuity of behaviour around dodecamer or higher multiples of

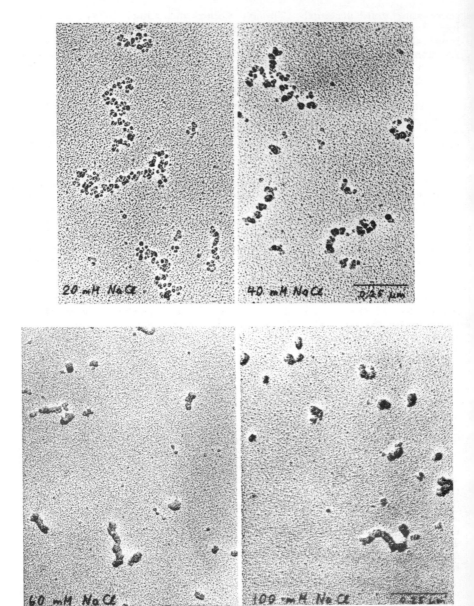

Fig. 12   Electron micrographs of chromatin containing H1, over a
          range of ionic strengths.   These show the steadily
          increasing compaction with ionic strength, through the
          intermediate "globular" structures to the relatively well
          ordered and compact 30 nm fibres (pictures kindly
          supplied by F. Thoma, cf. Thoma <u>et al</u>., 1979).

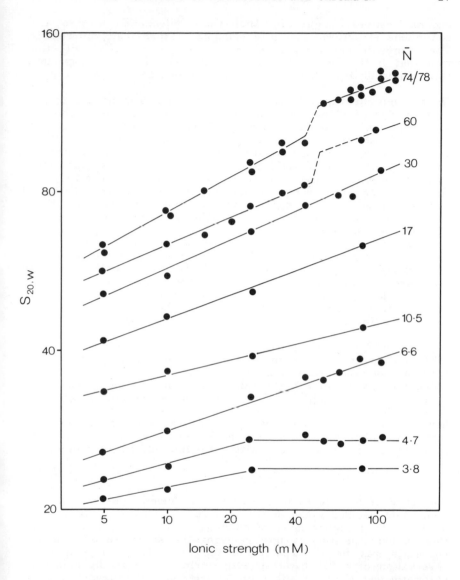

Fig. 13 Dependence of sedimentation rate upon ionic strength for
chromatin fractions of increasing size (Butler and Thomas,
1980).   The overall compaction is shown by the increasing
sedimentation coefficient with ionic strength in all cases:
the significance of the changes in behaviour is discussed
in the text. $\bar{N}$ is average size, in number of
nucleosomes.

six nucleosomes, suggesting that the condensation is into a continuous structure, that is a solenoid, rather than superbeads. The change in behaviour at 6 nucleosomes is identified with the formation of the first stable turn of the solenoid.   As expected, this folding is dependent on the presence of H1.

In rat liver chromatin, there is a further change in behaviour at an average size of 50 nucleosomes long.   Above this length there is an abrupt jump in sedimentation coefficient between ionic strengths 45 and 55 mM (Fig. 13).   This jump is probably due to the solenoid having only weak axial bonding at the lower ionic strengths and hence being susceptible to dynamic shearing.   Support for this interpretation comes from mild cross-linking, which mainly links H1.   This has no effect if carried out at low ionic strength but, if carried out at ionic strength above the jump, it stabilises the compact structure throughout the whole range of ionic strengths. The dynamic shearing would cause adjacent turns temporarily to open apart, accounting for the slower sedimentation and also the fluctuating helix observed by electron microscopy, since the fixation will preserve the gaps between unstacked turns as well as the tight structure.   This interpretation is further supported by the correspondence between the ionic strengths at which the jump in sedimentation coefficient is observed and that above which only continuous fibres are seen in the electron microscope.

Measurements by light scattering (Campbell, Cotter and Pardon, 1978) and neutron scattering (Suau, Bradbury and Baldwin, 1979) from solutions of chromatin give values of the mass per unit length which are consistent with the solenoidal model for the 30 nm fibres, at higher ionic strengths.   The smaller mass per unit length found at lower ionic strengths has been interpreted in terms of "open helices" still with about 6 subunits per turn, but the data do not in fact distinguish between such a model and one for helices with fewer subunits per turn, but with the subunits still in contact.

GENERAL DISCUSSION

Relevance of Structures to Chromatin in Nuclei.   We now have a moderately detailed model of the nucleosome and a description for at least the first higher level of folding, as it occurs in solution. One obvious question is how relevant the structure of soluble chromatin may be to its state in the nucleus.   The original observation of a 200 b.p. spacing during nuclease digestion of chromatin was made on intact nuclei (Hewish and Burgoyne, 1973) and direct lysis of nuclei on the specimen grids for the electron microscope readily yields the "beads-on-a-string" appearance of chromatin.   Since both of these features are easily reproduced from soluble chromatin, there is little doubt that the nucleosome structure is largely unperturbed on extraction of the chromatin.

The usual methods used for the preparation of long chromatin almost certainly disrupt the higher order structure and the question here becomes how accurately it is being reconstructed.   Fibres about 30 nm in diameter are seen in electron micrographs of

sections of fixed and embedded nuclei (Ris and Kubai, 1970) and
also spilling out of lysed nuclei (Labhart and Koller, 1981). These
30 nm fibres coming directly out of nuclei have a very similar
appearance to those formed by refolding of extracted chromatin at
the same ionic strength (Thoma, Koller and Klug, 1979). Low
angle X-ray diffraction has also been used to relate the structure
of chromatin in chromosomes, nuclei and intact cells to the models
derived for chromatin in solution (Langmore and Schutt, 1980).
This shows the presence of a weak 40 nm periodicity in intact cells
or nuclei and that this is directly related to the 30 nm side-by-side
packing of solenoidal chromosome fibres seen in the electron
microscope. While the change in spacing occurs because of
drying, it seems likely that the overall packing is similar in vivo to
that observed in solution, and hence that the models derived from
solution studies are relevant to chromatin in one of its natural
states.

Assembly of Chromatin.   The essential role of the $(H3)_2(H4)_2$
tetramer in the formation of the nucleosome has been discussed
above, and together with two H2A.H2B dimers this can fold the 165
base pairs of DNA into two complete superhelical turns. An H1
molecule could then bind onto the "outside" of the nucleosome, at
the point where the DNA enters and leaves, as it were "sealing off"
the nucleosome and in a position to interact with neighbouring H1
molecules to fold the higher order structure. Such a sequence of
events in time would provide a physical rationale for the temporal
order of assembly observed for histones onto newly replicated DNA
(Senshu, Fukuda and Ohashi, 1978; Worcel, Han and Wong, 1978;
Crémisi and Yaniv, 1980).   However, chromatin will not
"self-assemble" under physiological conditions because of the
overwhelmingly strong interaction between the histones and DNA,
and some additional mechanism is required to slow down this process
to allow time for proper assembly (reviewed by Laskey and
Earnshaw, 1980).   Probably the most biologically relevant
"chaperone" is the protein "nucleoplasmin" which is a highly acidic
polypeptide and catalyses the addition of histones from a pool onto
DNA (Laskey, Mills and Morris, 1977; Laskey et al., 1978).   It
has since been described in nuclei from many species (Krohne and
Franke, 1980).

While nucleoplasmin, or possibly some other similar proteins, can
assemble nucleosomes, these still are not spaced at the same
intervals as in naturally occurring chromatin and do not contain
correctly positioned H1.   The full assembly of chromatin can be
obtained with an extract from oocytes of Xenopus laevis (Laskey,
Mills and Morris, 1977) which contains nucleoplasmin and possibly
some other essential protein(s) yet to be described.   A further
factor affecting the assembly may be the state of modification of the
histones.   Many different modifications have been described
(reviewed by Bradbury and Matthews, 1981) and, while it is
frequently assumed that they are linked to the state of activity of
the chromatin, some of the modifications could well be transiently
present in nascent histones and essential for their assembly into
chromatin (Laskey and Earnshaw, 1980).

<u>Further Levels of Structure.</u>   Although the 30 nm fibre may
represent the state of much of the cellular chromatin throughout a
large part of the cell cycle, it must still be folded further into even
higher levels of structure, up to the chromosome in metaphase
cells.   One obvious possibility for the next level of structure is
helical coiling of the solenoid, which could give a fibre of thickness
compatible with chromosomes, but the evidence there is for it is
rather equivocal (Bak, Zeuthen and Crick, 1977).   Chromosomes
are clearly complicated structures and would certainly require
additional components to organise them.   Indications of how the
higher levels are organised come from several distinct lines of
evidence.   First there is the finding of independently supercoiled
"loops" of DNA containing about 80,000 b.p. in Drosophila
chromatin: these appear to be supercoiled even in the presence of
the core histones and are independently relaxed by nicking with
nuclease (Benyajati and Worcel, 1976).   The preparation containing
these loops, made in 0.9 M sodium chloride, 0.4% Nonidet P40, was
reported to contain RNA and the core histones, H2A, H2B, H3 and
H4, but no H1, which would be washed off by the salt.   Secondly,
detailed studies of the early stages of digestion of nuclei from rat
liver with both micrococcal nuclease and restriction endonucleases
have led to the suggestion that this chromatin may also be
organised into "domains" of between 34,000 and 75,000 b.p.
(Igó-Kemenes and Zachau, 1978).

These size estimates for the domains are interesting as they
correlate roughly with the size of the "chromomeres" described from
cytological studies, and this correlation has led to the suggestion
that each domain might correspond to a single staining band, or
cytogenetic unit, on polytene chromosomes.   While there is no

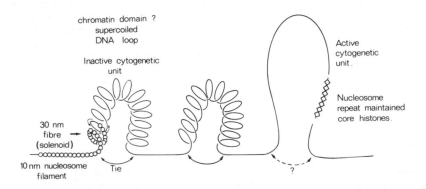

Fig. 14   Highly schematic picture of possible levels of organisation
           in chromatin.   The folding of the 10 nm nucleosome
           filament into the 30 nm fibre (or solenoid) is established,
           but the connection between this, the "radial loops" and the
           cytogenetic units and also the unfolding of the 30 nm fibre
           in active loops is speculative, but based upon images of
           lampbrush chromosomes.

direct proof of this correspondence or of that between a band and a single transcriptional unit, such assumptions allow a picture to be drawn which integrates the various known levels of structure for chromatin (Fig. 14).   In this picture, the nucleosome filament of inactive chromatin is folded solenoidally into the 30 nm fibre, lengths of which are folded back on each other and "tied" at their bases by special RNA or protein to give the loops.   When a given transcriptional unit becomes active, the solenoid would unwind concomitantly with modification of the nucleosomes (see below). This picture is also compatible with the further compaction of chromatin into the condensed metaphase chromosomes, as the "ties" could interact together, either directly or through other non-histone proteins, to give a helical array along the axis of the chromosome.   Such arrays have been observed and called a "scaffold" at the core of chromosomes from which all the histones had been removed (Fig. 15) (Adolph, Cheng and Laemmli, 1977; Paulson and Laemmli, 1977; Adolph, 1980).   Loops of DNA have been observed emanating from the scaffold and these are identical

Fig. 15   Electron micrographs of thin sections of mitotic HeLa cell
          fixed after swelling in hypotonic medium containing Mg[++]
          ions (from Adolph, 1980).   This shows a dense central
          region near the centre of the chromosome arms, with loops
          of chromatin projecting out from this, compatible with a
          model in which the overall morphology of the chromosome is
          determined by a "scaffold" located on its central axis
          (Adolph et al., 1977).   Scale bar represents 0.3 μm.

with the radial loops of the model just outlined (Marsden and Laemmli, 1979).

Transcriptionally Active Chromatin.   All of the foregoing discussion has been concerned with the structure of the bulk of the cellular DNA.   With the exception of highly actively transcribing ribosomal RNA genes (Franke and Scheer, 1978), even actively transcribed chromatin appears to possess a periodic nucleosome structure (Lacy and Axel, 1975; Mathis and Gorovsky, 1976; Bellard, Gannon and Chambon, 1978), but there is evidence that transcribing genes may exist in an altered conformation.   Thus transcribing DNA is preferentially susceptible to digestion by DNase I (Berkovitz and Doty, 1975; Weintraub and Groudine, 1976; Garel and Axel, 1976) and DNase II (Gottesfeld et al., 1974), and micrococcal nuclease cleaves the ovalbumin gene in hen oviduct more rapidly than the globin gene (Bellard, Gannon and Chambon, 1978). While the nucleosome repeat length appears to be inherited in somatic cells (Sperling, Tardieu and Weiss, 1980) and to correlate with the phenotypic expression (Sperling and Weiss, 1980), the average length appears shorter in cells with an increased amount of RNA synthesis (Morris, 1976; Thomas and Thompson, 1977; Lohr et al., 1977; Weintraub, 1978).   Moreover, it has recently been shown that the repeat on the actively transcribing DNA may be shorter than the average for the cell (Berkovitz and Riggs, 1981).

These changes in properties of transcribing chromatin correlate with an altered protein content.   Thus active chromatin appears to have a reduced content of H1 (Levy-Wilson and Dixon, 1979), which taken together with the requirement for H1 for formation of higher order structure would be compatible with a looser, less folded conformation for active regions.   Various non-histone proteins also appear to be concentrated on the active regions of the gene, in particular high mobility group proteins (HMG) 14 and 17 are not only found bound to actively transcribing chromatin, but also will selectively rebind such regions after being dissociated by raised salt concentrations (Weisbrod and Weintraub, 1979; Weisbrod, Groudine and Weintraub, 1980).

Besides the enhanced nuclease sensitivity of active regions of the chromatin, these may also contain "hypersensitive" regions (Wu et al., 1979; Stalder et al., 1980).   Unlike the general increased sensitivity, which is found both some time before and for a period after chromatin is transcribed, the hypersensitive sites appear to occur only during the active period for a region of chromatin, i.e. when a gene or set of genes is being expressed, and may well involve some regions of protein-free DNA within the hypersensitive site (McGhee et al., 1981).   A similar region of free DNA occurs in the minichromosomes of simian virus 40 and polyoma, around the origin of replication on the viral DNA (Varshavsky, Sundin and Bohn, 1979; Herbomel et al., 1981), but in this case the DNA appears not to be protected from nuclease at any time.   The absence of nucleosomes at specific sites on the DNA may thus come about in different ways, depending upon whether it is permanent, and could thus be due to a DNA sequence incompatible with packaging, or whether it occurs only at particular times, as required for the functioning of the genes. In this latter case, the uniformity of the histone complement suggests that the specificity will have to be directed by other proteins which are more variable and can therefore recognise both specific regions of the DNA and also signals for when they should act.

Conclusion. We are thus at an exciting stage in our understanding of the processes which take place in chromatin, where we at last have a picture of its underlying chemistry and structure. It is to be expected that the next stages will lead us to a more detailed knowledge so that we will begin to see the actual interactions bringing about the folding and how these are modified to change the structure as chromatin functions in the life of the cell.

ACKNOWLEDGEMENTS

We are grateful to Drs. K.W. Adolph and F. Thoma for kindly supplying us with prints of their electron micrographs.

We also thank the publishers for permission to reproduce prints of the following figures: Fig. 2 from Philosophical Transactions of the Royal Society, London; Fig. 3 from Journal of Molecular Biology and from Nature; Fig. 5 from Nature; Fig. 10 from Science and from Nucleic Acids Research; Fig. 15 from Experimental Cell Research.

REFERENCES

Adolph, K.W., 1980. Organization of chromosomes in mitotic HeLa cells. Exp. Cell Res. 125, 95-103.
Adolph, K.W., Cheng, S.M. and Laemmli, U.K., 1977. Role of nonhistone proteins in metaphase chromosome structure. Cell 12, 805-816.
Anderson, W.F., Ohlendorf, D.H., Takeda, Y. and Matthews, B.W., 1981. Structure of the cro repressor from bacteriophage λ and its interaction with DNA. Nature (Lond.) 290, 754-758.
Axel, R., 1975. Cleavage of DNA in nuclei and chromatin with staphylococcal nuclease. Biochemistry 14, 2921-2925.
Bak, A.L., Zeuthen, J. and Crick, F.H.C., 1977. Higher-order structure of human mitotic chromosomes. Proc. Nat. Acad. Sci. U.S.A. 74, 1595-1599.
Bakayev, V.V., Melnickov, A.A., Osicka, V.D. and Varshavsky, A.J., 1975. Studies on chromatin. II. Isolation and characterization of chromatin subunits. Nucleic Acids Res. 2, 1401-1419.
Bellard, M., Gannon, F. and Chambon, P., 1978. Nucleosome structure III: The structure and transcriptional activity of the chromatin containing the ovalbumin and globin genes in chick oviduct nuclei. Cold Spring Harb. Symp. Quant. Biol. 42, 779-791.

Bentley, G.A., Finch, J.T. and Lewit-Bentley, A., 1981. Neutron diffraction studies on crystals of nucleosome cores using contrast variation. J. Mol. Biol. 145, 771-784.

Benyajati, C. and Worcel, A., 1976. Isolation, characterization, and structure of the folded interphase genome of Drosophila melanogaster. Cell 9, 393-407.

Berkovitz, E.M. and Doty, P., 1975. Chemical and physical properties of fractionated chromatin. Proc. Nat. Acad. Sci. U.S.A. 72, 3328-3332.

Berkovitz, E.M. and Riggs, E.A., 1981. Characterization of rat liver oligonucleosomes enriched in transcriptionally active genes: Evidence for altered base composition and a shortened nucleosome repeat. Biochemistry 20, 7284-7290.

Bradbury, E.M. and Matthews, H.R., 1981. Histone variants, histone modifications and chromatin structure. Proc. 2nd SUNYA Conv. Disc. Biomolec. Stereodynamics (ed. R.H. Sarma) 2, 125-144.

Burgoyne, L.A., Hewish, D.R. and Mobbs, J., 1974. Mammalian chromatin substructure studies with the Ca-Mg endonuclease and 2-dimensional polyacrylamide-gel electrophoresis. Biochem. J. 143, 67-72.

Butler, P.J.G. and Thomas, J.O., 1980. Changes in chromatin folding in solution. J. Mol. Biol. 140, 505-529.

Camerini-Otero, R.D. and Felsenfeld, G., 1977a. Supercoiling energy and nucleosome formation: The role of the arginine-rich histone kernel. Nucleic Acids Res. 4, 1159-1181.

Camerini-Otero, R.D. and Felsenfeld, G., 1977b. Histone H3 disulfide dimers and nucleosome structure. Proc. Nat. Acad. Sci. U.S.A. 74, 5519-5523.

Camerini-Otero, R.D., Sollner-Webb, B. and Felsenfeld, G., 1976. The organization of histones and DNA in chromatin: Evidence for an arginine-rich histone kernel. Cell 8, 333-347.

Campbell, A.M. and Cotter, R.I., 1976. The molecular weight of nucleosome protein by laser light scattering. FEBS Lett. 70, 209-211.

Campbell, A.M., Cotter, R.I. and Pardon, J.F., 1978. Light scattering measurements supporting helical structures for chromatin in solution. Nucleic Acids Res. 5, 1571-1580.

Caron, F. and Thomas, J.O., 1981. Exchange of histone H1 between segments of chromatin. J. Mol. Biol. 146, 513-537.

Caspar, D.L.D. and Klug, A., 1962. Physical principles in the construction of regular viruses. Cold Spring Harb. Symp. Quant. Biol. 27, 1-24.

Chao, M.V., Gralla, J. and Martinson, H.G., 1979. DNA sequence directs placement of histone cores on restriction fragments during nucleosome formation. Biochemistry 18, 1068-1074.

Chung, S.-Y., Hill, W.E. and Doty, P., 1978. Characterization of the histone core complex. Proc. Nat. Acad. Sci. U.S.A. 75, 1680-1684.

Clark, R.J. and Felsenfeld, G., 1971. Structure of chromatin. Nature New Biol. 229, 101-106.

Crémisi, C. and Yaniv, M., 1980.   Sequential assembly of newly synthesized histones on replicating SV40 DNA.   Biochem. Biophys. Res. Commun. 92, 1117-1123.

Crick, F.H.C., 1976.   Linking number and nucleosomes.   Proc. Nat. Acad. Sci. USA 73, 2639-2643.

De Lange, R.J. and Smith, E.L., 1971.   Histones: Structure and function.   Ann. Rev. Biochem. 40, 279-314.

de Murcia, G. and Koller, Th., 1981.   The electron microscopic appearance of soluble rat liver chromatin mounted on different supports.   Biol. Cell 40, 165-174.

Dieterich, A.E., Eshaghpour, H., Crothers, D.M. and Cantor, C.R., 1980.   Effect of DNA length on the nucleosome low salt transition.   Nucleic Acids Res. 8, 2475-2487.

Eickbush, T.H. and Moudrianakis, E.N., 1978.   The histone core complex: An octamer assembled by two sets of protein - protein interactions.   Biochemistry 17, 4955-4964.

Engel, A., Sütterlin, S. and Koller, Th., 1980.   Estimation of the mass per unit length of soluble chromatin using the STEM. Proc. 7th Eur. Congr. Electron Microscopy (ed. P. Brederoo and W. de Priester) 2, pp. 548-549.   The Hague.

Finch, J.T. and Klug, A., 1976.   Solenoidal model for superstructure in chromatin.   Proc. Nat. Acad. Sci. U.S.A. 73, 1897-1901.

Finch, J.T., Lutter, L.C., Rhodes, D., Brown, R.S., Rushton, B., Levitt, M. and Klug, A., 1977.   Structure of nucleosome core particles of chromatin.   Nature (Lond.) 269, 29-36.

Finch, J.T., Lewit-Bentley, A., Bentley, G.A., Roth, M. and Timmins, P.A., 1980.   Neutron diffraction from crystals of nucleosome core particles.   Phil. Trans. Roy. Soc. Lond. B 290, 635-638.

Finch, J.T., Brown, R.S., Rhodes, D., Richmond, T., Rushton, B., Lutter, L.C. and Klug, A., 1981.   X-ray diffraction study of a new crystal form of the nucleosome core showing higher resolution.   J. Mol. Biol. 145, 757-769.

Franke, W.W. and Scheer, U., 1978.   Morphology of transcriptional units at different states of activity.   Phil. Trans. Roy. Soc. Lond. B 283, 333-342.

Garel, A. and Axel, R., 1976.   Selective digestion of transcriptionally active ovalbumin genes from oviduct nuclei. Proc. Nat. Acad. Sci. U.S.A. 73, 3966-3970.

Germond, J.E., Hirt, B., Oudet, P., Gross-Bellard, M. and Chambon, P., 1975.   Folding of the DNA double helix in chromatin-like structures from Simian Virus 40.   Proc. Nat. Acad. Sci. U.S.A. 72, 1843-1847.

Goodwin, G.H., Walker, J.M. and Johns, E.W. 1978.   The high mobility group (HMG) nonhistone proteins.   In "The Cell Nucleus" (ed. H. Busch), 6, 182-219.   Academic Press, New York.

Gordon, V.C., Knobler, C.M., Olins, D.E. and Schumaker, V.N., 1978.   Conformational changes of the chromatin subunit.   Proc. Nat. Acad. Sci. U.S.A. 75, 660-663.

Gottesfeld, J.M., Garrard, W.T., Bagi, G., Wilson, R.F. and Bonner, J., 1974. Partial purification of the template-active fraction of chromatin: A preliminary report. Proc. Nat. Acad. Sci. U.S.A. 71, 2193-2197.

Hagerman, P.J., 1981. Investigation of the flexibility of DNA using transient electric birefringence. Biopolymers 20, 1503-1535.

Herbomel, P., Saragosti, S., Blangy, D. and Yaniv, M., 1981. Fine structure of the origin-proximal DNAase I - hypersensitive region in wild-type and EC mutant polyoma. Cell 25, 651-658.

Hewish, D.R. and Burgoyne, L.A., 1973. Chromatin substructure. The digestion of chromatin DNA at regularly spaced sites by a nuclear deoxyribonuclease. Biochem. Biophys. Res. Commun. 52, 504-510.

Hjelm, R.P., Kneale, G.G., Suau, P., Baldwin, J.P. and Bradbury, E.M., 1977. Small angle neutron scattering studies of chromatin subunits in solution. Cell 10, 139-151.

Hozier, J., Renz, M. and Nehls, P., 1977. The chromosome fiber: Evidence for an ordered superstructure of nucleosomes. Chromosoma 62, 301-317.

Huntley, G. and Dixon, G.H., 1972. Histone T sequence. J. Biol. Chem. 247, 4916-4919.

Igó-Kemenes, T., Omori, A. and Zachau, H.G., 1980. Non-random arrangement of nucleosomes in satellite I containing chromatin of rat liver. Nucleic Acids Res. 8, 5377-5390.

Igó-Kemenes, T. and Zachau, H.G., 1978. Domains in chromatin structure. Cold Spring Harb. Symp. Quant. Biol. 42, 109-118.

Isenberg, I., 1979. Histones. Ann. Rev. Biochem. 48, 159-191.

Jorcano, J.L., Meyer, G., Day, L.A. and Renz, M., 1980. Aggregation of small oligonucleosomal chains into 300-Å globular particles. Proc. Nat. Acad. Sci. U.S.A. 77, 6443-6447.

Klug, A. and Lutter, L.C., 1981. The helical periodicity of DNA on the nucleosome. Nucleic Acids Res. 9, 4267-4283.

Klug, A., Rhodes, D., Smith, J., Finch, J.T. and Thomas, J.O., 1980. A low resolution structure for the histone core of the nucleosome. Nature (Lond.) 287, 509-516.

Kornberg, R.D., 1974. Chromatin structure: A repeating unit of Histones and DNA. Science 184, 868-871.

Krohne, G. and Franke, W.W., 1980. A major soluble acidic protein located in nuclei of diverse vertebrate species. Exptl. Cell Res. 129, 167-189.

Labhart, P. and Koller, Th., 1981. Electron microscope specimen preparation of rat liver chromatin by a modified Miller spreading technique. Eur. J. Cell Biol. 24, 309-316.

Lacy, E. and Axel, R., 1975. Analysis of DNA of isolated chromatin subunits. Proc. Nat. Acad. Sci. U.S.A. 72, 3978-3982.

Langmore, J.P. and Schutt, C., 1980. The higher order structure of chicken erythrocyte chromosomes in vivo. Nature (Lond.) 288, 620-622.

Laskey, R.A. and Earnshaw, W.C., 1980. Nucleosome assembly. Nature (Lond.) 286, 763-767.

Laskey, R.A., Mills, A.D. and Morris, N.R., 1977.   Assembly of SV40 chromatin in a cell-free system from Xenopus eggs.   Cell 10, 237-243.

Laskey, R.A., Honda, B.M., Mills, A.D. and Finch, J.T., 1978. Nucleosomes are assembled by an acidic protein which binds to histones and transfers them to DNA.   Nature (Lond.) 275, 416-420.

Levy-Wilson, B. and Dixon, G.H., 1979.   Limited action of micrococcal nuclease on trout testis nuclei generates two mononucleosome subsets enriched in transcribed DNA sequences. Proc. Nat. Acad. Sci. U.S.A. 76, 1682-1686.

Lippard, S.J. and Hoeschele, J.D. 1979.   Binding of cis- and trans-dichlorodiammineplatinum(II) to the nucleosome core. Proc. Nat. Acad. Sci. USA 76, 6091-6095.

Lohr, D., Corden, J., Tatchell, K., Kovacic, R.T. and Van Holde, K.E., 1977.   Comparative subunit structure of HeLa, ycast, and chicken erythrocyte chromatin.   Proc. Nat. Acad. Sci. U.S.A. 74, 79-83.

Lutter, L.C., 1977.   Deoxyribonuclease I produces staggered cuts in the DNA of chromatin.   J. Mol. Biol. 117, 53-69.

Lutter, L.C., 1978.   Kinetic analysis of deoxyribonuclease I cleavages in the nucleosome core: Evidence for a DNA superhelix.   J. Mol. Biol. 124, 391-420.

Lutter, L.C., 1979.   Precise location of DNase I cutting sites in the nucleosome core determined by high resolution gel electrophoresis.   Nucleic Acids Res. 6, 41-56.

Lutter, L.C., 1981.   DNase II digestion of the nucleosome core: Precise locations and relative exposures of sites.   Nucleic Acids Res. 9, 4251-4265.

McGhee, J.D. and Felsenfeld, G., 1980.   The number of charge-charge interactions stabilizing the ends of nucleosome DNA.   Nucleic Acids Res. 8, 2751-2769.

McGhee, J.D., Wood, W.I., Dolan, M., Engel, J.D. and Felsenfeld, G., 1981.   A 200 base pair region at the 5' end of the chicken adult β-globin gene is accessible to nuclease digestion.   Cell 27, 45-55.

McKay, D.B. and Steitz, T.A., 1981.   Structure of catabolite activator protein at 2.9 Å resolution suggests binding to left-handed B-DNA.   Nature (Lond.) 290, 744-749.

Marsden, M.P.F. and Laemmli, U.K., 1979.   Metaphase chromosome structure: Evidence for a radial loop model.   Cell 17, 849-858.

Martinson, H.G., True, R.J. and Burch, J.B.E., 1979.   Specific histone-histone contacts are ruptured when nucleosomes unfold at low ionic strength.   Biochemistry 18, 1082-1089.

Mathis, D.J. and Gorovsky, M.A., 1976.   Subunit structure of rDNA-containing chromatin.   Biochemistry 15, 750-755.

Mirzabekov, A.D., Shick, V.V., Belyavsky, A.V. and Bavykin, S.G., 1978.   Primary organization of nucleosome core particle of chromatin: Sequence of histone arrangement along DNA.   Proc. Nat. Acad. Sci. U.S.A. 75, 4184-4188.

Morris, N.R., 1976.   A comparison of the structure of chicken erythrocyte and chicken liver chromatin.   Cell 9, 627-632.

Noll, M., 1974.   Subunit structure of chromatin.   Nature (Lond.) 251, 249-251.

Noll, M. and Kornberg, R.D., 1977.   Action of micrococcal nuclease on chromatin and the location of histone H1.   J. Mol. Biol. 109, 393-404.

Olins, A.L. and Olins, D.E., 1974.   Spheroid chromatin units (ν bodies).   Science 183, 330-332.

Olins, D.E. and Wright, E.B., 1973.   Glutaraldehyde fixation of isolated eukaryotic nuclei.   Evidence for histone-histone proximity.   J. Cell Biol. 59, 304-317.

Oudet, P., Gross-Bellard, M. and Chambon, P., 1975.   Electron microscopic and biochemical evidence that chromatin structure is a repeating unit.   Cell 4, 281-300.

Pardon, J.F. and Wilkins, M.H.F., 1972.   A super-coil model for nucleohistone.   J. Mol. Biol. 68, 115-124.

Pardon, J.F., Wilkins, M.H.F. and Richards, B.M., 1967. Super-helical model for nucleohistone.   Nature (Lond.) 215, 508-509.

Pardon, J.F., Cotter, R.I., Lilley, D.H.J., Worcester, D.L., Campbell, A.H., Wooley, J.C. and Richards, B.H., 1978. Scattering studies of chromatin subunits.   Cold Spring Harb. Symp. Quant. Biol. 42, 11-22.

Paulson, J.R. and Laemmli, U.K., 1977.   The structure of histone-depleted metaphase chromosomes.   Cell 12, 817-828.

Peck, L.J. and Wang, J.C., 1981.   Sequence dependence of the helical repeat of DNA in solution.   Nature (Lond.) 292, 375-378.

Prunell, A. and Kornberg, R.D., 1982.   Variable center to center distance of nucleosomes in chromatin.   J. Mol. Biol. 154, 515-523.

Prunell, A., Kornberg, R.D., Lutter, L., Klug, A., Levitt, M. and Crick, F.H.C., 1979.   Periodicity of deoxyribonuclease I digestion of chromatin.   Science 204, 855-858.

Renz, M., 1979.   Heterogeneity of the chromosome fiber.   Nucleic Acids Res. 6, 2761-2767.

Renz, M., Nehls, P. and Hozier, J., 1977a.   Histone H1 involvement in the structure of the chromosome fiber.   Cold Spring Harb. Symp. Quant. Biol. 42, 245-252.

Renz, M., Nehls, P. and Hozier, J., 1977b.   Involvement of histone H1 in the organisation of the chromosome fiber.   Proc. Nat. Acad. Sci. U.S.A. 74, 1879-1883.

Rhodes, D., and Klug, A., 1980.   Helical periodicity of DNA determined by enzyme digestion.   Nature (Lond.) 286, 573-578.

Rhodes, D. and Klug, A., 1981.   Sequence-dependent helical periodicity of DNA.   Nature (Lond.) 292, 378-380.

Richards, B., Pardon, J., Lilley, D., Cotter, R. and Worcester, D., 1977.   The sub-structure of nucleosomes.   Cell Biol. Int. Rep. 1, 107-116.

Ris, H. and Kubai, D.F., 1970.   Chromosome structure.   Ann. Rev. Genet. 4, 263-294.

Ruiz-Carrillo, A., Puigdomènech, P., Eder, G. and Lurz, R., 1980.   Stability and reversibility of higher ordered structure of interphase chromatin: Continuity of deoxyribonucleic acid is not required for maintenance of folded structure.   Biochemistry 19, 2544-2554.

Senshu, T., Fukuda, M. and Ohashi, M., 1978.   Preferential association of newly synthesized H3 and H4 histones with newly replicated DNA.   J. Biochem (Japan) 84, 985-988.

Shaw, B.R., Herman, T.M. Kovacic, R.T., Beaudreau, G.S. and Van Holde, K.E., 1976.   Analysis of subunit organization in chicken erythrocyte chromatin.   Proc. Nat. Acad. Sci. U.S.A. 73, 505-509.

Shick, V.V., Belyavsky, A.V., Bavykin, S.G. and Mirzabekov, A.D., 1980.   Primary organization of the nucleosome core particles.   Sequential arrangement of histones along DNA. J. Mol. Biol. 139, 491-517.

Shore, D., Langowski, J. and Baldwin, R.L., 1981.   DNA flexibility studied by covalent closure of short fragments into circles.   Proc. Nat. Acad. Sci. U.S.A. 78, 4833-4837.

Simpson, R.T., 1978.   Structure of the chromatosome, a chromatin particle containing 160 base pairs of DNA and all the histones. Biochemistry 17, 5524-5531.

Sollner-Webb, B. and Felsenfeld, G., 1975.   A comparison of the digestion of nuclei and chromatin by staphylococcal nuclease. Biochemistry 14, 2915-2920.

Sollner-Webb, B. and Felsenfeld, G., 1977.   Pancreatic DNAase cleavage sites in nuclei.   Cell 10, 537-547.

Sperling, L. and Klug, A., 1977.   X-ray studies on "native" chromatin.   J. Mol. Biol. 112, 253-263.

Sperling, L. and Tardieu, A., 1976.   The mass per unit length of chromatin by low-angle X-ray scattering.   FEBS Lett. 64, 89-91.

Sperling, L. and Weiss, M.C., 1980.   Chromatin repeat length correlates with phenotypic expression in hepatoma cells, their dedifferentiated variants, and somatic hybrids.   Proc. Nat. Acad. Sci. U.S.A. 77, 3412-3416.

Sperling, L., Tardieu, A. and Weiss, M.C., 1980.   Chromatin repeat length in somatic hybrids.   Proc. Nat. Acad. Sci. U.S.A. 77, 2716-2720.

Stalder, J., Larsen, A., Engel, J.D., Dolan, M., Groudine, M. and Weintraub, H., 1980.   Tissue-specific DNA cleavages in the globin chromatin domain introduced by DNAase I.   Cell 20, 451-460.

Stein, A., Bina-Stein, M. and Simpson, R.T., 1977.   Cross-linked histone octamer as a model of the nucleohistone core.   Proc. Nat. Acad. Sci. U.S.A. 74, 2780-2784.

Steinmetz, M., Streeck, R.E. and Zachau, H.G., 1978.   Closely spaced nucleosome cores in reconstituted histone.DNA complexes and histone-H1-depleted chromatin.   Eur. J. Biochem. 83, 615-628.

Stellwagen, R.H. and Cole, R.D., 1969.   Chromosomal proteins. Ann. Rev. Biochem. 38, 951-990.

Stockley, P.G. and Thomas, J.O., 1979.   A nucleosome-like particle containing an octamer of the arginine-rich histones H3 and H4.   FEBS Lett. 99, 129-135.

Strätling, W.H., 1979.   Role of histone H1 in the conformation of oligonucleosomes as a function of ionic strength.   Biochemistry 18, 596-603.

Strätling, W.H., Müller, U. and Zentgraf, H., 1978. The higher order repeat structure of chromatin is built up of globular particles containing eight nucleosomes. Exptl. Cell Res. 117, 301-311.

Suau, P., Kneale, G.G., Braddock, G.W., Baldwin, J.P. and Bradbury, E.M., 1977. A low resolution model for the chromatin core particle by neutron scattering. Nucleic Acids Res. 4, 3769-3786.

Suau, P., Bradbury, E.M. and Baldwin, J.P., 1979. Higher-order structures of chromatin in solution. Eur. J. Biochem. 97, 593-602.

Thoma, F., Koller, Th. and Klug, A., 1979. Involvement of histone Hl in the organization of the nucleosome and of the salt-dependent superstructures of chromatin. J. Cell Biol. 83, 403-427.

Thomas, J.O. and Butler, P.J.G., 1977. Characterization of the octamer of histones free in solution. J. Mol. Biol. 116, 769-781.

Thomas, J.O. and Butler, P.J.G., 1978. The nucleosome core protein. Cold Spring Harb. Symp. Quant. Biol. 42, 119-125.

Thomas, J.O. and Butler, P.J.G., 1980. Size-dependence of a stable higher-order structure of chromatin. J. Mol. Biol. 144, 89-93.

Thomas, J.O. and Khabaza, A.J.A., 1980. Cross-linking of histone Hl in chromatin. Eur. J. Biochem. 112, 501-511.

Thomas, J.O. and Kornberg, R.D., 1975a. An octamer of histones in chromatin and free in solution. Proc. Nat. Acad. Sci. U.S.A. 72, 2626-2630.

Thomas, J.O. and Kornberg, R.D., 1975b. Cleavable cross-links in the analysis of histone-histone associations. FEBS Lett. 58, 353-358.

Thomas, J.O. and Thompson, R.J., 1977. Variation in chromatin structure in two cell types in the same tissue: A short DNA repeat length in cerebral cortex neurons. Cell 10, 633-640.

Varshavsky, A.J., Sundin, O. and Bohn, M., 1979. A stretch of "late" SV40 viral DNA about 400 bp long which includes the origin of replication is specifically exposed in SV40 minichromosomes. Cell 16, 453-466.

Walker, J.M., Hastings, J.R.B. and Johns, E.W., 1977. The primary structure of a non-histone chromosomal protein. Eur. J. Biochem. 76, 461-468.

Wang, J.C., 1979. Helical repeat of DNA in solution. Proc. Nat. Acad. Sci. U.S.A. 76, 200-203.

Weintraub, H., 1978. The nucleosome repeat length increases during erythropoiesis in the chick. Nucleic Acids Res. 5, 1179-1188.

Weintraub, H. and Groudine, M., 1976. Chromosomal subunits in active genes have an altered conformation. Science 193, 848-856.

Weintraub, H., Palter, K. and Van Lente, F., 1975. Histones H2A, H2B, H3, and H4 form a tetrameric complex in solutions of high salt. Cell 6, 85-110.

Weintraub, H., Worcel, A. and Alberts, B., 1976.   A model for chromatin based upon two symmetrically paired half-nucleosomes. Cell 9, 409-417.

Weisbrod, S. and Weintraub, H., 1979.   Isolation of a subclass of nuclear proteins responsible for conferring a DNase I-sensitive structure on globin chromatin.   Proc. Nat. Acad. Sci. U.S.A. 76, 630-634.

Weisbrod, S., Groudine, M. and Weintraub, H., 1980.   Interaction of HMG 14 and 17 with actively transcribed genes.   Cell 19, 289-301.

Weischet, W.O., Tatchell, K., Van Holde, K.E. and Klump, H., 1978.   Thermal denaturation of nucleosomal core particles. Nucleic Acids Res. 5, 139-160.

Whitlock, J.P. and Stein, A., 1978.   Folding of DNA by histones which lack their NH_2-terminal regions.   J. Biol. Chem. 253, 3857-3861.

Wilkins, M.H.F., Zubay, G. and Wilson, H.R., 1959.   X-ray diffraction studies of the molecular structure of nucleohistones and chromosomes.   J. Mol. Biol. 1, 179-185.

Worcel, A., Han, S. and Wong, M.L., 1978.   Assembly of newly replicated chromatin.   Cell 15, 969-977.

Wu, C., Bingham, P.M., Livak, K.J., Holmgren, R. and Elgin, S.C.R., 1979.   The chromatin structure of specific genes: I. The evidence for higher order domains of defined DNA sequence.   Cell 16, 797-806.

Genes: *Structure and Expression*
Edited by A. M. Kroon
© 1983 John Wiley & Sons Ltd.

ACTIVATION AND FUNCTION OF CHROMATIN

P. N. Bryan
Institut für Molekularbiologie II
der Universität Zürich
Zürich, Switzerland

O. H. J. Destree
Anatomisch - Embryologisch Laboratorium
Universiteit van Amsterdam
Amsterdam, The Netherlands

The packaging of eucaryotic DNA into the nucleus involves successive levels of folding mediated by the histones as well as other proteins (1). In addition to compacting the considerable length of DNA to manageable dimensions, chromatin appears also to provide an orderly filing system permitting selective retrieval of information by the transcriptional machinery. Structural changes which occur in chromatin in the process of establishing a transcriptionally active template will be the focal point of this review. A complete understanding of chromatin activation, however, is limited from the beginning since only the first level of chromatin structure, the nucleosome, is understood in any detail (2). The preponderance of recent work has concerned itself with such questions as what happens to nucleosome conformation during activation, and, are distinctive chromatin structures associated with DNA sequences regulatory in transcription? We emphasize these recent studies in our coverage of the topic. Chromatin activation, of course, involves changes in higher order structure as well as on the nucleosomal level though both are probably interrelated. In an attempt to provide a semblence of balance to our discussion, we will begin by mentioning several long-established correlations of highly condensed chromatin with transcriptional inactivity.

HETEROCHROMATIN

Cycles of condensation and decondensation during the cell cycle coincide with periods of relative transcriptional inactivity and activity (3). Following telophase most but not all of the chromatin returns to a dispersed state, called euchromatic, as RNA synthesis increases. Some regions of chromosomes do not decondense after mitosis resulting in a pattern of heterochromatin always occurring at characteristic locations in homologous chromosomes. This type of condensed chromatin is never activated in either chromosome of a pair in any daughter cell and is therefore called constitutive heterochromatin. It has been found that when chromosome segments are translated from euchromatic to heterochromatic regions, the genes involved become inactivated (position effect variegation (4)). It is thought that a structural "effect" is spreading from the heterochromatic to the euchromatic material. This effect is usually inherited by the daughter cells, but, in some instances the inactivated chromatin region is gradually reactivated.

Another example of the correlation between condensation and inactivity is the phenomenon of X-chromosome inactivation (5). At an early stage in mammalian development one of the two X-chromosomes does not return to the decondensed state and will never be reactivated in any of the somatic daughter cells. In germ line cells, however, the heterochromatic X-chromosome decondenses and is reactivated at sometime during the onset of meiosis. The mechanisms of chromosome condensation and reactivation are unknown.

Nuclei of mature erythrocytes of birds and amphibians have highly condensed chromatin which is inactive in replication and transcription (3). The histone composition of these nuclei is clearly different from that of other cell types especially for its lysine-rich histones. A long-standing speculation is that the specific lysine-rich histone (H5) is involved with the inactivation of the chromatin. This is not a simple relationship as H5 is already present in progenitor erythroblasts of chicken before cessation of nucleic acid synthesis (6). It is thought that a sequential phosphorylation and dephosphorylation of the H5 histone during erythrogenesis determines the condensation of the chromatin and thereby its activation (7).

When chicken erythrocytes are fused with HeLa cells the red cell nucleus becomes reactivated in transcription and replication while it starts to swell. This reactivation is dependent on the synthesis and transport of specific nuclear proteins from the cytoplasm to the nucleus (8). It is not known what the early events in the reactivated erythrocyte chromatin are and what their influence on chromatin structure and transcription might be.

BIOCHEMICAL CHARACTERIZATION

Activation of chromatin by decondensation of heterochromatin is a course level of gene regulation since only a fraction of euchromatin is active in transcription. The transcribable subset of chromatin must possess other distinguishing characteristics. Two decades ago molecules regulating transcription were thought to act on essentially naked DNA as in bacteria and regions of DNA complexed with histones were thereby inactivated (9). In the latter part of the seventies, however, it was established that almost all eukaryotic DNA, including at least some sequences active in transcription are packaged into nucleosomes. Evidence for this point comes both from biochemical and electron microscopic evidence (3,10). The most notable examples of non-nucleosomal DNA are ribosomal cistrons during periods of intense transcription (3,10). This class of genes will be discussed later in more detail. The fact that in most cases a conserved nucleosome structure is shared by both regions of DNA, active and inactive in transcription, raises the question of what additional components and structures are specific to active regions which identify them for selective transcription. Since the transcribed portion of the genome is generally only a small fraction of the total, its characterization depends upon methods which can selectively detect components and structures associated with "active regions."

Determination of the structure of active chromatin has been approached both by electron microscopic examination of transcription complexes and by partial separation of chromatin into transcribed and nontranscribed fractions. Both these approaches have been comprehensively reviewed (10,11) so we will only summarize a few points about the composition of active chromatin.

Most biochemical studies of chromatin enriched in transcribed sequences indicate that the stoichiometry of the four core histones and the histone:DNA ratio are similar in active and total chromatin (10). More controversial is how the Hl class of histones is associated with active chromatin. Because Hl histones are particularly sensitive to proteolysis and exchange, measurement of Hl stoichiometry in active fractions is difficult. Since Hl histones are required for higher order folding, their absence or lower content in active chromatin could explain the apparent lack of higher order folding seen by electron microscopy of transcription complexes.

It has been conjectured that postsynthetic modifications of histones known to occur in chromatin (acetylation, phosphorylation, methylation, ribosylation and ubiquitination) are involved in the modulation of chromatin architecture and therefore in the regulation of gene expression. Most of the evidence for this concept is corollary. Acetylation and ubiquitination have gained most recent attention and will therefore be briefly mentioned.

Acetylation of histones occurs in vivo on lysine residues in the basic N-terminal regions of all four nucleosomal histones. Acetylation lessens the positive charge of histones, and may result in loosening electrostatic histone:DNA interactions. Correlations between an increase of histone acetylation and increased RNA synthesis have been numerous (10). Recent work has shown that active globin gene-containing nucleosomes are enriched in acetylated H3 and H4 although not stoichimetrically (12). If acetylation is involved in establishing active chromatin structure, then apparently not every nucleosome need contain acetylated histones.

Ubiquitin is a 76 residue protein present ubiquitously in all organisms tested. In the chromatin of eucaryotes it can covalently bind to approximately 5 to 15% of the nucleosomal H2A (13). One or both H2A molecules in a nucleosome can be ubiquitinated and less frequently H2B has been found ubiquitinated. The binding of ubiquitin to H2A is cell-cycle dependent. It is removed from nucleosomes shortly before metaphase and rebound in G1 phase. Ubiquibinated H2A is almost missing in nontranscribed satellite chromatin (1 in 25 nucleosomes) while one in two nucleosomes is ubiquitinated in heat shock genes of Drosophila (14). The consequence of ubiquitination might be a weakening of the interactions between neighboring nucleosomes as a step in unfolding higher order chromosomal structures prior to transcription.

Among the nonhistone proteins of chromatin, the "high-mobility group" proteins (HMG's) are the best characterized (15). These proteins are small (about 30,000 MW), found in many different eucaryotic orgainisms and lack substantial species and tissue specificity. The relative amounts of these proteins differ from tissue to tissue but no more than 10% of the nucleosomes in a particular cell contain HMG's. The HMG's have an unusual amino-acid composition with about 30% acidic and 25% basic residues. HMG proteins are extracted with 5% PCA or 0.35 M NaCl from nuclei or chromatin.

Recently a method has been devised (16) which
demonstrates that two of the HMG proteins interact
specifically with active chromatin. HMG 14 and 17 can
be cross-linked to agarose (or glass beads) and used as
an affinity column for isolation of active globin
gene-containing nucleosomes. Mononucleosomes (free of
HMG proteins and Hl by washing in 0.55 M NaCl) were
isolated using this procedure from different cell types
of chicken and frog (12,16). In both animals,
nucleosomes containing globin coding sequences can be
selectively fish out from total mononucleosomes if they
are isolated from erythroid cells, active or previously
active in globin synthesis.

What the effects of HMG 14 and 17 are on the nucleosome
structure is not clear. Studies on HMG 14 and 17
binding to total nucleosomes show that they bind to two
specific sites on the nucleosome core (17,18). The
packing of polynucleosomes into the higher order
chromatin fibers is not interfered by HMG 14 and 17,
suggesting that they do not induce an extended
configuration of chromatin (19).

STUDIES ON SPECIFIC GENES

A quantum jump in the understanding of active chromatin
structure has been attained using nucleases as probes
of chromatin containing specific DNA sequences. We
will discuss in detail some of the specific gene
systems which have been studied to date. This approach
involves straightforward extension of the nuclease
digestion techniques used to analyse total chromatin
but depends on the availability of radioactively
labeled hybridization probes allowing selective
detection of specific sequences. The most commonly
used nuclease probes are micrococcal nuclease and
DNAase I (20,21). Micrococcal nuclease makes double-
stranded cuts specifically in linker DNA between
nucleosome cores and can therefore provide information
about nucleosomal organization. DNAase I introduces
single-stranded nicks with only slight preference for
linker over core DNA, showing the relative
accessibility of DNA in chromatin to a macromolecular
probe.

The first studies of specific genes began in 1976 when
using labeled cDNA probes made from globin and
ovalbumin mRNA it was established: 1) globin and
ovalbumin sequences can be isolated in nucleosomes from
tissue producing these proteins, and 2) active globin
and ovalbumin genes are degraded to non-hybridizable
material by DNAase I several fold faster than the bulk
of the DNA (22,23). Preferential sensitivity to DNAase
I was soon shown to be a property of other genes in the

active state and indicated that nucleosomes associated
with transcribed sequences exist in a conformation
different from inactive nucleosomes. As the complement
of genes which have been cloned and characterized on
the DNA level has grown, studies of the chromatin of
specific genes has grown in parallel. The use of
nuclease digestion, Southern blotting and filter
hybridization to specific probes has been invaluable in
detecting chromatin structures associated with unique
chromosomal regions. This method is illustrated
(Figure 1) by comparing the total pattern of DNA
fragments produced by micrococcal nuclease in Xenopus
chromatin with the pattern of tRNA gene sequences
visualized by hybridization to a labeled probe. It can
be easily determined that tRNA genes are organized into
a very regular nucleosomal structure on the basis of
the ladder of fragments produced by micrococcal
nuclease.

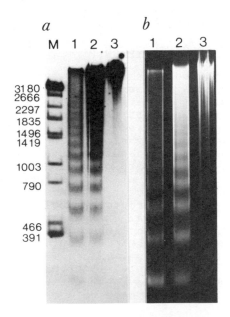

Figure 1. Micrococcal nuclease digestion of Xenopus
laevis cultured kidney cell nuclei. a) M. size marker;
1-3. decreasing amounts of nuclease digestion. DNA
blotted and hybridized to a ^{32}P-labeled tDNA probe. b)
1-3. same DNA as a) stained with Ethidium bromide.

A clever adaptation of the Southern blotting technique,
called indirect end-labeling (24,25), permits mapping
of nuclease sensitive regions of chromatin in relation
to a defined DNA sequence. In indirect end-labeling,
chromatin is partially digested with a nuclease and the
DNA is purified and then cleaved to completion with a
restricion enzyme. After gel electrophoresis and
Southern blotting the DNA fragments cut on one end with
the restriction enzyme and the other with the nuclease
can be selectively detected by hybridization to a short
radioactively labeled probe abutting the chosen
restriction enzyme site. The technique is analogous in
many ways to the footprinting of specific DNA binding
proteins on an isolated end-labeled restriction
fragment (26).

Using this technique it has been demonstrated that in
general not all sequences along the chromatin fiber are
homogeneously accessible to nucleases, but rather a
number of defined regions exist which are extremely
sensitive to nuclease attack (27). These
hypersensitive sites reflect an addition structural
feature presumably not directly related to the general
DNAaseI sensitivity of activated chromatin. General
sensitivity is reflected in the rapid reduction of
active sequences to oligonucleotides whereas
hypersensitive sites are the result of rapid production
of double stranded breaks over relatively small regions
30-300 bp in length. Some times regions will be
hypersensitive to only DNAase I and not micrococcal
nuclease. The preferred mode of DNAase I attack is to
make single stranded nicks in double-stranded DNA
templates (21). One might assume therefore that at
least two cleavage events must occur very close
together in order for the cut to be detectable on
native gels, which have been used exclusively to map
hypersensitive sites. Exactly how close together two
nicks must be to cause a double-stranded break will
certainly depend on how the DNA is isolated and also on
the local base composition around the cutting sites.
These sorts of considerations must be taken into
account in speculations about the physical nature of
hypersensitive regions. The general conclusion that
chromatin around genes has interesting structural
features superimposed on the basic nucleosome structure
remains firm, however, and invites speculations about
their role in controlling gene activity.

DNAase I hypersensitive regions are particularly
interesting because they have a nonstatistical tendency
to occur in regions upstream of genes active or
activated for transcription (27). Two possible
explanations of the structural basis for
hypersensitivity are the following: 1)

hypersensitivity is due to a small region of DNA being
more or less naked and therefore highly susceptible to
nuclease attack, or 2) the sequence-specific binding of
a macromolecule could result in particular
phosphodiester bonds becoming very vulnerable to
nuclease. This second phenomenon is seen in
footprinting protein-DNA interactions with DNAase I.
Protein binding results in not only protection of
certain bonds but also in rapid cleavage at others
(26). Both of these structures probably occur in 5'
flanking regions of genes.

Another mechanism by which particular DNA sequences can
be preferentially exposed is through a specific
register of nucleosomes with respect to DNA sequence,
often referred to as nucleosome phasing (28,29).
Studies on bulk populations have shown that nucleosomes
are arrayed regularly for the most part along the DNA,
but no specific register with regard to particular DNA
sequences is detectable (30). Several studies on
specific gene regions in the past two years have shown
evidence of some preferential positioning (28,29). If
nucleosome phasing is a property of chromatin which can
be altered during cell differentiation then the
availability of different sets of regulatory sequences
could be modulated through shifts in phasing. As of
yet, however, no conclusive experimental evidence
exists correlating nucleosome phasing with active
chromatin.

A slightly different application of probing specific
gene sequences involves mapping sites of DNA
methylation, a modification often correlated with
inactive gene regions (31,32). Most methylations in
eukaryotic DNA occur in the C^5 position of cytosine at
CG sequences. A subset of the potentially methylatable
positions can be probed with the restriction enzymes
isoschizomers Msp I and Hpa II. Msp I cuts all CCGG
sequences irrespective of methylation while Hpa II cuts
only the unmethylated sites. Using these enzymes the
pattern of methylation of a particular region of DNA
can be determined.

At the present a substantial catalogue of genes has
been studied by the types of methods just discussed.
Although many large gaps in an overall picture still
remain, a flavor of what active chromatin is like is
beginning to emerge. Representatives of genes
transcribed by all three eukaryotic polymerases have
now been studied as well as genes expressed in a
variety of situations. As might be expected there are
some structural features common to most active genes as
well as features unique to genes under a particular

type of regulation. We will attempt to review data on
representatives of different classes of genes and try
to point out some unifying principles, as well as
differences.

<u>Globin Genes</u>. Globin genes of vetebrates are multigene
families whose expression follows a developmental
program during embryogenesis (33). The α and β globin
gene families of chicken are undoubtedly the best
studied system on the chromatin level. The β-globin
cluster of chicken contains four genes, two of which (ρ
and ε) code for embryonic globins produced in primitive
red cells and two (β and $β^H$) for adult globins
expressed in the definitive red cell line (Figure 2)
(34). The α-family characterized thus far consists of
three genes: $α^π$, expressed only in primitive red cells
and $α^A$ and $α^D$, believed expressed in both primitive and
definitive red cells (35). The change from the
embryonic to the adult pattern of expression occurs
between days five and nine of development. Chromatin
structural changes concomitant with developmental
switching have been carefully studied and document
events in both activation and inactivation of globin
genes (36,37,38,39,40).

In presumptive red cell precursors of first day chicken
embryos, prior to the start of globin synthesis, the
globin genes are relatively insensitive to DNAase I
digestion and the DNA is extensively methylated (36).
By the onset of embryonic globin synthesis a large
region extending many Kb around both embryonic and
adult globin genes has become moderately sensitive to
DNAase I. The coding regions of the embryonic genes,
contained within this large domain, are very sensitive
to DNAase I during the next several days when
exclusively embryonic globin chains are being
synthesized in the primitive erythroblasts. With the
appearance of the definitive red cell line producing
adult globins, the adult globin genes become very
sensitive to DNAase I while the embryonic genes assume
a conformation moderate in sensitivity. The large
domain of moderate accessibility to nuclease is
hypothesized to result from chromatin decondensation of
all globin genes within a cluster. The regions of
highest accessibility correspond roughly to the globin
transcription unit. The nucleosomes of these regions
have been shown to be associated with the non-histone
proteins HMG 14 and 17 which are, at least in part,
responsible for the sensitive conformation (41).
Enhanced DNAase I sensitivity of coding regions
persists even in the mature erythrocyte, which has shut
off globin synthesis. This continued sensitivity in
the transcriptionally inert erythrocyte nucleus would
seem to indicate that globin genes retain some aspects

of their active conformation during subsequent shut
down of globin synthesis and chromatin condensation.
Despite the consistency of DNAase I sensitivity data in
chicken, curiously the globin genes of adult duck are
preferentially sensitive neither in immature nor mature
erythroid cells (42).

Globin switching in the chicken is accompanied by
changes in the pattern of DNAase I hypersensitive sites
around the gene clusters (Figure 2) (37,38,39,40,43).
These sensitive regions occur often but not exclusively
in the 5' flanking sequences, close to the start of
transcription of the active genes. The best studied
hypersensitive region is one occurring 60-260 bp 5' to
the transcription start of the adult β globin gene
(40). This site becomes sensitive in definitive
erythroid cells from 14 day chicken embryos. The site
has been shown to be accessible to a variety of
endonucleases and can be released as a protein free
fragment in 50% yield, indicating histones may be
absent from this region. The accessible region
includes the conserved CCAAT sequence motif at -70 bp
from the start of transcription but not the Goldberg-
Hogness box at -30. It is known that sequences -70 to
-110 are needed for efficient transcription of the
rabbit β globin gene transfected into mouse tissue
culture cells (44) but unfortunately similar data is
not available for chicken globin genes.

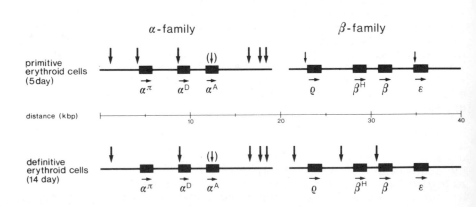

Figure 2. Map of chicken α and β-globin gene families.
Vertical arrows denote sites of DNAase I
hypersensitivity in primitive erythroid cells (5 day)
and definitive erythroid cells (14 day). Large arrows
show major hypersensitive sites, small arrows minor
sites.

One might imagine that other globin genes expressed in definitive erythroblasts would have DNAase I hypersensitivity sites at analogous positions to β , but this is not strictly the case. The hypersensitivity site 5' to the β^H gene is more than 1 Kb away from the transcription start site (37). The α^D gene has a very weak site close to its 5' end and the α^A gene may have no site. All three embryonic globin genes have at least a weak site in primitive erythroblasts (38). Because of the qualitative differences in intensity and positioning of hypersensitive sites, it is difficult to formulate specific models of how these structures are involved in the pattern of globin expression. As eukaryotic regulatory sequences for transcription become better understood perhaps the role of hypersensitive sites will become clearer.

Additional data on globin gene expression is available from two types of inducible erythroid cell lines. The first is derived from chicken bone marrow cells infected with a temperature sensitive mutant of the retrovirus, avian erythroblastosis virus (ts AEV) (43). It is believed that viral infection arrests the differentiation process at a particular stage. At elevated temperatures the products of the viral transforming gene are inactivated, and after several days the cell lines begin to synthesize mRNA for both adult and embryonic hemoglobins. The fact that both adult and embryonic genes are activated indicates regulation is not entirely normal since uninfected erythroid cells activate either one or the other sets of globin genes. One particular cell line acquires characteristic hypersensitive sites on both adult and embryonic globin genes upon inactivation of the virus and subsequent globin gene activation. Another cell line, hypothesized to be farther along in red cell differentiation before ts AEV infection, already has the pattern of hypersensitivity even before globin gene activation. The hypersensitive sites on globin genes are therefore neither totally the result of or sufficient for the transcription process. Exactly when the hypersensitivity pattern is established during erythropoesis is not known. Interestingly globin production in the virus infected cells is not accompanied by changes in methylation as it is in normal erythroid cells. The β-globin genes remain extensively methylated even during globin synthesis. Clearly demethylation of most CG groups is not an absolute prerequisite of transcription. There may nevertheless be a few key positions from which methyl groups must be removed to permit transcription.

Murine erythroleukemia cells can be induced to
differentiate by a number of chemical agents (45). The
mouse α and β major genes have been probed with DNAase
I both before and after induction (46). The globin
gene sequences in uninduced cells are already
preferentially sensitive to DNAase I as compared to the
inactive ovalbumin gene. After induction the β major
gene acquires a hypersensitive site very close to the
transcription initiation site. The α globin gene also
acquire several hypersensitive sites but they have not
yet been mapped. No changes in the methylation pattern
were detectible around either α or β globin genes
during the chemical induced differentiation process.

Both of the inducible transformed cell lines seem to
establish chromatin structures (as assayed by DNAase I)
similar to normal erythroid cells but interestingly do
not exhibit any dramatic changes in methylation
pattern. This would indicate that methylation has no
causal relationship to the DNAase I sensitive
structures.

Heat Shock Genes of Drosophila. The heat-induced
activation of specific chromosomal loci was first
detected cytologically in the polytene chromosomes of
Drosophila salivary glands (47). It provides an
unusual and particularly convenient system in which to
study activation and inactivation of genes. Elevation
of Drosophila from 25°C to 37°C induces the rapid
production of a small number of specific proteins which
alter the cellular metabolic machinery and seem to
allow cells to survive the environment insult. The
heat shock products appear to be induced in all tissues
both adult and embryonic as well as in several stable
Drosophila cell lines. Synthesis of most other mRNAs
is suppressed during heat shock. Normal patterns of
RNA synthesis return within several hours after the
animals are returned to 25°C.

The most abundant heat shock product is a 70,000 dalton
protein denoted hsp 70. Prior to activation the hsp 70
gene is organized into regularly spaced nucleosomes as
assayed with M. nuclease and shows little or no
preferential susceptibility to DNAase I degradation
(48-50). Within minutes after heat shock the
nucleosomal ladder becomes blurred and disappears
completely after ~30 minutes. This apparent disruption
or displacement of nucleosomes occurs on coding regions
and appears not to extend into the flanking regions.
Likely, transient disruption of nucleosomal structure
is a result of polymerases traversing the gene at an
increasing frequency as the heat shock response
progresses. The nucleosomal pattern gradually is
reestablished after return to lower temperature.

The hsp 70 genes also become increasingly sensitive to
DNAase I roughly in parallel to the disruption of the
micrococcal nuclease ladder suggesting that DNAase I
sensitivity is also a manifestation of the actual
transcription process (49,50). Is the structural basis
for DNAase I sensitivity in heat shock genes therefore
different than for activated globin genes? Sensitivity
of globin coding regions to DNAase I precedes mRNA
synthesis and has been shown to be, at least in part,
caused by HMG 14 and 17 binding. Drosophila also have
HMG-like proteins and conceivably nucleosomes in hsp
coding regions could bind HMG's upon heat shock and
thereby acquire DNAase I sensitivity (51). It is
perhaps more likely that not all aspects of active
chromatin structure are shared between genes as
different as globin genes, which are activated slowly
in a process requiring cell division, and heat shock
genes, which display an almost instantaneous response
to induction.

All of the major heat shock loci have hypersensitive
sites near the 5' end of the transcription unit
(24,52). The sites are very clearly seen in uninduced
cells. In the induced state the increase in DNAase I
sensitivity of the entire coding region masks the
specific cutting sites. The Hsp 70 gene has a major
site covering ~300 base pairs which extends into the 5'
edge of the protein coding region. In the Hsp 83 gene
three hypersensitive sites fall within a 700 bp region
again including the site of RNA transcription. A
cluster of five sites also occur in the 3' flanking
sequences. Strong sites occur at the 5' end of each of
the small heat shock genes hsp's 22, 23, 26, 28, which
may also extend through their sites of initiation and
TATA boxes (52). More accurate mapping of the sites
around the small hsp will be needed to verify whether
TATA box and initiation site sensitivity is indeed a
common denominator in the heat shock process. With the
data at hand it is tempting to speculate that
accessibility of the promoter regions is related to the
potential for rapid gene activation.

We mention in passing that changes in DNA methylation
pattern have frequently been correlated with gene
expression possibly by marking genes for activation by
the removal of methyl groups. Drosophila (and other
insects) have very little if any methylation so they
may employ different strategies in their developmental
programming (31).

Egg White Proteins. Ovalbumin, conalbumin, lysozyme
and ovamucoid are the major proteins produced by the
tubular gland cells of chicken oviduct (53,54). The
differentiation of the oviduct and subsequent

transcription of these proteins is under control of sex
steroid hormones. Because the egg white proteins are
produced in widely different amounts they represent a
good opportunity to study differential gene expression
in response to a common stimulus. Comparison of
sequences flanking these four genes shows that each has
a TATA region approximately 30 bp upstream from the
start of transcription but reveal no other features of
the DNA sequence suggesting a possible common mechanism
of hormonal response (55).

Several aspects of the chromatin structure of
conalbumin and ovalbumin genes (which account for 10%
and 50%, respectively, of the total protein produced in
induced oviduct) have been studied over the last
several years (56,57,58). These genes, in stimulated
oviduct tissue, resemble active globin genes in that
their coding regions are highly sensitive to DNAase I
and undermethylated compared to tissues not producing
egg white proteins. The ovalbumin gene is linked to
two related genes X and Y which are expressed at low
levels in oviduct. This entire locus containing about
100 Kb of DNA is preferentially DNAase I sensitive in
oviduct (59). This sensitivity is not a direct result
of transcription because in the absence of estrogen
stimulation the sensitive state remains although
synthesis of ovalbumin mRNA ceases (53).

Early solution hybridization studies demonstrated that
ovalbumin coding sequences can be isolated as monomer
and short oligomer nucleosomes after micrococcal
nuclease digestion of active oviduct (23). More recent
analysis of micrococcal nuclease digests on Southern
blots shows however, that ovalbumin coding sequences
and 5' flanking region do not produce a typical pattern
of oligomer nucleosomes (58). Taken together those
results could mean that nucleosomes are present but the
distance between cores is somewhat variable. The
ladder of DNA fragments is least clear when hybridized
to coding probes, suggesting RNA polymerases may
contribute in disrupting regular nucleosomal
organization. At least 2.5 Kb of 5' flanking DNA also
displays a somewhat irregular packaging (58). An
interesting possibility would be that the binding of
receptor-hormone complex results in the reorganization
of the 5' chromatin structure. Unfortunately similar
fine mapping studies for conalbumin or the other egg
white proteins have not been reported so the generality
of this observation for other steriod hormone inducible
genes is unknown.

Neither conalbumin or ovalbumin have DNAase I
hypersensitive sites anywhere in the vicinity of the
transcription start sites making them notable

exceptions to the rule that such sites are
prerequisites of Pol II transcription (57). That genes
under tight hormonal control may not need 5'
hypersensitive structures may be a point to consider in
devising general models of chromatin activation.

Ovalbumin and conalbumin mRNAs are produced at
different rates in oviduct and the two genes also
differ greatly in the speed of their response to
secondary estrogen induction (53). Conalbumin mRNA
synthesis is stimulated in tubular gland cells almost
immediately after estrogen administration while
ovalbumin mRNA production begins after a lag of about 3
hours. Studies of chromatin up to now have not
detected any profound differences between the two
genes. Hopefully more detailed examination will
provide clues to their differential response to
estrogen stimulation.

Histone Genes. The chromatin structure of histone
genes has been studied in Drosophila and the sea urchin
(60,61). The major class of histone genes in both
organisms are organized on the DNA level into highly
repeated tandem units with each unit containing one
each of the five histone genes (62). Histones are, of
course, required in any dividing cell but unless
division is very rapid as in the early cleavage stages
of embryogenesis then demand for histones could
conceivably be met by a few sets of genes or
alternatively, very low level transcription of all gene
repeats. This inherent uncertainty must kept in mind
in making conclusions about the structure of active
chromatin. Histone genes are nontheless extremely
interesting because their expression can be cell cycle
dependent, developmentally regulated and in certain
instances tissue specific.

Histone gene sequences from Drosophila tissue culture
cells can be demonstrated by Southern blotting to
reside within more or less regularly arrayed
nucleosomes (60). Mapping nuclease sensitive regions
along the repeat unit by indirect end-labeling reveals
that the 5' ends of all five histone genes are highly
sensitive to attack. Both micrococcal nuclease and
DNAase I cut at similar sites falling very close to the
5' termini of the genes although a number of other
sites occur elsewhere, particularly in the long
nontranscribed spacer region upstream of the Hl gene.
Unlike the situation for most other pol II genes, the
5' hypersensitive areas do not appear to include the
TATA sequence or other upstream regions. The
hypersensitive pattern is established by the first hour
of embryonic development and has been found in all
tissues examined to date.

In contrast the chromatin structure of sea urchin
histone genes undergoes numerous changes in nuclease
sensitivity pattern during early development (61).   The
sea urchin, P. miliaris, contains several different
classes of histone repeating units.   Shortly after
fertilization the majority class of genes is switched
on and the rate of mRNA production per gene reaches a
peak of one transcript/gene min about the 128 cell
stage of development (63).   The rate/gene from this
particular cluster then gradually decreases and is no
longer detectable after the hatching stage of
development, as other sets of histone genes become
active.   The chromatin structure of this majority class
of genes undergoes developmental changes correlated to
transcriptional activity.   At the peak of synthetic
activity in early blastula stage, some nucleosomal
organization is detectable on both coding regions and
non-transcribed spacer regions as evidenced by a
indistinct ladder of micrococcal nuclease fragments on
Southern blots with only the first 4-5 bands in the
pattern resolvable.   Regions of ~300 bp on the 5' end
of each of the five genes are hypersensitive to both M.
nuclease and DNAase I.   These regions stretch from
about -100 bp before the transcription start site to
+200 bp into mRNA coding sequences (Figure 3).   This
pattern contrasts with the relatively narrow
hypersensitive areas flanking Drosophila histone genes.

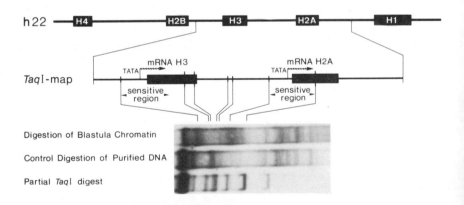

Figure 3.   Micrococcal nuclease cutting pattern of
embryonically expressed histone genes (h22), mapped by
indirect end-labeling (p.   ).   Positions of Taq I
restriction sites shown by vertical lines allow
accurate positioning of nuclease sensitive areas.   Free
DNA control shows pattern of bands produced by
micrococcal nuclease on histone DNA due to its sequence
specificity.

Cutting of both nucleases is relatively evenly
distributed throughout the hypersensitive regions in
the sea urchin. Analysis of DNA sequences important in
specifiying rate and accuracy of transcription
initiation of the H2A gene has shown that, in addition
to the TATA sequence at -30 bp, an upstream sequence -
100 to -400 bp (termed modulator) is necessary for
efficient transcription (64). While the TATA box is
exposed in the active chromatin structure the
"modulator" sequence is apparently not preferentially
accessible. The hypersensitive regions seem not the
result of rapid polymerase initiation because they
remain accessible until midblastula stages, after
transcription rate/gene has decreased by an order of
magnitude. After hatching, when detectable
transcription from this cluster has ceased, all
preferentially sensitive sites disappear.

Other Pol II Genes. The correlation of gene activity
with the presence of 5' hypersensitive sites has been
further documented by studies on several other Pol II
transcribed genes. The Preproinsulin II gene of rat
has a site spanning from the transcription start site
to 300 bp upstream in insulin producing pancreatic β
cell tumors (65). Tissues producing no insulin do not
contain this site. A ribosomal protein 49 gene in
Drosophila, which is constituatively expressed at low
levels in virtually all cells, has a striking array of
five hypersensitive sites spaced at fairly regular
intervals of 250 bp near the 5' end of the coding
sequences (66).

Combining Drosophila genetics with chromatin mapping
studies has provided further information on the
interrelationship among DNA sequence, hypersensitivity
and gene expression (67). The glue protein gene is
expressed in salvary glands of wild type third instar
larvae, whose chromatin has two major DNAase I
hypersensitive sites; one at -330 bp one at -405 bp
from the 5' end of the gene. These sites are not
hypersensitive in chromatin of embryos,in which the
glue gene in inactive. The Drosophila strain BER1 has
a deletion of ~100 bp which eliminates sequences around
-405. In BER1 neither of the major hypersensitive
sites is visible in larval salavary glands even though
the -330 sequence is still present. The sequences
around -405 therefore seem required for establishing
the pattern of hypersensitivity and the activity of the
gene.

Animal Viruses. The papovaviruses, SV40 and polyoma,
have yielded important information on the relationship
between DNA sequences required for transcription and
chromatin structure. Both SV40 and polyoma DNA are

packaged into minichromosomes by cellular histones and
are transcribed by cellular polymerase II. Sequences
important for the transcription of both early and late
viral genes lie close to the viral origin of
replication. The regions from the origin to several
hundred bps into the "late" side of the DNA have been
demonstrated to be accessible to various nucleases in
approximately 20-25% of viral minichromosomes for both
polyoma and SV40 (69,70). Electron micrographs reveal
that in a fraction of SV40 minichromosomes several
hundred bps near the origin lack the typical beaded
structure seen throughout the rest of the molecule
(71). This would indicate that nucleosomes in this
region, if present at all, are labile and disrupted
during sample preparation. It has not actually been
shown that the fraction of minichromosomes with the gap
are transcriptionally active.

A two times tandemly repeat sequence of 72 bp occurs
within the "gap" region of SV40 minichromosomes. One
complete copy of this sequence is required for SV40 to
be viable (68). Interestingly the 72 bp sequence can
stimulate the transcription of some linked genes by two
orders of magnitude in recombinant molecules introduced
into cultured cells (72). The level of stimulation is
not very sensitive to the distance and orientation of
the SV40 sequence in relation to the gene. There are
many possible mechanisms by which the viral DNA could
enhance transcription. Reorganizing the chromatin
structure of the gene-containing molecule by creating a
gap region analogous to the situation in viral
minichromosomes is one possibility often mentioned.

The DNA of the gap region in polyoma is also able to
enhance transcription similarly to the SV40 sequence
(73).

Active and inactive endogenous retroviral genes in
chicken differ in their sensitivity to DNAase I (74).
A RAV-0 virus designated ev 3 is integrated into
chromosomal DNA of certain strains of chicken. The ev 3
provirus contains deletions and cannot produce
infection virus, but does produce several viral encoded
transcripts. This locus is DNAase I sensitive and
contains a hypersensitive site in each of the long
terminal repeated sequences (LTR's) flanking the
provirus. The LTR's contain the viral promoter and RNA
initiation site as well as other sequences important in
RNA production. Interestingly, there is considerable
sequence homology between some LTR sequences of
retroviruses and the nuclease sensitive region of SV40.
Inactive endogenous chicken retrovirus lacks both
general sensitivity to DNAase I and hypersensitive
sites in the LTR's.

The activity and timing of expression of retrovirus in
mouse embryos has been shown to depend on the site of
chromosomal integration (75). This documents the
importance of the local chromosomal environment in
determining the pattern of viral expression during
development and differentiation.

Pol III Transcribed Genes. 5s and tRNA genes are a
class of genes distinct from those coding for mRNA.
They are transcribed by RNA pol III and have promoters
specific to this polymerase residing entirely within
the coding regions (76). 5s RNA is associated with the
eukaryotic ribosome and the 5s genes are generally
highly reiterated and clustered in tandem units
separate from the 18 and 28S RNA genes. Because the
coding region is only ~130 bp, an entire gene could be
contained within a single core nucleosome. Chromatin
structure of 5s genes has been studied in both Xenopus
and Drosophila (77,78). The 5s genes are organized
into regularly spaced nucleosomes, as assayed by
micrococcal nuclease, with the repeat distance the same
as the bulk repeat distance for a particular tissue.
The relationship between the nucleosomal repeat
distance and the gene repeat unit is interesting in
Drosophila. The 5s repeat unit is only 375 bp in
length which is within experimental error, the size of
a dimer nucleosome. The coincidence between
periodicity in the DNA sequence and chromatin
organization means that, once established, a phase
relationship between DNA sequence and nucleosomes could
be propagated for long distances resulting in a long
array of 5s genes occupying a common orientation on
nucleosomes. This organization is suggestive of
mechanisms to activate or inactivate a long battery of
genes through shifts in nucleosome phasing.
Nucleosomes are reported to have two preferred
locations on 5s genes of Drosophila, but correlation of
one or the other with gene activity is lacking because
the genes are highly repeated (78).

5s genes in Xenopus are about 20,000 times repeated per
haploid genome with the vast majority used only in
developing oocytes. Multiple but distinct phase
relationships of oocytes-type genes have been reported
in somatic cell types in which these genes are silent
(77).

The transcription machinery of 5s genes is the best
characterized of any eukaryotic gene. In addition to
pol III, correct initiation of 5s transcription
requires a 40,000 dalton transcription faction which
binds to a DNA sequence in the center of the gene (79).
Chromatin reconstituted from purified histones and 5s
plasmid DNA in the absence of cellular factors is

poorly transcribed in vitro. The presence of the 5s transcription factor during the reconstitution results in efficient in vitro transcription (80). Binding of the factor may have to precede nucleosome formation in order to form an active template. Whether the factor contributes to the active chromatin structure by excluding nucleosomes from the control region of the gene or by a more indirect effect is not known.

Chromatin structure of tRNA genes, another class of pol III transcribed genes has been examined in chicken embryos and various tissues of Xenopus (81,82). They have been found to be packaged into nucleosomes in the situations thus far examined. The first reported instance of specific nucleosome phasing with respect to a gene was on tRNA genes in chicken embryos (81). The majority of genes coding for three species of tRNA have a common location within nucleosomes with the 70 bp of coding sequence lying slightly off center within the core structure. A phased relationship around the genes is maintained only for the next several nucleosomes. In this arrangement the transcription promoter elements, which lie within the coding sequences, are located near the center of the core nucleosome structure. Again, it is not known whether this organization for the majority of the tRNA genes is representative of the activated state of embryonic chicken genes.

In Xenopus, the tRNA genes occur in very high copy number. The chromatin structure of a 100 times tandemly repeated unit containing seven different tRNA genes has been studied in both metabolically active and inactive tissues (82). In the condensed and transcriptionally inactive nucleus of mature erythrocytes a phase relationship between nucleosome and DNA sequence is detectable with nucleosomes of the majority of repeating units occupying identical positions (Figure 4). In this instance phasing of nucleosomes may result from or facilitate folding of nucleosome fiber into a condensed heterochromatic state. No phasing is detectible in metabolically active tissues.

The tRNA genes in active tissues of Xenopus are not preferentially degraded by DNAase I (unpublished observations). Whether this is because active tRNA genes are not DNAase I sensitive or because only a small fraction of the genes are active is not known.

rRNA Genes. Ribosomal genes are the third class of eukaryotic genes and are transcribed by pol I. They are typically arranged in tandemly repeat units containing a transcribed portion coding for a common

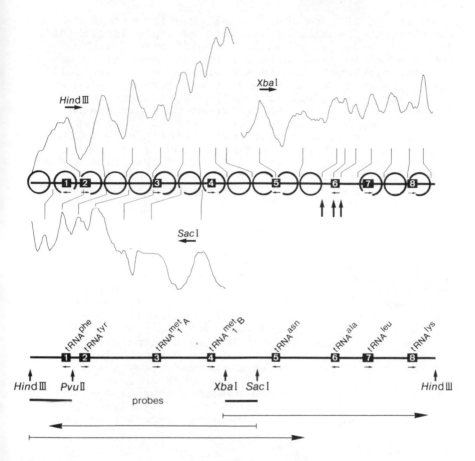

Figure 4. Positions of nucleosomes on tRNA gene repeat unit in Xenopus laevis erythrocytes. Micrococcal nuclease cutting sites were mapped by indirect end-labeling (p.   ). Densitometric tracings show positions of relative accessibility (peaks) and protection (troughs). Micrococcal nuclease cutting sites were mapped from 3 restriction enzyme sites as indicated.

precursor of 18S and 28S rRNA alternating with a nontranscribed spacer of somewhat variable length (83). Characterization of ribosomal gene chromatin with nuclease probes also suffers from the uncertainty of how many ribosomal repeats are active in a given cell type. The point of most intense controversy is to what extent active nucleolar chromatin is packaged into nucleosomes. Most of the pertinent data concerning this question has already been discussed in review articles so it will be dealt with briefly here (3,10). During intense ribosomal transcription as in midvitellogenic stages of oogenesis, polymerases are so tightly packed on the transcribed fiber that the presence of intact nucleosomes particles would be precluded. Whether nucleosomes particles occur in nontranscribed spacer in this situation is still debated (84,85).

Activated ribosomal genes in Tetrahymena and Physarum have been shown to be digested faster by DNAase I than bulk chromatin reminiscent of pol II transcribed genes (86,87). Studying interspecies hybrids of Xenopus laevis and borealis also supports this correlation (88). Hybrid tadpoles express almost exclusively the laevis type rRNA genes and the laevis genes are preferentially degraded with DNAase I compare to borealis genes. Examination of the DNA methylation pattern reveals no differences between laevis and borealis genes. The laevis type genes are, in fact, both heavily methylated and DNAase I sensitive - an argument that undermethylation is not tightly coupled to DNAase I sensitivity or ribosomal gene activation.

CONCLUSIONS

Cataloging observations about chromatin structure is naturally much easier than fitting available information into a coherent picture. Many activated genes certainly share common characteristics in their sensitivity to nuclease but very few generalizations hold for all genes thus far examined. We will attempt to summarize by mentioning several features of structure which seem to apply to many activated genes (Table 1).

TABLE 1.   Changes in chromatin structure during
           during activation.

inactive

nucleosomal.
often condensed.
few nonhistones.
DNA often highly methylated.

potentially active   transcribed
transcribed (low level)   (high level)

nucleosomal   nucleosomal structure
"altered" conformation.   "disrupted."

association of HMG 14,17 (globin).
various levels general
DNAase I sensitivity.
5' hypersensitivity.
histone acetylation,
other modifications.
DNA often undermethylated.

During the activation process most genes retain at
least some semblance of regular nucleosomal packaging.
Complete disappearance of normal nucleosomal
organization is observed in several instances where the
transcription rate is particularly high (e.g.,
ribosomal genes in oocytes, induced heat shock genes)
suggesting nucleosome particles may be transiently
disrupted or displaced as polymerases read through.
The activation of chromatin in most instances seems to
involve a punctuation rather than a disruption of
regular nucleosomal organization.

Pol II transcribed genes become preferentially
sensitive to DNAase I during gene activation. (10,89).
This phenomenon is complex in that different levels of
sensitivity are observed and it is unlikely that DNAase
I sensitivity has a single common structural basis in
different genes. In the case of the chicken globin
genes, the general sensitivity results at least in part
from the binding of HMG 14,17 to the nucleosomes
containing globin sequences and every nucleosome in the
transcribed region appears preferentially sensitive
(12,16). Learning the generality of this observation
in other systems will be an extremely important step in
understanding gene activation.

In most cases general DNAase I sensitivity seems to precede transcription rather than result from it (10,89). One exception to this generalization are Drosophila heat shock genes which acquire sensitivity roughly at the onset of transcription (50).

General DNAase I sensitivity has been suggested to reflect an "accessible" chromatin structure which is able to serve as a template for phlymerase. Likely in various genes, the "accessible" structure is built in different ways.

Sites hypersensitive to DNAase I also show considerable diversity in size, intensity and positions with respect to transcribed sequences. Well established is that hypersensitivity sites tend to appear in 5' flanking sequences of pol II transcribed genes prior to activation (27,89). Hypersensitive regions also occur elsewhere in chromatin, the 5' sites being a subset of the total given special attention because they may be associated with DNA sequences having a regulatory role in transcription. Comparisons within a single class of genes of same organism often reveals considerable consistency in the pattern of hypersensitive sites. For example, each heat shock gene has a hypersensitive region which appears to encompass the TATA box and point of RNA initiation (27). Comparison among different types of genes reveals variation in the intensity and positioning of hypersensitive regions. The relationship of hypersensitivity to regulatory sequences will remain obscure until a better understanding of both the molecular basis of hypersensitivity and the nature of eukaryotic regulatory sequences is reached.

One approach which promises to help sort out the interrelationship between DNA sequence and specific chromatin structures is to study the packaging of cloned genes reintroduced into eukaryotic cells. There are already several cases in which the expression of foreign genes is regulated in the new host (90-93).

Such systems allow the simultaneous assessment of DNA sequence manipulations on transcription and chromatin structure.

At present the understanding of active chromatin structure resembles an unfinished puzzle with only the border pieces assembled. The interior of the picture is more difficult to put together and will probably require not only careful characterization of more gene systems by the techniques we have described here, but also application of new approaches which allow

regulatory DNA sequences to be studied within the
context of chromatin structure.

ACKNOWLEDGEMENTS

We are grateful to Dr. W. Folk for critically reading
the manuscript and providing useful advice. We thank
B. Bryan for preparing the manuscript and F. Ochsenbein
for photography and skillful drawings. Acknowledged
also are Drs. M. L. Birnstiel and H. Hofstetter for
helpful discussions and suggestions. P. N. B. is a
fellow in cancer research supported by a grant from the
Daymon Runyon - Walter Winchell Cancer Fund and the
Canton of Zurich.

REFERENCES

1.   Lewis, C. D. and Laemmli, V. K. (1982). Cell 29,
     171-181.
2.   McGhee, J. D. and Felsenfeld, G. (1989). Ann.
     Rev. Biochem. 49, 1115-1156.
3.   Lewin,B. (1980). Gene Expression II, Eucaryotic
     Chromosomes, 2nd ed., John Wiley and Sons, New
     York.
4.   Lewis, E. B. (1950). Adv. Genet. 3, 75-115.
5.   Johnson, M. H. (1982). Nature 296, 493-494.
6.   Appels, R., Wells, J. R. E. (1972). J. Mol. Biol.
     70, 425-434.
7.   Sung, M. T. (1977). Biochemistry 16, 286-290.
8.   Appels, R., Tallroth, E., Appels, D. M. and
     Ringertz, N. R. (1975). Exp. Cell Res. 92, 70-78.
9.   Watson, J. D. (1965). Molecular Biology of the
     Gene, W. A. Benjamin, Inc., New York.
10. Mathis, D., Oudet, P. and Chambon, P. (1980).
     Progress in Nucleic Acid Research and Molecular
     Biology 24, 1-55.
11. Gottesfeld, J. M. (1977). Methods in Cell biology
     16, 421-436.
12. Weisbrod, S. (1982). Nucleic Acids Res. 10, 2017-
     2042.
13. Goldknopf, I. A. and Busch, H. (1976). The Cell
     Nucleus, part C (H. Busch, ed.), pp. 149-180,
     Academic Press, New York.
14. Levinger, L. and Varshavski, A. (1982). Cell 28,
     375-385.
15. Johns, e.W. (1982). the HMG Chromosomal Proteins,
     Academic Press, London (in press).
16. Weisbrod, S. T. and Weintraub, H. (1981). Cell
     23, 391-400.
17. Mardian, J., Paton, A. E., Bunick, G. and Olins,
     D. E. (1980). Science 209, 1534-1536.
18. Sandeen, G., Wood, W. I. and Felsenfeld, G.
     (1980). Nucleic Acids Res. 8, 3757-3778.

19. McGhee, J. D., Rau, D. C. and Felsenfeld, G. (1982). Nucleic Acid Res. 10, 2007-2016.
20. Anfinsen, C. B., Cuatrecasas, P. and Taniuchi, H. (1971). The Enzymes, P. D. Boyer, Ed., Vol. IV 3rd Edition, pp. 177-204, Academic Press, New York.
21. Laskowski, Sr., M. (1971). The Enzymes, P. D. Boyer, Ed., Vol. IV, 3rd Edition, pp. 289-311, Academic Press, New York.
22. Weinbraub, H. and Groudine, M. (1976). Science 193, 848-856.
23. Garel, A. and Axel, R. (1976). Proc. Nat. Acad. Sci. 73, 3966-3970.
24. Wu, Carl (1980). Nature 286, 854-860.
25. Nedospasov, S. A. and Georgiev, G. P. (1980). Biochem. Biophys. Res. Commun. 92, 532-539.
26. Galas, D. J. and Schmitz, A. (1978). Nucleic Acids Res. 5, 3157-3170.
27. Elgin, S. C. R. (1981). Cell 27, 413-415.
28. Kornberg, R. (1981). Nature 292, 579-580.
29. Zachau, H. G. and Igo-Kemenes, T. (1981). Cell 24, 597-598.
30. Prunell, A. and Kornberg, R. D. (1977). Cold Spring Harbor Symp. Quant. Biol. 42, 103-108.
31. Felsenfeld, G. and McGhee, J. (1982). Nature 296, 602-603.
32. Razin, A. and Riggs, A. D. (1980). Science 210, 604-610.
33. Groudine, M., Holzer, H., Scherrer, K. and Therwath, A. (1974). Cell 3, 243-247.
34. Dolan, M., Sugarman, B. J.,Dodgson, J. B. and Engel, J. D. (1981). Cell 24, 669-678.
35. Dodgson, J. B. and Engel, J.D. (1980). Proc. Nat. Acad. Sci. USA 77, 2596-2599.
36. Stalder, J., Groudine, M., Dodgson, J. B., Engel, J. D. and Weintraub, H. (1980). Cell 19, 973-980.
37. Stalder, J., Larsen, A., Engel, J. D., Dolan, M., Groudine, M. and Weintraub, H. (1980). Cell 20, 451-460.
38. Weintraub, H., Larsen, A. and Groudine, M. (1981). Cell 24, 333-344.
39. Groudine, M. and Weintraub, H. (1981). Cell 24, 393-401.
40. McGhee, J. D.,Wood, W. I., Dolan, M., Engel, J. D. and Felsenfeld, G. (1981). Cell 27, 45-55.
41. Weisbrod, S., Groudine, M. and Weintraub, H. (1980). Cell 19, 289-301.
42. Camerini-Otero, R. D.,Sollner-Webb, B., Simon, R. H., Williamson, P., Zasloff, M. and Felsenfeld, G. (1977). Cold Spring Harbor Symp. Quant. Biol. 42, 57-74.
43. Weintraub, H., Beug, H., Groudine, M. and Graf, T. (1982). Cell 28, 931-940.

44. Dierks, P., van Ooyen, A., Dobkin, C., Reiser, J., Weber, H. and Weissman, C. (1982). J. of Cell Biochem. Supplement 6, 344.

45. Marks, P. A. and Rifkind, R. A. (1978). Ann. Rev. Biochem. 47, 419-448.

46. Sheffery, M., Rifkind, R. A. and Mark, P. A. (1982). Proc. Natl. Acad. Sci. USA 79, 1180-1184.

47. Ashburner, M. and Bonner, J. J. (1979). Cell 17, 241-254.

48. Wu, C., Bingham, P. M., Livak, K. J., Holmgren,R. and Elgin, S. C. R. (1979). Cell 16, 797-806.

49. Wu, C., Wong, Y. C. and Elgin, S. C. R. (1979). Cell 16, 807-814.

50. Levy, A. and Noll, M. (1981). Nature 289, 198-203.

51. Bassuk, J. A. and Mayfield, J. E. (1982). Biochem. 21, 1024-1027.

52. Keene, M. A., Corces, V., Lowenhaupt, K. and Elgin, S. C. R. (1981). Proc. Nat. Acad. Sci. 78, 143-146.

53. Palmiter, R. D., Mulvihill, E. R., McKnight, G. S. and Senear, A. W. (1977). Cold Spring Harbor Symp. Quant. Biol. 42, 639-648.

54. Schuetz, G., Nguyen, M. C., Giesecke, K., Hynes, N. E., Groner, B., Wurtz, T. and Sippel, jA. E. (1977). Cold Spring Harbor Symp. Quant. Biol. 42, 617-624.

55. Breathnach, R. and Chambon, P. (1981). Ann. Rev. Biochem. 50, 349-383.

56. Mandel, J. L. and Chambon, P. (1979). Nucl. Acids Res. 7, 2081-2103.

57. Kuo, M. T., Mandel, J. L. and Chambon, P. (1979). Nucleic Acids Res. 7, 2105-2114.

58. Bellard, M., Dretzen, G., Bellard, F., Oudet, P. and Chambon, P. (1982). EMBO Journal 1, 223-230.

59. O'Malley, B. W., Stumph, W. E., Lawson, G. M. and Tsai, M. J. (1982/. J. Cell Biochemistry, Supplement 6, 262.

60. Samal, B., Worcel, A., Louis, C. and Schedl, P. (1981). Cell 23, 401-409.

61. Bryan, P. N. and Birnstiel, M. L. Manuscript in preparation.

62. Hentschel, C. C. and Birnstiel, M. L. (1981). Cell 25, 301-313.

63. Maxson, R. E. and Wilt, F. H. (1981). Dev. Biol. 83, 380-386.

64. Grosschedl, R. and Birnstiel, M. L. (1982). Proc. Nat. Acad. Sci. 79, 297-301.

65. Wu,C. and Gilbert, W. (1981). Proc. Nat. Acad. Sci. 78, 1579-1580.

66. Wong, Y-C,O'Connell, P., Rosbash, M., Elgin and S. C.R. (1981). Nucleic Acids Res. 9, 6749-6762.

67. Shermoen, A. W. and Beckendorf, S. K. Cell, in press.

68. Griffith, J. D. (1975). Science 187, 1202-1203.
69. Varshavsky, A. J., Sundin, O. H. and Bohn, M. J.
     (1979). Cell 16, 453-466.
70. Herbomel, P., Saragosti, S., Blangy, D. and Yaniv,
     M. (1981). Cell 25, 651-658.
71. Jakobovits, E. B., Bratosin, S. and Aloni, Y.
     (1980). Nature 285, 263-265.
72. Banerji, J., Rusconi, S. and Schaffner, W. (1981).
     Cell 27, 299-308.
73. de Villier, J. and Schaffner, W. (1982). Nucleic
     Acids Res. 9, 6251-6264.
74. Groudine, M., Eisenmann, R. and Weintraub, H.
     (1981). Nature 292, 311-317.
75. Jaenisch, R., Jaehner, D., Hobis, P., Simon, I.,
     Loehler, J., Harbers, K. and Grotkapp, D. (1981).
     Cell 24, 519-529.
76. Hall, B. D., Clarkson, S. G. and Tocchini-
     Valentini, G. (1982). Cell 29, 3-5.
77. Gottesfeld, J. M. and Bloomer, L. S. (1980). Cell
     21, 751-760.
78. Louis, C., Schedl, P., Samal, B. and Worcel, A.
     (1980). Cell 22, 387-392.
79. Pelham, H. R. B. and Brown, D. D. (1980). Proc.
     Nat. Acad. Sci. 77, 4170-4174.
80. Gottesfeld, J. and Bloomer, L. S. (1982). Cell
     28, 781-791.
81. Wittig, B. and Wittig, S. (1979). Cell 18, 1173-
     1183.
82. Bryan, P. N., Hofstetter, H. and Birnstiel, M. L.
     (1981). Cell 27, 459-466.
83. Brown, D. D. and Sugimoto, K. (1973). J. Mol.
     Biol. 78, 397-416.
84. Pruitt, S. C. and Grainger, R. M. (1981). Cell
     24, 711-
85. Labhart, P. and Koller, T. (1982). Cell 28, 279-
     292.
86. Reeves,R. and Jones, A. (1976). Nature 260, 495-
     500.
87. Giri, C. P. and Gorovsky, M. A. (1980). Nucl.
     Acids Res. 8, 197-214.
88. Macleod, D. and Bird, A. (1982). Cell 29, 211-
     218.
89. Weisbrod, S. (1982). Nature 297, 289-295.
90. Hynes, N. E., Kennedy, N., Rahmsdorf, V. and
     Broner, B. (1981). Proc. Nat. Acad. Sci. 78,
     2038-2042.
91. Lee, F., Mulligan, R., Berg, P. and Ringold, G.
     (1981). Nature 294, 228-232.
92. Kurtz, D. T. (1981). Nature 291, 629-631.
93. Hauser, H., Gross, G., Bruns, W., Hochkeppel, H.,
     Mahr, V. and Collins, J. (1982). Nature 297,
     650-654.

Genes: *Structure and Expression*
Edited by A. M. Kroon
© 1983 John Wiley & Sons Ltd.

STRUCTURE AND FUNCTION OF RIBOSOMAL RNA

Harry F. Noller and Peter H. van Knippenberg

INTRODUCTION

The present day student of biochemistry would have little trouble in describing the functions of two important classes of cellular ribonucleic acids, i.e., messenger RNA and transfer RNA. The essence of the interaction of these molecules in translation of the genetic message is indeed easily understood by everyone who is familiar with RNA–RNA interaction through base pairing, although the rules must be bent somewhat to explain the genetic code (1). For most biochemists this knowledge has been sufficient to satisfy one's mind as far as RNA function is concerned. One should realize, however, that the fraction of cellular RNA of this type constitutes only 5 to 10 percent of the total RNA in most cells. Roughly 90 percent of total cellular RNA is present in particles, called ribosomes, that are specifically designed to facilitate the interaction between mRNA and tRNA during protein synthesis.

In biochemical terms, the ribosome is a huge complex, an assembly of some 50 to 60 different protein molecules and at least three RNA molecules with a total mass of several million daltons (2). Between one half to two thirds of this mass is RNA. For "ribosomologists" it is a challenge to establish the structure and function of the individual molecules of which the ribosome is composed. Although studies on the function and on the structure of ribosomal components have gone on alongside for many years, it is fair to state that at the moment our knowledge of the structure of the proteins and the nucleic acid is far more advanced than our understanding of their function.

For a variety of historical reasons, earlier work on ribosome structure and function emphasized the ribosomal proteins. In

recent years, advances in nucleic acid biochemistry have permitted
elucidation of the primary and secondary structures of the rRNAs
from a variety of organisms; at the same time, a great deal of
evidence has been obtained to support the notion that rRNA is
involved directly in ribosome function, and does not serve simply
as a scaffold for assembly of ribosomal proteins. In an
evolutionary sense, it can be argued that ribosomal RNA
represents the essence of ribosome structure, since it is hard to
imagine how the primordial ribosome could have been made from the
very molecules which it evolved to produce (3).

THE RIBOSOMAL NUCLEIC ACIDS

Ribosomal particles are not only found in all cells,
prokaryotic and eucaryotic, but also in subcellular organelles
(mitochondria and chloroplasts). The active ribosome consists of
a small and a large subunit. Traditionally, the approximate
sedimentation coefficient has been taken as a crude measure to
indicate the size of the ribosome, its subunits and the ribosomal
RNAs.

Since the overwhelming majority of published information on
ribosomes deals with E. coli ribosomes, we shall direct our focus
to the ribosomes of this organism, with reference to other
organisms, chloroplasts and mitochondria where appropriate.
Bacterial ribosomes (called 70S ribosomes) are composed of about
60 percent RNA and 40 percent protein by mass. The 70S ribosome
can be dissociated into a 30S and a 50s subunit, with respective
molecular weights of 0.85×10^6 and 1.75×10^6, respectively (2).
The 30S subunit contains 16S RNA (1542 nucleotides, in E. coli)
and a single copy of each of 21 different proteins. The 50S
subunit contains 23S RNA (2904 nucleotides), 5S RNA (120
nucleotides) and 31 different proteins. These various
macromolecular components assemble with one another by means of
non-covalent interactions.

The ribosomes of chloroplasts and their subunits are very
similar in size to bacterial ribosomes and also sediment at 70S,

50S and 30S. They also contain 23S and 5S RNA in their 50S
subunit and a 16S RNA in the 30S subunit, but in addition the
large subunit contains small RNA molecules of variable length (4).
Mitochondrial ribosomes are structurally diverse, varying in gross
properties from one organism to the other (5). Their
sedimentation values range from 55 to 80S for the intact ribosome,
40 to 60S for the large and 30 to 55S for the small subunit.
However, these values are of course influenced by the shape of the
particles and their relative content of protein and RNA. In
general, mitochondrial ribosomes from animal cells are much
smaller than those from other sources (yeast, Neurospora, plant
cells). The same holds true for the RNAs: the large subunit of
animal mitochondria contains a 16S RNA while the corresponding RNA
in yeast is 21S and in Neurospora 25S; the small subunit has a 12S
RNA in animals, a 15S RNA in yeast and a 19S RNA in Neurospora.
Typically in all these ribosomes a 5S RNA (or a RNA that
corresponds to bacterial 5S RNA) is lacking, but plant
mitochondrial ribosomes may have such an RNA. The ribosomes in
the cytoplasm of all eucaryotic cells sediment at a value close to
80S. Their subunits have a sedimentation coefficient of roughly
60S and 40S. Although this does by no means exclude variability
among the cytoplasmic ribosomes they all contain a 5S RNA, a 5.8S
RNA and a 28S RNA molecule in the 60S particle and a 18S RNA in
the small subunit. Recent evidence shows that the eucaryotic 5.8S
RNA corresponds to the 5' end of prokaryotic 23S RNA (6, 7). Also
a 2S RNA found in ribosomes of certain insects (8) and a 4.5S RNA
in some plant ribosomes (9) are strongly homologous to regions in
23S RNA of E. coli (7).

It is only because of the rapid progress in nucleotide
sequence analysis of the last few years that we are able to make
comparisons that do not rely on such crude parameters as a
sedimentation coefficient. The conservation of sequence and/or
secondary structure in corresponding RNA molecules from different
sources makes it possible to relate molecules with one another,
even if their resemblance is very weak on the basis of physical
parameters alone (i.e., size).

We shall now proceed with a discussion of the structure and function of the three major classes of ribosomal RNA, i.e., "5S RNA" from the large subunit, "16S RNA" from the small subunit, and "23S RNA" from the large subunit.

5S RNA Structure

Many dozens of 5S RNA sequences have now been determined, from a variety of sources (for a partial listing, see refs. 10 and 11). In spite of numerous attempts, a satisfactory secondary structure model for this relatively small molecule (120 nucleotides) was not deduced until 1975, when Fox and Woese brought the powerful concept of comparative sequence analysis to bear on 5S RNA (12). This approach is based on the assumption that the 5S RNAs of widely divergent organisms are likely to have the same basic secondary /structure, in spite of numerous differences in their nucleotide sequences. Thus, a mutation in a base involved in a helix usually (but not always) demands a compensating mutation in the complementary base with which it is paired. When several of these compensating changes can be found for a given helix, a compelling case can be made for the biological existence of the helix. Conversely, incorrect helices (of which there are very many) can usually be disproven by the existence of non-compensated mutations, that give rise to non-pairs in a potential helix. The greater the number of sequences at one's disposal, the greater is one's ability to arrive at the correct structure. A refined version (13) of the Fox and Woese model for 5S RNA is shown in figure 1a. There are four helical elements, each of which is now supported by extensive comparative sequence evidence. Similar models have been derived for the secondary structures of eucaryotic and archaebacterial 5S RNAs (14-16).

5S RNA illustrates an important concept relating to RNA structure that has emerged from the sequence work: in related RNA molecules secondary structure dominates conservation of primary structure. This point is nicely demonstrated in Figure 1. The nucleotide sequences of E. coli and human 5S RNA show very little

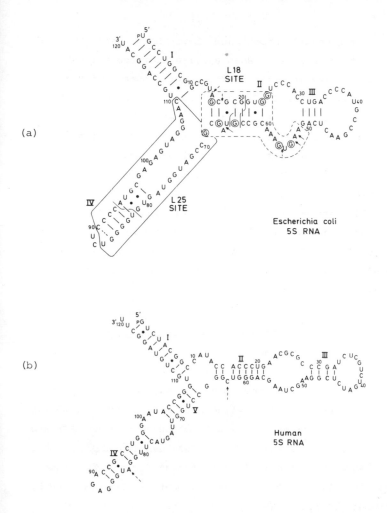

Figure 1.   Secondary structure models for 5S RNA (a) Eubacterial
(E. coli) 5S RNA (12-14) showing binding regions for ribosomal
proteins L18 and L25. Arrows indicate RNase A or T$_2$ cuts which
occur in the free RNA but not in the protein-RNA complexes (25);
curved lines show cobra venom RNase cuts that are inhibited by the
bound   proteins.   Circled   guanines   are   protected against
modification by kethoxal (24) or dimethylsulfate (13) by protein
L18.   (b) Eucaryotic (human) 5S RNA (14).   The arrows indicate the
bulged bases which are putative protein recognition sites.

homology. (No one would predict them to be related.) However, the
molecules look very similar when folded into secondary structure
models (Figure 1).

Several experimental approaches have been used to test
secondary structure models. When applied to ribosomal RNA, one
must interpret such results with caution, for a number of reasons:
(1) The biologically active conformation can be altered during the
course of isolation of the RNA; (2) in the absence of proteins,
the RNA structure may be altered; (3) the probes themselves (e.g.,
nucleases or chemical reagents) can perturb RNA conformation; (4)
the ionic environment (Mg^{++}, monovalent cations, pH, etc.) can
affect RNA conformation drastically. (5) There may be more than
one biologically active conformation for ribosomal RNAs. This
last point is aptly illustrated by the observation of Crothers and
co-workers (17) that 5S RNA in solution can undergo a
conformational transition under ionic conditions very close to
physiological.

Among the more useful conformational probes currently
available are structure-specific reagents and enzymes, tritium
exchange, chemical crosslinking and a variety of spectroscopic
methods. With the exception of nuclear magnetic resonance (NMR),
the latter methods usually do not provide sufficiently detailed
information to test specific structural models; NMR has so far
been successfully applied only to small molecules such as tRNA and
specific fragments of large RNAs. The availability of the
bacteriocins has enabled the isolation of a 49 nucleotide fragment
from E. coli 16S RNA (18, 19). This fragment has been studied by
physical methods (19-21). The proposed helix at the 3' end
(Figure 2) could indeed be visualized by proton NMR (20). The
stability of the hairpin against thermal denaturation is decreased
by methylation of the adenines (21), but whether this effect has
anything to do with the in vivo function of the methyl groups is
unknown. Thus, the presently available detailed data on rRNA
conformation come mainly from chemical and enzymatic approaches.
In all studies where sufficient care has been taken to ensure that
the 5S RNA was in its native conformation, the experimental

results are in close accord with the proposed structure. The only discrepancies are those cases where the experimental data indicate the existence of higher order structure in regions not base-paired in the model. This is seen, for example, around positions 44-47, 70-76 and 100-106 (Fig. 1). Very likely, these regions of the molecule are organized in some kind of tertiary structure. Further refinements of the secondary structure have recently been proposed by Studnickna et al. (22), and involve pairing of residues 28-30 with 53-55. Crosslinking of position 41 to position 72 by the bivalent reagent phenyldiglyoxal suggests possible tertiary folding that would bring these residues into mutual proximity (23). Ultimate elucidation of the 5S RNA structure awaits single crystal x-ray diffraction analysis; for the meanwhile, these molecules, as for ribosomes and their subunits, have proven difficult to crystallize.

5S RNA-Protein Interactions

For molecular biologists interested in the mechanism of protein-RNA recognition, the ribosome is a rich source of such interactions. At least 15 of the 52 r-proteins of E. coli are known to interact directly and specifically with rRNA, and three of these bind to 5S RNA. Because of its small size and relative ease of preparation, 5S RNA is an attractive model system for studying RNA-protein recognition. The binding of two of its cognate proteins, L18 and L25, has been studied in some detail. Use of chemical and enzymatic probes (13, 24) has permitted localization of the L18 binding region to positions #16-24 and #51-69. These sequences interact to form the central helix shown in figure 1. This helix nearly always contains an interesting irregularity consisting of a single "bulged" nucleotide at position 66. There is evidence to support the notion that this irregularity, in addition to other structural features, is a recognition signal which enables the protein to distinguish this helix from the multitude of other cellular RNA structures that it must confront during the course of ribosome assembly (13).

Protein L25 binds to a region of 5S RNA that has been localized to about positions 70–106 (14, 25, 26). At least part of this site is a helix (Fig. 1), which may extend to form a highly irregular helical structure (16). Again, it is attractive to imagine that the structural irregularity may help to provide some sort of recognition signal for the protein.

Functional role of 5S RNA

It has often been suggested that 5S RNA provides a binding site for tRNA (e.g., refs. 27 and 28) in which the conserved $G_{44}AAC$ sequence would pair with the constant GTψC of tRNA. Although attractive, this model has received criticism; for example, these sequences appear to be relatively inaccessible to various single strand–specific probes (29), and the tRNA sequence has been found to be intricately involved in maintaining tertiary structure in that molecule (30, 31).

More recently, Pace and coworkers have tested the latter hypothesis directly and elegantly (32). In these experiments, a sequence including the putative tRNA binding region of 5S RNA was deleted by a series of specific ribonuclease cleavages. The deleted 5S RNA was then incorporated into 50S ribosomal subunits by in vitro reconstitution, and the protein synthesis activity of the particles tested by a standard poly U-directed synthesis of polyphenylalanine. Ribosomes containing deleted 5S RNA were active in this assay, demonstrating that the $G_{44}AAC$ sequence is certainly not an obligatory binding site for the tRNA in this system. It cannot yet be ruled out that some aspect of protein synthesis not tested by this in vitro system (e.g., initiation via f·Met-tRNA) may depend on this sequence in 5S RNA. In any event, it is fair to say that the functional role(s) of 5S RNA is presently unknown.

Primary structures of 16S and 23S RNAs

Development of DNA cloning and rapid sequencing methods has permitted elucidation of the nucleotide sequences of several 16S and 23S RNAs, and their eucaryotic analogues. Currently available complete large rRNA sequences and their respective chain lengths are listed in Table 1. The E. coli rRNAs are representative with a 16S RNA of 1542 residues (33–35) and a 23S RNA of 2904 (36).

The 16S-like RNAs vary in size by almost a factor of two between human mitochondrial 12S RNA (37) and Xenopus cytoplasmic 18S RNA (38); the corresponding 23S-like RNAs exhibit a similar size diversity (Table 1). In spite of this variation, there is a strong family resemblance between rRNAs from all organisms and their organelles.

Modified Nucleotides

In the 16S RNA of E. coli the position of 10 modified (methylated) nucleotides with a total of 13 methyl groups has been established (Table 2)(39, 40).

Table 2. Positions of methylated nucleotides in the
sequence of 16S RNA of E. coli

Position	Modified nucleotide
527	m_2^7G
966,1207,1516	m_2^5G
967,1407	m_4^5C
1402	m^4Cm
1498	m^5U
1518,1519	m_2^6A

Since originally a larger number of modified nucleotides was reported (41, 42) it remains possible that the list of Table 2 will have to be extended. The 16S RNA of P. vulgaris, which shows 93% homology with the sequence of E. coli (43) contains all the modifications of Table 2 at similar positions, except $_m^5C$ (at position 1407, which instead is an unmodified C.

The small ribosomal subunit RNA of hamster mitochondria contains one residue each of m^5U, m^4C, m^5C and two residues of m_2^6A (44, 45). 18S RNA of Xenopus laevis contains many more modified

Table I

Sequenced 16S-like rRNAs:

	organism	residues	reference
Eubacteria			
	E. coli	1542	33
	B. brevis	(1540)	53
	P. vulgaris	1544	107
Chloroplasts			
	Maize chloroplast	1490	55
	Tobacco chloroplast	1485	108
Archaebacteria			
	Halobacterium volcanii	1469	118
Eukaryotes			
	Yeast cytoplasmic (18S)	1799	109
	Xenopus laevis (18S)	1825	38
	Rat liver (18S)	1874	110
Mitochondria			
	Human mitochondria	954	37
	Bovine mitochondria	958	111
	Mouse mitochondria	956	112
	Rat mitochondria	953	113
	Yeast mitochondria	1686	114,115
	Aspergillus mitochondria	1437	116
	Paramecium mitochondria	(1607)	117

Sequenced 23S-like rRNAs:

	organism	residues	reference
Eubacteria			
	E.coli	2904	36
	B. stearothermophilus	2931	54
Chloroplasts			
	Maize chloroplast	2903	56
	Tobacco chloroplast	2904	119
Eukaryotes			
	Yeast cytoplasmic (26S)	3393	112,123
Mitochondria			
	Human mitochondria	1559	37
	Bovine mitochondria	1571	111
	Mouse mitochondria	1582	112
	Rat mitochondria	1584	120
	Paramecium mitochondria	2380+	121

nucleotides of which some 40 have been placed in the sequence (38). Most of the methylations in eucaryotic cytoplasmic ribosomal RNAs are 2'-O-methylribose and are found in the 3' parts of the RNAs (47, 48). In E. coli 16S RNA, 9 of the 13 methyl groups (Table 2) are located within 150 nucleotides from the 3' end.

Little is known about the function of the modified nucleotides in ribosome function with the possible exception of the two adjacent dimethylated adenines near the 3' end (see below).

Secondary structures of the large rRNAs

Computer searches for potential base-paired sequences in E. coli 16S RNA (49) showed that there are on the order of 10,000 plausible helices; only about 60 to 70 of these can coexist in a single structure, however. The problem thus became one of elimination of a vast number of incorrect helices. Although free energy prediction rules (50) and results of biochemical probing experiments (51, 52) were of great help in the task, confident identification of the biologically relevant structures depended on comparative sequence analysis. For this purpose, the corresponding rRNAs from Bacillus brevis (53) or B. Stearothermophilus (54) were sequenced. Additionally, the maize chloroplast rRNA sequences became available (55, 56) and were also used in the analysis. Secondary structure models for the 16S and 23S rRNAs based on these data were deduced (49, 57) and are shown in Figures 2 and 3. Two other groups have independently derived secondary structure models for these RNAs (58-61), and there is now reasonably good agreement between the 16S RNA models.

It is useful to point out a few of the properties of the secondary structure models. In addition to local hairpin (or stem-loop) structures, similar to those found in tRNA, base-pairing in rRNA also occurs between sequences that are hundreds or even thousands of nucleotides apart in the primary structure. A typical example is the pairing between sequences at positions 30 and 550 in 16S RNA (Fig. 2). This interaction sequesters about

500 nucleotides into a 5' terminal domain. Interestingly, a fragment corresponding almost exactly to this domain has been isolated after gentle ribonuclease treatment of 16S RNA (62). Similarly, interaction between positions 570 and 880 and 930 and 1390 define the central and 3' major domains. 23S RNA is organized into 6 domains, again having average sizes of about 500 nucleotides per domain (Fig 3). Additionally, the 5' and 3' termini of 23S RNA are base paired, as is found in tRNA and 5S RNA, but not 16S RNA.

After the secondary structure models had been derived, primary structures of large rRNAs from several additional sources were completed (Table 1). These are all consistent with the general secondary structure models (Figs. 2, 3), allowing for insertions and deletions where appropriate (63). This argues strongly for the general validity of the proposed models. Interestingly, deletions and insertions tend always to occur in the same positions in the structure, irrespective of the phylogenetic relatedness of the organisms. This suggests that there are "variable" regions of the structures which can tolerate gross alteration during evolution, perhaps by analogy to the variable loop of tRNA. Conversely, other parts of these molecules show extreme conservation of primary and/or secondary structure. The pattern of conservation undoubtedly provides important clues to crucial aspects of ribosome structure and function. As we shall see below, there is experimental evidence to suggest what the nature of some of these may be.

Three-dimensional organization of 16S RNA

Certain constraints on the 3-dimensional organization of rRNA are dictated by their secondary structures. This, however, is not sufficient to begin attempts to deduce their structural arrangement in the ribosome. For example, the 16S RNA model (Fig. 2) must be folded approximately in half in both dimensions in order to reduce its overall size to that of the 30S ribosomal subunit (64). One can, however, extrapolate from other information to arrive at a low resolution assignment of the positions of specific regions of 16S RNA in the ribosome.

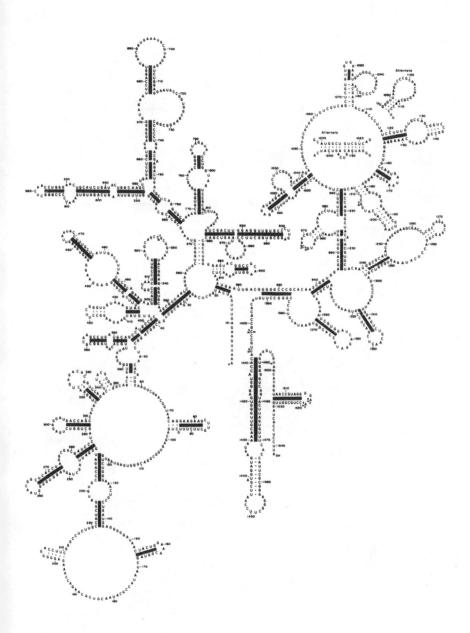

Figure 2. Secondary structure model for 16S ribosomal RNA (E. coli) (49, 63, 64). Helices considered to be proven by phylogenetic comparative evidence are shaded.

84

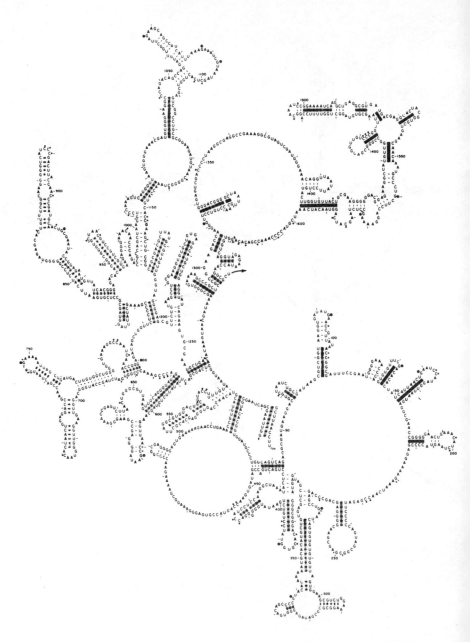

Figure 3. Secondary structure model for 23S ribosomal RNA (E. coli)(7). Symbols are as for fig. 2 and as described in ref 7.

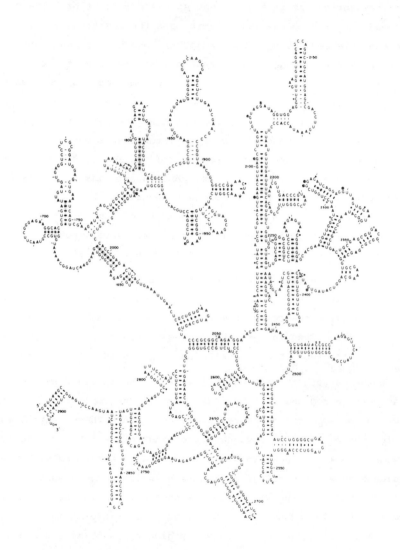

Positions of ribosomal proteins in the 30S subunit have been studied by a number of ingenious approaches (65-69). One of these utilizes antibodies raised against specific r-proteins. These are allowed to bind to the 30S subunit, and the positions of specific proteins are deduced by visualizing the location of bound antibody with respect to recognizable morphological features of the subunit by electron microscopy (68, 69). A second method is to measure protein-protein distances by neutron scattering (67). This approach uses pairs of deuterated proteins reconstituted into an otherwise normal ribosomal subunit. Interference patterns created by scattering from the deuterated protein pairs yield measurements of the distance between their centers of mass. Quite good agreement is observed between protein positions obtained by these two methods. A consensus arrangement of some of the proteins based on these data is shown in Figure 4, superimposed on the electron microscopy model for the 30S subunit (64, 67).

Specific regions of the 16S RNA can then be assigned to positions in the e.m. model using direct or indirect evidence. The direct evidence comes from the location of antibodies directed at specific sites in the rRNA. These include the $_{m}^{7}G$ at position 527 (70), the two $_{m_{2}}^{6}A$'s at positions 1518 and 1519 (71), and the 3' terminus of the RNA chain (72-74). The indirect evidence is that certain regions of the RNA have been implicated in the binding of specific ribosomal proteins. Since these regions must be in contact with their cognate proteins in the ribosome, we can assign them to positions in the model accordingly. These assignments are summarized in Figure 4, which correlates sites in the secondary structure model with positions in the structural model of the 30S subunit. Clearly, much more information of this kind will be required before anything approaching a structural representation of the RNA in the 30S subunit will be forthcoming.

Functional role of rRNA

It was mentioned earlier that the functional role of 5S RNA is presently unknown; this is not the case for the large rRNAs,

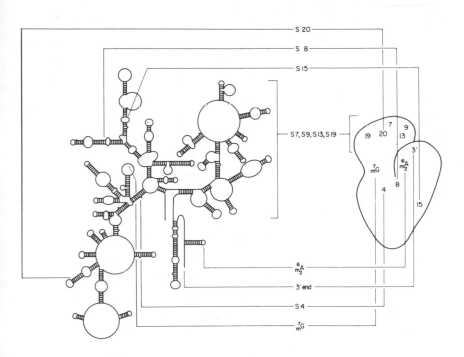

Figure 4: Assignment of specific regions of 16S ribosomal RNA to positions in the 3-dimensional model for the 30S ribosomal subunit. The shape of the 30S subunit is from electron microscopy (68). Positions of specific nucleotides and proteins are from immuno-electron microscopy (68-73) and from neutron scattering (67). Protein binding sites on the 16S RNA are summarized in ref. 124.

where a number of lines of evidence have clearly implicated them in crucial roles in protein synthesis. These include sensitivity or resistance to a variety of antibiotics, mRNA selection, and association of 30S and 50S subunits. Also likely is the possibility that rRNA is somehow directly involved in binding tRNA.

Some mitochondria carry only a single copy of the rRNA genes; using these organelles in genetic experiments, alterations involving point mutations in rRNA have been shown to affect the antibiotic sensitivity of mitochondrial ribosomes. Because of the close relationship between the mode of action of certain antibiotics and specific ribosomal functions (e.g., chloramphenicol and peptidyl transferase (75)), it is not unreasonable to imagine that some of these antibiotics may act by directly binding to functional sites that are at least in part composed of portions of the rRNA.

An interesting structure near the 3' ends of the RNAs of virtually all small ribosomal subunits is the sequence of $m_2^6Am_2^6A$ (reviewed in 76). In a mutant of E. coli that is resistant to the antibiotic kasugamycin, the methyl groups on the adenines are lacking (77). This is due to the absence of a specific methylase, the product of the ksgA gene (78). The effect of kasugamycin on initiation of protein synthesis as well as the effect of methyl groups in the absence and the presence of the antibiotic has been studied extensively (78-82). Kasugamycin inhibits initiation of protein biosynthesis by interfering with the messenger-dependent binding of initiator tRNA to 30S ribosomes (79). 70S-bound initiator tRNA is not affected. Curiously, however, 30S ribosomes from resistant bacteria (and thus lacking the methyl groups on the adjacent adenines) are also sensitive in this in vitro reaction, and only display resistance in the presence of the 50S subunit (either from sensitive or resistant bacteria)(79). Since the methyl groups also influence the interaction of the subunits (82), it was hypothesized that resistance is due to an enhanced rate of interaction of the mutant 30S initiation complex with the 50S ribosome (83). The 30S·mRNA·fMet-tRNA complex could escape from kasugamycin inhibition by a faster reaction with the 50S subunits.

Shine and Dalgarno (84-86) made a very specific proposal for recognition of translational initiation sites on mRNA by a sequence near the 3' end of 16S RNA. They observed that the AUG codons in initiation sites were preceded approximately 10 nucleotides to the 5' side by a purine-rich sequence that could potentially base pair with the pyrimidine cluster at the 3' end of 16S RNA. This "Shine and Dalgarno" model has been supported experimentally. For example, treatment of a complex consisting of ribosomes, a phage RNA fragment containing an initiation site and initiator tRNA, with colicin E3 (which cleaves the 16S RNA near its 3' end; see below) yielded a complex containing phage RNA fragment and a large oligonucleotide from the 3' end of 16S RNA (52).

A mutation in the 0.3 mRNA of bacteriophage T7 that greatly diminishes translation of this gene was found to be due to a single base alteration 11 nucleotides to the 5' side of the initiator codon (88). This resulted in a change from GAGGU (complementary to the Shine-Dalgarno sequence ACCUC in 16S RNA) to GAAGU. A suppressor mutation was then isolated that restored translation of the message. The sequence of the suppressor was found to be GGAGU, restoring complementarity with the 16S RNA, but in a sequence that overlaps the original complementarity. This result constitutes a compelling case for the in vivo utilization of the Shine-Dalgarno mechanism in initiation of translation. Other evidence (mainly based on the use of oligonucleotides complementary to the 3' end of 16S RNA to block initiation) lends support to the Shine and Dalgarno proposal (89-92). Much of the evidence for this theory has been extensively reviewed (97, 94).

The importance of the 3' region of 16S RNA in initiation of protein synthesis and in the interaction of ribosomal subunits has also become clear from other observations. First, cleavage of the 16S RNA in situ at 49 nucleotides from the 3' end by the bacteriocin colicin E3 (95) and cloacin DF13 (96) strongly inhibits protein biosynthesis (96, 97). Intactness of the 16S RNA at this position is required for proper function of the initiation factor IF-1 in initiation and in subunit interaction (98). It has also been reported that the cleavage strongly reduces the elongation rate of protein synthesis and increases the fidelity of codon reading (99).

Another approach that has proven useful in identifying possible functional sites is protection of the RNA from chemical modification or nuclease digestion by functional ligands. It was shown in early experiments that kethoxal modification of certain guanosine residues of 16S RNA in 30S subunits causes inactivation of tRNA binding (100). Subunits could be protected from inactivation by bound tRNA, and ribosome reconstitution experiments demonstrated that modification of ribosomal proteins had no effect on tRNA binding. Kethoxal modification was also shown to cause loss of 30S-50S subunit association albeit at a somewhat slower rate (101).

Localization of the kethoxal-reactive sites responsible for loss of these functions has been carried out in several stages. First, a diagonal electrophoresis method was employed to identify sites of kethoxal modification (51). Second, protection experiments were done, in which ^{32}P labelled 30S subunits were bound with unlabelled 50S subunits to create "tight couple" 70S ribosomes (102). Many sites in 16S RNA were shown to have a greatly reduced reactivity in 70S compared with 30S subunits. Nearly all the protected sites are clustered in the central and 3' terminal domains (Fig. 5). This finding is independently corroborated by studies using protection against ribonuclease T_1 (103) and cobra venom ribonuclease (104), where 50S subunits again protect 16S RNA at specific sites in the central and 3' terminal domains. Additionally, protection of a site in the 5' domain is observed in the cobra venom experiments.

Evidence that the protected 16S RNA sites may actually make contact with 50S subunits comes from modification-selection experiments (101). 30S subunits are modified for a short time

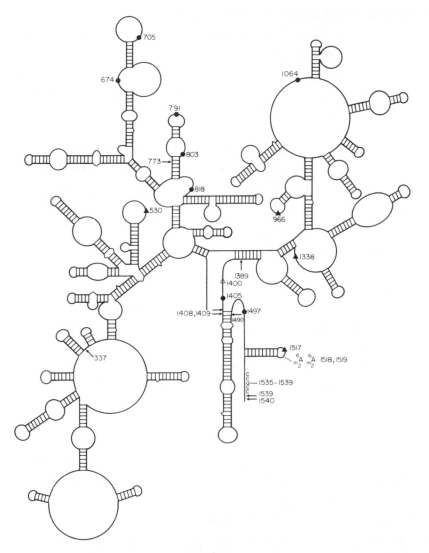

Figure 5. Sites in 16S RNA implicated in specific ribosomal functions. (●), protected from kethoxal by 50S subunits (102) or sites whose modification causes loss of 50S subunit binding (101); (▲) protected from kethoxal in polysomes but not in vacant 70S ribosomes (46); (→) protected from cobra venom RNase by 50S subunits (104); (△) site of crosslinking of wobble base in tRNA to 16S RNA (105); ($m_2^6Am_2^6A$), sites of post-transcriptional methylation affecting Kasugamycin sensitivity (77-78); (CCUCC), the "Shine-Dalgarno" sequence, important for mRNA selection (84-88).

RNA function and phylogenetic conservation

Phylogenetic comparison of rRNA structures reveals a wide spectrum of conservation of primary and secondary structural features (49, 57, 63, 64, 106). At one extreme is the deletion or insertion of entire helical stems; at the other are invariant sequences on the order of 10 to 20 nucleotides long, and universally conserved base pairing features. There is a "core" of secondary structure that is virtually invariant among the presently known rRNAs, and the universal sequences appear to be included within this core. This doubtless reflects an extremely high precision of structure, deviation from which would be lethal.

An early observation was that many of the kethoxal-reactive sites in 16S RNA are in highly conserved sequences (102, 106). It is significant that the conserved sites are all found to be protected, either in 70S ribosomes or in polysomes. This correlation is consistent with the interpretation that the protected sites directly contact their respective ligands during protein synthesis. It must be kept in mind, however, that there is an alternate interpretation for the protection experiments. Binding of a ligand to 30S ribosomal subunits could also bring about an induced conformational change affecting the reactivity of these sites. While no less interesting a conclusion than simple protection, the implications of the two interpretations are quite different. In the case of direct protection, the fact of high sequence conservation implies a site specifically evolved to interact with the ligand. In contrast, the allosteric interpretation in its simplest form would imply that the sequence has been conserved to interact with some other part of the 30S subunit, either RNA or protein. It will be difficult to distinguish between these possibilities; fresh thinking combined with newly developed methods need to be brought to bear on this and the other demanding problems raised here.

Future directions

Some of the main questions that need to be considered in future studies on ribosomal RNA include: What is the nature of with kethoxal, such that about half the population has lost the ability to bind 50S subunits. The competent 30S particles are then selected out of the total population by adding excess 50S subunits and isolation of the resulting 70S ribosomes by sucrose gradient centrifugation. The kethoxal modification pattern of the competent 30S fraction is then compared with that of the total 30S population. Sites whose modification interferes with subunit association will be missing from the pattern obtained from competent 30S subunits. The results of these studies show that the missing sites correspond almost precisely to those identified by the protection experiments.

In an attempt to probe the sites protected by tRNA (and mRNA), analogous kethoxal protection studies were performed on polysomes (46). The reactivity of each site was measured in both polysomes and in "vacant" or "run-off" 70S ribosomes. Here, the only expected difference is the presence of mRNA and tRNA in polysomes, but not in 70S monosomes. The tRNA occupancy was monitored directly by isolation of bound tRNA on denaturing gels and quantitation by means of its ^{32}P radioactivity. In addition to the sites previously shown to be protected in 70S, protection of several new sites was observed (Fig. 5); these were either only marginally protected or unprotected in vacant 70S ribosomes.

A compelling demonstration of the close proximity of tRNA and rRNA is the direct crosslinking of the 5' (wobble) base of the anticodon of tRNA to position 1400 of 16S RNA (105). Irradiation of certain ribosome-bound tRNAs with long wavelength ultraviolet light causes a photochemical reaction which covalently links the two bases in a cyclobutane dimer adduct with yields of more than 50 percent. This result strongly suggests intimate association of the anticodon region of tRNA with the 16S RNA. The site of crosslinking in 16S RNA is in a region shown to be strongly protected from kethoxal and nucleases in 70S ribosomes, supporting the prevailing notion that tRNA is bound somewhere between the two ribosomal subunits.

protein–RNA interaction in the ribosome? How do the RNA and protein moieties cooperate mechanistically during the translation process? Does the ribosome contain moving parts? If so, what are they and how do they work? What is the nature of rRNA–tRNA interaction? What kinds of tertiary structure does rRNA form (and what is its 3-dimensional structure?)? Can we conceptualize how the first ribosome may have evolved?

References

1. Crick, F. H. C. (1966) J. Mol. Biol. _19_, 548–555.

2. "Ribosomes," Chambliss, G., Craven, G., Davies, J., Davis, K., Kahan, L. and Nomura, M., eds.) Univ. Park Press, Baltimore, 1980.

3. Woese, C. R. (1980) ref. 2, pp. 357–373.

4. Dyer, T. A., Bowman, C. M. and Payne, P. I. (1976) in Nucleic Acids and Protein Synthesis in Plants," pp. 121–123, Plenum Press, N.Y.

5. Boynton, J. E., Gillham, N. W. and Lambowitz, A. M. (1979) ref. 2, pp. 903–950.

6. Nazar, R. N. (1980) FEBS Lett. _119_, 212–214.

7. Noller, H. F., Kop, J., Wheaton, V., Brosius, J., Gutell, R. R., Kopylov, A. M., Dohme, F., Herr, W., Stahl, D. A., Gupta, R. and Woese, C. R. (1981) Nucleic Acids Res. _9_, 6167–6189.

8. Pavlakis, G. N., Jordan, B. R., Wurst, R. M. and Vournakis, J. N. (1979) Nucleic Acids Res. _7_, 2213–2238.

9. Mackay, R. M. (1981) FEBS Lett. _123_, 17–18.

10. Erdmann, V. A. (1982) Nucleic Acids Res. _10_, r93–r115.

11. Hori, H. and Osawa, S. (1979) Proc. Nat. Acad. Sci. USA _76_, 381–385.

12. Fox, G. E. and Woese, C. R. (1975) Nature _256_, 505–507.

13. Peattie, D. A., Douthwaite, S., Garrett, R. A. and Noller, H. F. (1981) Proc. Nat. Acad. Sci. USA _78_, 7331–7335.

14. Garrett, R. A., Douthwaite, S., and Noller, H. F. (1981) Trends Biochem. Sci. _6_, 137–139.

15. Luehrsen, K. R. and Fox, G. E. (1981) Proc. Nat. Acad. Sci. USA _78_, 2150–2154.

16. Stahl, D. A., Luehrsen, K. R., Woese, C. R., Pace, N. R. (1981) Nucleic Acids Res. _9_, 6129–6137.

17. Kao, T. A. and Crothers, D. M. (1980) Proc. Natl. Acad. Sci. USA _77_, 3360–3364.

18. Baan, R. A., van Charldorp, R., van Leerdam, E. van Knippenberg, P., Bosch, L., de Rooij, J. F. M. and van Boom, J. H. (1976) FEBS Lett. _71_, 351–355.

19. Yuan, R. C., Steitz, J. A., Moore, P. B. and Crothers, D. M. (1979) Nucleic Acids Res. 7, 2399-2418.

20. Baan, R. A., Hilbers, C. W., van Charldorp, R., Van Leerdam, E., Van Knippenberg, P. A. and Bosch, L. (1977) Proc. Nat. Acad. Sci. USA 74, 1028-1031.

21. Van Charldorp, R., Heus, H. A., van Knippenberg, P. H., Joordens, J., de Bruin, S. H. and Hilbers, C. W. (1981) Nucleic Acids Res. 9, 4413-4422.

22. Studnicka, G. M., Eiserling, F. A. and Lake, J. A. (1981) Nucleic Acids Res. 9, 1885-1904.

23. Hancock, J. and Wagner, R. (1982) Nucleic Acids Res. 10, 1257-1269.

24. Garrett, R. A. and Noller, H. F. (1979) J. Mol. Biol. 132, 637-648.

25. Douthwaite, S., Christensen, A. and Garrett, R. A. (1982) Biochemistry 21, 2313-2320.

26. Douthwaite, S., Garrett, R. A., Wagner, R. and Feunteun, J. (1979) Nucleic Acids Res. 6, 2453-2470.

27. Erdmann, V. A. (1977) Prog. Nucleic Acid Res. Mol. Biol. 18, 45-90.

28. Forget, B. G. and Weissman, S. M. (1967) Science 158, 1695-1699.

29. Noller, H. F. and Garrett, R. A. (1979) J. Mol. Biol. 132, 637-648.

30. Rich, A. and Raj Bhandary (1976) Ann. Rev. Biochem. 45, 805.

31. Clark, B. F. C. (1977) Prog. Nucleic Acid Res. Mol. Biol. 20, 1.

32. Pace, B., Matthews, E. A., Johnson, K. D., Cantor, C. R. and Pace, N. R. (1982) Proc. Nat. Acad. Sci. USA 79, 36-40.

33. Brosius, J., Palmer, M. L., Kennedy, P. J., and Noller, H. F. (1978) Proc. Nat. Acad. Sci. USA 75, 4801-4805.

34. Carbon, P., Ehresmann, C., Ehresmann, B. and Ebel, J. P. (1979) Eur. J. Biochem. 100, 399-410.

35. Brosius, J., Dull, T. J., Sleeter, D. D. and Noller, H. F. (1981) J. Mol. Biol. 148, 107-127.

36. Brosius, J., Dull, T. J. and Noller, H. F. (1980) Proc. Nat. Acad. Sci. USA 77, 201-204.

37. Eperon, I., Anderson, S. and Nierlich, D. P. (1980) Nature 286, 460-467.

38. Salim, M. and Maden, B. E. H. (1981) Nature 291, 205-208.

39. Ehresmann, C., Stiegler, P., Fellner, P. and Ebel, J. P. (1972) Biochimie 54, 901-967.

40. van Charldorp, R., Heus, H. A., and van Knippenberg, P. H. (1981) Nucleic Acids Res. 9, 2717-2725.

41. Fellner, P. (1969) Eur. J. Biochem. 11, 12-27.

42. Dubin, D. T. and Gunalp, A. (1967) Biochim. Biophys. Acta 2134, 106-123.

43. Carbon, P., Ebel, J. P. and Ehresmann, C. (1981) Nucleic Acids Res. 9, 2325-2333.

44. Dubin, D. T., Taylor, R. H. and Davenport, L. W. (1978) Nucleic Acids Res. 5, 4385-4397.

45. Baer, R. J. and Dubin, D. T. (1981) Nucleic Acids Res. 9, 323-337.

46. Brow, D. A. and Noller, H. F. (1982) J. Mol. Biol. (in press).

47. Khan, M. S. N. and Maden, B. E. H. (1978) Eur. J. Biochem. 84, 241-250.

48. Khan, M. S. N., Salim, M. and Maden, B. E. H. (1978) Biochem. J. 169, 531-542.

49. Woese, C. R., Magrum, L. J., Gupta, R., Siegel, R. B., Stahl, D. A., Kop, J., Crawford, N., Brosius, J., Gutell, R., Hogan, J. J. and Noller, H. F. (1980) Nucleic Acids Res. 8, 2275-2294.

50. Tinoco, I., Borer, P. N., Dengler, B., Levine, M. D., Uhlenbeck, O. C., Crothers, D. M. and Gralla, J. (1973) Nature New Biology 246, 40-41.

51. Noller, H. F. (1974) Biochemistry 13, 4694-4703.

52. Ehresmann, C., Stiegler, P., Carbon, P., Ungewickell, E. and Garrett, R. A. (1980) Eur. J. Biochem. 103, 439-446.

53. Kop, J., Kopylov, A. M., Noller, H. F., Siegel, R., Gupta, R. and Woese, C. R. (unpublished).

54. Kop, J., Wheaton, V., Gupta, R., Woese, C. R. and Noller, H. F. (unpublished).

55. Schwartz, Z. and Kössel, H. (1980) Nature 283, 739-742.

56. Edwards, J. and Kössel, H. (1981) Nucleic Acids Res. 9, 2853-2869.

57. Noller, H. F., Kop, J., Wheaton, V., Brosius, J., Gutell, R. R., Kopylov, A. M., Dohme, F., Herr, W., Stahl, D. A., Gupta, R. and Woese, C. R. (1981) Nucleic Acids Res. 9, 6167-6189.

58. Zwieb, C., Glotz, C. and Brimacombe, R. (1981) Nucleic Acids Res. 9, 3621-3640.

59. Stiegler, P., Carbon, P., Ebel., J.-P. and Ehresmann, C. (1981) Eur. J. Biochem. 120, 487-495.

60. Glotz, C., Zwieb, C., Brimacome, R., Edwards, K. and Kössel, H. (1981) Nucleic Acids Res. 9, 3287-3306.

61. Branlant, C., Krol, A., Machatt, M. A., Pouyet, J., Ebel, J.-P., Edwards, K. and Kössel, H. (1981) Nucleic Acids Res. 9, 4303-4324.

61. Branlant, C., Krol, A., Machatt, M. A., Pouyet, J., Ebel, J.-P., Edwards, K. and Kössel, H. (1981) Nucleic Acids Res. 9, 4303-4324.

62. Garrett, R. A., Ungewickell, E., Newberry, V., Hunter, J. and Wagner, R. (1977) Cell Biol. Int. Rep. 1, 487-502.

63. Woese, C. R., Gutell, R. R. and Noller, H. F. (in preparation)

64. Noller, H. F. and Woese, C. R. (1981) Science 212, 403-411.

65. Traut, R. R., Lambert, J. M., Boileau, G. and Kenny, J. W. (1980) ref. 2, pp. 89-110.

66. Huang, K. H., Fairclough, R. H. and Cantor, C. R. (1975) J. Mol. Biol. 97, 443-470.

67. Moore, P. B. (1980) ref. 2, pp. 111-133.

68. Lake, J. A. (1980) ref. 2, pp. 207-236.

69. Stöffler, G., Bald, R., Kastner, B., Lührmann, R., Stöffler-Meilicke, M. and Tischendorf, G. (1980) ref. 2, pp. 171-205.

70. Trempe, M. R. and Glitz, D. G. (1982) J. Biol. Chem. (in press).

71. Politz, S. M. and Glitz, D. G. (1977) Proc. Nat. Acad. Sci. USA 74, 1468-1472.

72. Olson, H. M. and Glitz, D. G. (1979) Proc. Nat. Acad. Sci. USA 76, 3769-3773.

73. Shatsky, I. N., Mochalova, L. V., Kojouharova, M. S., Bogdanov, A. A. and Vasiliev, V. D. (1979) J. Mol. Biol. 133, 501-515.

74. Lührmann, R., Stöffler-Meilicke, M. and Stöffler, G. (cited in ref. 69).

75. Cundliffe, E. (1980) ref. 2, pp. 555-581.

76. Van Charldorp, R. and van Knippenberg, P. H. (in press).

77. Helser, T. L., Davies, J. E. and Dahlberg, J. E. (1971) Nature New Biol. 233, 12-14.

78. Helser, T. L., Davies, J. E. and Dahlberg, J. E. (1973) Nature New Biol. 235, 6-9.

79. Poldermans, B., Goosen, N. and van Knippenberg, P. H. (1979) J. Biol. Chem. 254, 9085-9089.

80. Poldermans, B., van Buul, C. and van Knippenberg, P. H. (1979) J. Biol. Chem., 254, 9090-9093.

81. Poldermans, B., Roza, L. and van Knippenberg, P. H. (1979) J. Biol. Chem. 254, 9094-9100.

82. Poldermans, B., Bakker, H. and van Knippenberg, P. H. (1980) Nucleic Acids Res. 8, 143-151.

83. Poldermans, B. (1980) Ph.D. Thesis, State University of Leiden, The Netherlands.

84. Shine, J. and Dalgarno, L. (1974) Proc. Nat. Acad. Sci. USA 71, 1342-1346.

85. Shine, J. and Dalgarno, L. (1975) Nature 254, 34-38.

86. Shine, J. and Dalgarno, L. (1975) Eur. J. Biochem. 57, 221-230.

87. Steitz, J. A. and Jakes, K. (1975) Proc. Nat. Acad. Sci. USA 72, 4734-4738.

88. Dunn, J. J., Buzash-Pollert, E. and Studier, W. F. (1978) Proc. Nat. Acad. Sci. USA 75, 2741-2745.

89. Taniguchi, T. and Weissman, C. (1978) Nature 275, 770-772.

90. Eckhardt, H. and Lührmann, R. (1979) J. Biol. Chem. 22, 11185-11188.

91. Backendorf, C., Overbeek, G. P., van Boom, J. H., van der Marel, G., Veeneman, G. and van Duin, J. (1980) Eur. J. Biochem. 110, 599-604.

92. Backendorf, C., Ravensbergen, C. J. C., van der Plas, J., van Boom, J. H., Veeneman, G. and van Duin, J. (1981) Nucleic Acids Res. 9, 1425-1444.

93. Steitz, J. A. (1979) in "Biological Regulation and Develoment" (Goldberger, R. F., ed.) 349-399, Plenum Press, New York.

94. Gold, L., Pribnow, D., Schneider, T., Shinedling, P., Singer, B. S. and Stormo, G. (1981) Ann. Rev. Microbiol. 35, 365-403.

95. Senior, B. W. and Holland, I. B. (1971) Proc. Nat. Acad. Sci. USA 68, 959-963.

96. de Graaf, F. K., Niekus, H. G. D. and Klootwijk, J. (1973) FEBS Lett. 35, 161-165.

97. Bowman, C. M., Dahlberg, J. E., Ikemura, T., Konisky, J. and Nomura, M. (1971) Proc. Nat. Acad. Sci. USA 68, 964-968.

98. Baan, R. A., Duijfjes, J. J., van Leerdam, E., van Knippenberg, P. H. and Bosch, L. (1976) Proc. Nat. Acad. Sci. USA 73, 702-706.

99. Twilt, J. C., Overbeek, G. P. and van Duin, J. (1979) Eur. J. Biochem. 94, 477-484.

100. Noller, H. F. and Chaires, J. B. (1972) Proc. Nat. Acad. Sci. USA 69, 3115-3118.

101. Herr, W., Chapman, N. M. and Noller, H. F. (1979) J. Mol. Biol. 130, 433-449.

102. Chapman, N. M. and Noller, H. F. (1977) J. Mol. Biol. 109, 131-149.

103. Santer, M. and Shane, S. (1977) J. Bacteriol. 130, 900-910.

103. Santer, M. and Shane, S. (1977) J. Bacteriol. 130, 900–910.

104. Vassilenko, S. K., Carbon, P., Ebel, J. P. and Ehresmann, C. (1981) J. Mol. Biol. 152, 699–721.

105. Prince, J. P., Taylor, B. H., Thurlow, D. L., Ofengand, J. and Zimmermann, R. A. (1982) Proc. Nat. Acad. Sci. USA (in press).

106. Woese, C. R., Fox, G. E., Zablen, L., Uchida, T., Bonen, L., Pechman, K., Lewis, B. J. and Stahl, D. (1975) Nature 254, 83–86.

107. Carbon, P., Ebel., J.-P., and Ehresmann, C. (1981) Nucleic Acids Res. 9, 2325–2333.

108. Tohdoh, N. and Sugiura, M. (1982) Gene 17, 213–218.

109. Rubtsov, P. M., Musakhanov, M. M., Zakharyev, V. M., Krayev, A. S., Skryabin, K. G. and Bayev, A. A. (1980) Nucleic Acids Res. 8, 5779–5794.

110. Chan, Y.-L. and Wool, I. G. (personal communication).

111. Anderson, S., de Bruijn, M. H. L., Coulson, A. R., Eperon, I. C., Sanger, F. and Young, I. G. (1982) J. Mol. Biol. 156, 683–717.

112. van Etten, R. A., Walberg, M. W. and Clayton, D. A. (1980) Cell 22, 157–170.

113. Kobayashi, M., Seki, T., Katsuyuki, Y. and Koike, K. (1981) Gene 16, 297–307.

114. Li, M., Tzagoloff, A., Underbrink-Lyon, K. and Martin, N. C. (1982) J. Biol. Chem. 257, 5921–5928.

115. Sor, F. and Fukuhara, H. (1980) C.R. Acad. Sci. Paris 291D, 933–936.

116. Köchel, H.-G. and Küntzel, H. (1981) Nucleic Acids. Res. 9, 5689–5696.

117. Seilhammer, J. J. and Cummings, D. J. (personal communication).

118. Gupta, R. and Woese, C. R. (personal communication).

119. Takaiwa, F. and Sugiura, M. (1982) Eur. J. Biochem. 124, 13–19.

120. Saccone, C., Cantatore, P., Gadaleta, G., Gallerani, R., Lanave, C., Pepe, G. and Kroon, A. M. (1981) Nucleic Acids Res. 9, 4139–4148.

121. Seilhammer, J. J. and Cummings, D. J. (1981) Nucleic Acids Res. 9, 6391–6406.

122. Veldman, G. M., Klootwijk, J., de Regt, V. C. H. F., and Planta, R. J. (1981) Nucleic Acids Res. 9, 6935–6952.

123. Georgiev, O. I., Nikolaev, N., Hadjiolov, A. A., Skryabin, K. G ., Zacharyev, V. M. and Bayev, A. A. (1981) Nucleic Acids Res. 9, 6953–6958.

124. Zimmermann, R. A. (1980) ref. 2, pp. 135–169.

Genes: *Structure and Expression*
Edited by A. M. Kroon
© 1983 John Wiley & Sons Ltd.

STRUCTURE AND ROLE OF EUBACTERIAL RIBOSOMAL PROTEINS

Roger A. Garrett

INTRODUCTION

The eubacterial ribosome with its fifty or so proteins and three RNA molecules is a highly complex enzyme; its main function is to polymerize amino acids into proteins. It differs from the other DNA and RNA polymerase systems chiefly in that it contains a large proportion of RNA. In an adjoining chapter, by Noller and Van Knippenberg, the structural organization of the RNA and its special functional roles are considered. The examples they provide of important functional roles for the RNA are supported by the evolutionary view that a primitive ribosome consisted mainly of RNA (1,2). Defining the respective functional contributions of the proteins and RNA in the highly sophisticated eubacterial ribosome is not easy, however. There have been various arguments for one or other component playing a dominant role at different stages of protein biosynthesis; it is possible, for example, that during mRNA binding to the ribosome the RNA is mainly involved, and that proteins effect the catalysis of peptidyl transfer, but there are strong indications that both macromolecular components contribute directly to both processes and it is likely that at all steps of protein biosynthesis the actions of the two macromolecular components are concerted.

This article starts by summarizing the present knowledge on the structures of the ribosomal proteins and their interactions amongst themselves and with the RNA, and continues by considering the phenomenum of the coordinated action of proteins and RNA during ribosome assembly, synthesis and function.

PROTEIN STRUCTURE

There is now general agreement that the ribosomal proteins exhibit fairly stable and compact tertiary structures. A comparison of evidence from several approaches including limited proteolysis, hydrodynamic methods, nuclear magnetic resonance and X-ray and neutron scattering methods indicates that although many of the proteins have irregular shapes, they have substantial amounts of organized tertiary structure (3,4). This has only recently become clear, however. Localization of antigenic determinants of proteins on the ribosome by immuno electron microscopy yielded evidence for highly extended protein structures. For example, protein S4 (M. Wt. -23,137) and the small proteins S15 (M. Wt. -10,001) and S18 (M. Wt. -8,896) were each assigned widely separated antibody binding sites on the ribosome surface, giving them minimum lengths of 150-200 Å and leaving them little opportunity to form a tertiary structure (5,6,7). It has only recently been established that the antibodies were contaminated with other ribosomal protein-specific antibodies such that many of the data, especially from the Berlin group, are unreliable (8,9). Physical-chemical studies on ribosomal proteins, purified under strongly denaturing conditions, also lent support to elongated shapes for some of the proteins; both low angle X-ray and neutron scattering and hydrodynamic measurements yielded abnormally large gyration radii for some proteins giving axial ratios in the range 5:1 to 10:1 (reviewed in 3). There were two main weaknesses in the latter approaches, however. First, most of the proteins exhibited low solubility when dialysed into ribosomal reconstitution buffer (30 mM Tris-Cl, pH 7.8, 20 mM $MgCl_2$, 300 mM KCl and 6 mM 2-mercaptoethanol) which resulted in large error limits for the measurements and, second, it was unclear to what extent the proteins had assumed their "native" structure.

The best evidence for the highly structured state of the proteins derives from high resolution NMR studies (270 MHz) initiated by Morrison and colleagues; they first examined proteins fractionated under strongly denaturing conditions and, subsequently, proteins prepared under much milder conditions (10,11). Although several of the former group of proteins exhibited evidence of tertiary structural interactions in the form of ring current shifted apolar methyl resonances and perturbation of aromatic ring proton resonances resulting from the close proximity of methyl groups and aromatic residues, these effects were strongly enhanced for the latter group of proteins implying that a greater fraction of these protein molecules were in the "native" state. Moreover, in measurements at higher resolution (470 MHz) Moore and colleagues selected two RNA-binding proteins, L11 and L25, which were prepared in urea and then carefully renatured; they concluded that the spectra closely resemble those of globular proteins (12,13). Shape measurements on ribosomal proteins either prepared under mild fractionation conditions, or carefully renatured, indicate that whereas a few protein structures approximate to a spherical shape (eg. 14), many exhibit irregular shapes in the free state and that axial ratios in the range 3:1 to 6:1 are fairly common (reviewed in 3). Higher values have consistently been found for proteins S1 and S4 (15,16); whether these shapes persist within the ribosome structure is unclear. It may be that the flexible domains of some of the RNA binding proteins such as S4 and L18, which exist in the free protein, only become structured in the assembled ribosome through additional protein-protein or protein-RNA interactions (reviewed in 17). Certainly, radius of gyration estimates computed from neutron scattering data from the whole 30S subunit indicate that, with the possible exception of S1 and S4, the proteins have fairly compact structures (4).

The ribosomal proteins of *E. coli* have proved exceedingly difficult to crystallize. While this may partly reflect the difficulty of renaturing urea-denatured proteins, it has also been the experience with proteins fractionated under milder conditions in LiCl. The main success with the *E. coli* proteins has been the crystallisation of the C- and N-terminal halves of protein L12. The former has been analyzed to 2.6 Å by X-ray diffraction by Liljas and colleagues at Uppsala University (18). The dimensions of the fragment are 20 x 20 x 35 Å. It is composed of three α-helical regions and three β-strands all aligned with the long axis of the fragment as indicated in Figure 1. The helices form one layer and the antiparallel β-strands form another. Apart from its relative simplicity, the structure has other special features. For example, the β-sheet side of the molecule has a largely hydrophobic surface, and hydrocarbon side chains are found exclusively between the two layers. Moreover, there is an anion bind-

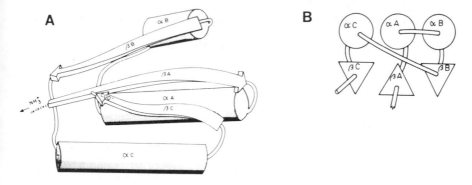

Fig. 1. Idealized representations of the L12 C-terminal domain.
In A. helices are represented by cylinders and β-structures
as arrows pointing in the direction from N- to C-terminus.
B. The structure seen from one edge. Δ, β-strand pointing
into the paper; ∇,β-strand pointing out of the paper; 0,α-
helix. This simple two layer structure is composed of three
helix-strand couples in the sequence $(\beta_\alpha)_A$ $(\alpha\beta)_B$ $(\alpha\beta)_C$.
Reproduced from reference 18 in *Nature* with permission of
A. Liljas.

ing site near the turn between the β_A-strand and the α_A-helix which

is analogous to that found in a large family of enzymes that bind

nucleotides or other anions at the junction between the C-terminal

side of a β-strand and the N-terminus of a helix. It occurs close to

residues Ala 61, Gly 62, and Lys 65, all of which are conserved in

the known sequences. This observation may correlate with the finding

that the C-terminal region of L12 mediates the hydrolysis of GTP

which is dependent on elongation factors EF-Tu and EF-G, and the

authors speculate that EF-Tu binds to the ribosome with phosphates of

the GTP molecule at the anion binding site of this C-terminal domain.

Recently, a protein complex from *Bacillus stearothermophilus*

ribosomes, which is equivalent to the $(L12)_4$-L10 pentamer, has also

been crystallized (19), as have a few other ribosomal proteins in-

cluding L29 from *E. coli* and BL10, BL17 and BL34 from *B. stearothermo-*

philus (20). It will be interesting to see whether the very unusual structure presented in Figure 1 is exclusive to the L12 protein or whether further exceptional structures will be found amongst the ribosomal proteins.

PROTEIN-PROTEIN INTERACTIONS

At present, there is minimal insight into the nature and extent of protein-protein interactions in the ribosome, mainly because the proteins have traditionally been extracted from the RNA, and fractionated, under conditions which dissociate protein-protein complexes (low pH treatment and 6 M urea). Evidence for the existence of protein complexes derives mainly from attempts to isolate proteins under non-denaturing conditions. One complex consisting of four copies of protein L12 and protein L10 has been isolated from both *E. coli* and *B. stearothermophilus* (22); the former gave rise to an additional spot on the standard 2-D gel electrophoresis system and was incorrectly defined as a protein L8. A stable complex of proteins S13 and S19 was also isolated from the 30S subunit and shown by various criteria to be a stable protein-protein complex. Both of the *E. coli* complexes form equimolar and site-specific complexes with the ribosomal RNA's (23,24). Various other putative complexes have been isolated by these mild fractionation methods including S2-S3, S3-S4-S5, S13-S20, L3-L23, L13-L19 and L16-L27 (25). Although no rigorous criteria could be found for their specificity, there is circumstantial evidence from other ribosomal studies that some of them, and other protein pairs, exist as complexes in the ribosome.

Aune and co-workers (26,27) have investigated mixtures of

proteins under equilibrium conditions, in the analytical ultracentri-
fuge and have detected complex formation between S3-S4, S4-S5, S3-S5
and S5-S8; they have managed to circumvent the common problem of protein self aggregation in the measurements. Some of the putative
protein complexes have also been chemically cross-linked in the ribosome in high yield using short cross-linking reagents; S13-S19, S5-S8,
S6-S18 and S7-S9 gave particularly high yields (28, reviewed in 17).
Pairs S5-S8 and S13-S19 were also cross-linked on the ribosome and
then the cross-linked proteins were extracted and reconstituted into
active 30S subunits (29). Distances measured between the centres of
mass of several proteins, within the 30S subunit, provide additional
evidence for possible protein-protein contacts; pairs S4-S5, S4-S12,
S5-S8, S6-S8 and S7-S9 all have centre of mass separations of ≤ 40 Å
(4). A final pointer to possible protein-protein contacts in the
ribosome is the assembly interdependence of proteins; several of the
proteins implicated in complex formation stimulate one another during
ribosomal assembly; these include S2-S3, S13-S20, S5-S8, S6-S18,
S7-S9, S7-S13-S19 and L3-L23 (30,31, see also Figure 5); moreover,
the interaction of L5 and L18 has been proposed as an essential step
in the 5S RNA-23S RNA interaction (32).

PROTEIN-RNA RECOGNITION

Considerable effort has gone into establishing the basis of
protein-RNA recognition in ribosomes. At an early stage many of the
purified proteins were found to form stable complexes with the ribosomal RNA's (reviewed in 17,33) and these single protein-RNA complexes seemed ideal systems for study.

Using the approach of limited ribonuclease digestion of reconstituted complexes of single proteins and RNA, several of the
primary RNA binding proteins, including S4, S8, S15, S20, L1, L11,
L23, L24, and L25 were shown to protect their RNA sites against
digestion and could be isolated as protein-RNA fragment complexes.
However, the expectation that these sites might contain readily dis-
cernible common structural features was not fulfilled. In particular,
the size and, therefore, the structural complexity of the RNA regions
varied widely from around 35-40 nucleotides for the binding sites of
proteins S8 and L25 on 16S RNA and 5S RNA, respectively, to around
450 nucleotides for the sites of proteins S4 and L24 located at the
5'-ends of the large ribosomal RNA's (reviewed in 17,33,34). The
significance of these results became clearer when it was shown that
the binding sites of some of the proteins, in particular those of S4,
S20, L23 and L25, constituted stable RNA domains which could also be
isolated from the renatured free RNA in the absence of protein,
albeit at lower ribonuclease concentrations (eg. 35,36). Another
group of protein binding sites including those of S7, L2, L3, L4, L20
and the $(L12)_4$-L10 complex on the large RNA's, and that of L18 on 5S
RNA, did not yield discrete protected fragments and it is tempting to
conclude that they may attach to more than one RNA domain; the site
of the latter protein, L18, which has been investigated in some detail appears to extend over a large part of the 5S RNA (37).

The corresponding experiments were also performed on the RNA-
bound ribosomal proteins. Proteolytic digestion of reconstituted
protein-RNA complexes demonstrated that protection of the protein
structural domains tends to occur; for example, exhaustive digestion

of both S4-16S RNA and L18-5S RNA complexes yields the same resistant protein fragment as was obtained from the free protein again under milder digestion conditions. For the S4 protein, an N-terminal fragment of 46 amino acids was excised and for the L18 a highly basic 17 amino acid N-terminal fragment was deleted (reviewed in 17). Although similar protected fragments have been reported from proteins L1, L3 and L23 complexed with 23S RNA (38), these digestion characteristics were not a general phenomenum. Other RNA-binding proteins, including S8, S15, S20 and L25, exhibited high resistance to proteolytic digestion when complexed with RNA and, compatible with this result, resistant fragments of the free protein did not reassociate specifically with the RNA (32,39). These results suggest a fairly complex and extensive protein site interacting with the RNA.

Apart from facilitating the localisation of the interacting regions of both protein and RNA, the above mentioned studies yielded little insight into the chemical specificity of the interactions. Some tentative leads did arise, however, from studies on the secondary structure of 16S RNA. Noller (40) reported that some of the putative protein binding regions contained double helical regions with single looped out nucleotides; one of these looped out nucleotides, in the S4 binding site, was accessible to kethoxal modification in the free RNA but partly protected when the protein S4 was added; it was concluded that it might constitute a protein recognition site. Structural studies on 5S RNA revealed that it also contained such a putative helix with a bulged nucleotide within the L18 binding site (41). This is illustrated in the secondary structural model of 5S RNA presented in Figure 2. A combination of ribo-

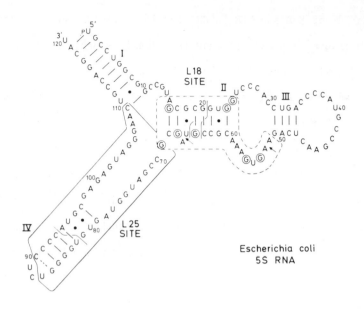

Fig. 2. Secondary structural model of *E. coli* 5S RNA showing the binding regions of proteins L18 and L25. Arrows indicate ribonuclease A or T$_2$ cuts which are protected by the bound protein. Lines drawn into the helix show double helix-specific cobra venom ribonuclease cuts which are affected by the bound proteins (37). The encircled guanosones were protected against modification by kethoxal or dimethyl sulphate by L18 (34,41).

nuclease digestion and chemical modification methods were employed to investigate its possible involvement in the protein interaction (37, 41). It was found that the protein protected the internucleotide bond U_{65}-A_{66} preceding the putative bulged nucleotide against ribonuclease hydrolysis; it also protected the adjacent cuts within helix II which are drawn in Figure 2. Moreover, when the bulged nucleotide A_{66}, which is the most reactive adenosine in the 5S RNA, was carbethoxylated with diethyl pyrocarbonate, those RNA molecules containing a modified A_{66} exhibited a much reduced binding affinity for the L18 protein (41). It is likely, therefore, that such short

helices, with bulged nucleotides, are important for protein-RNA
recognition. Possibly, the nature and position of the bulged nucleotides acts as an indicator so that the protein can distinguish between
the many short helical regions in the RNA. Another type of interaction
is exemplified by the L25 binding site drawn in Figure 2 (and by the
S8 site shown in Figure 3). These RNA structures also consist of
double helical regions but they contain weak or mismatched base pairs
which would lead to an unstable and distorted helix which might be
readily recognised by a protein.

During the above studies several examples of protein-induced
conformational changes have been detected in the RNA structure, particularly in 5S RNA (reviewed in 34), but also in the large RNA's.
It is still too early, though, to assess to what extent these effects
are important for the assembly and functioning of the ribosome.

DUAL NUCLEIC ACID BINDING SITES FOR PROTEINS INVOLVED IN CONTROL OF PROTEIN AND RNA SYNTHESIS

The rate of ribosome production in *E. coli* controls the rate
of protein synthesis and therefore the growth rate. The complex
manner in which the synthesis of RNA and proteins are coordinated
and controlled has been lucidly reviewed recently by Gausing (42).
Although the majority of control occurs at the transcription level,
a mechanism of translational control of ribosomal protein synthesis
has recently been elucidated (reviewed in 43). When the proteins are
in excess they bind to their own mRNA's and inhibit further translation of the message. The proteins which feed back are all primary
binding proteins which bind strongly to the 16S, 23S or, possibly,

5S RNA. The evidence suggests that they recognise a binding site on the mRNA that exhibits similar structural features to the ribosomal RNA site but has a weaker binding affinity for the protein.

The protein genes are arranged in groups in transcriptional units (or polycistronic operons) in the eubacterial chromosome. One of these, the spc operon, is shown below; the order of the protein genes from the promotor P is given. Protein S8, when in excess,

$$\text{L15 L30 S5 L18 L6 } \boxed{\text{S8}} \text{ S14 L5 L24 L14 P}$$
$$\longleftarrow$$

can inhibit translation of the mRNA from this transcriptional unit. The ribosomal RNA binding site for S8 is shown in Figure 3A and the putative binding site on the mRNA, which occurs in the intercistronic region between L24 and L5, is shown in Figure 3B. The latter is the likely binding region because S8 inhibits translation of L5 and the distal cistrons, but not those of L14 and L24.

Several other "feed back" proteins have been characterized. They are given here with the protein genes that they control enclosed in brackets: S4 (S13 + S11 + α-subunit of RNA polymerase + L17), S7 (S7 + EF-G + EF-Tu), L1 (L11 + L1), L4 (S10, L3, L4, L23, L2, (S19, L22), S3, L16, L29, L17) and L10 (L10, L12, β and β' subunits of RNA polymerase). Apart from the site for L1 (44), we still have insufficient knowledge of the exact binding regions of the proteins on the ribosomal RNA's for a meaningful search to be made for related structures in the mRNA. There is also no clear idea of the mechanism by which translation is inhibited: whether the protein-mRNA complex is simply inactive in translation and/or rapidly degraded or whether

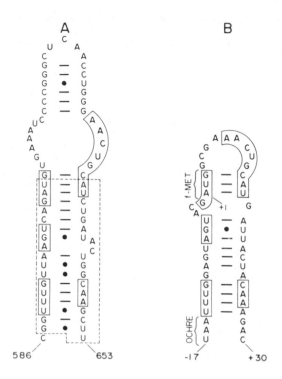

Fig. 3. Model of secondary structure of the S8 binding site on (A)
16S RNA and (B) the putative site on mRNA. Homologies which
are considered significant are boxed. In (A) the ribosomal
binding site of S8 is indicated by a broken line. In (B) the
L24 coding region ends at -18 and the L5 coding region begins
at +4. Reproduced from reference 43 in *Trends in Biochemical
Sciences* with permission of M. Nomura.

further production of the mRNA is also inhibited.

The mechanism by which ribosomal RNA synthesis is controlled,
and coordinated with protein synthesis in eubacteria, is unknown.
There is, however, a eukaryotic system that has recently been studied
extensively. In oocytes of *Xenopus laevis* and *borealis* 5S RNA is
transcribed separately from the large rRNA's, and is accumulated
prior to the large ribosomal RNA synthesis and cytoplasmic ribosome
assembly. Transcription of the 5S RNA genes is controlled by a factor

(TF III A) which can associate with both the 5S RNA gene and with the 5S RNA (45-47). When the 5S RNA is transcribed it complexes with the factor. When sufficient 5S RNA has been synthesized, there is no more factor available to effect further RNA transcription. Immunological studies have detected small amounts of a related but larger transcription factor in somatic cells (48). Circumstantial evidence suggests that both factors may have evolved from a protein which binds to 5S RNA in ribosomes by divergent evolution. Moreover, the region of the 5S RNA gene which is protected by the protein closely parallels the region of the RNA which is protected by a ribosomal protein of the same molecular weight (49).

RIBOSOMAL SUBUNIT RECONSTITUTION

The assembly of the ribosome *in vitro* has been studied not only with a view to establishing the "order" of protein assembly and the interrelationships of the proteins, but also as a means of determining the functions of the proteins. In their classic studies of omitting one protein at a time from the 30S subunit reconstitution mixture, Traub and Nomura investigated the effects on both ribosome assembly and function (50). They introduced the concept of "assembly" proteins and "functional" proteins. It was considered that the former would fold and maintain the RNA structure and that the latter were responsible for function. In this model the RNA itself was assigned a predominantly structural role in forming a framework for the functional proteins. Although the concept is still used, it has become dated. Many studies have demonstrated that the RNA is able to fold into a tertiary structure in the absence of proteins if subjected to a suitable renaturation procedure (eg. 36,51), and while the proteins may help to select for one of different possible conformations, they do not appear to organise the RNA-structure as originally conceived.

Equally, these early studies revealed that very few of the "functional" proteins were essential for any of the assayed functions, such as the initiator tRNA binding or viral mRNA binding, although this could have been due to the technical difficulty of ensuring that any one protein was completely absent from the reconstitution mixture.

There are two pertinent questions about such studies. First, what do they tell us about ribosome assembly *in vivo*? *In vivo*, the ribosomes assemble from their 5'-ends. It seems very likely that each RNA domain (averaging about 500 nucleotides) folds up as it is formed into a tertiary structure. It has been suggested that the 5'-domain which forms first, and binds protein S4 in 16S RNA and L24 in 23S RNA, acts as a nucleus for the assembly of the remainder of the RNA (36). The proteins then assemble with the structured domains, and possibly earlier if the RNA site is recognisable before the domain structure is complete as it may be, for example, for proteins such as S20 in domain I and S8 and S15 in domain II which have relatively simple RNA sites (17,33; see Figure 3); these proteins may assemble prior to completion of the domain structure. Subsequently, the other domains fold up and accrue their protein component. Presumably, then these protein-RNA domains interact to form a depleted ribosomal subunit and then the last proteins assemble. Some intermediate assembly particles have been characterised both *in vivo* in assembly defective mutants and during *in vitro* reconstitution experiments (30,31,52); although some have similar, but not identical, protein compositions it is still unclear to what extent they are related structurally. Evidence for the assembly of the proteins during the transcription of RNA is provided in electron micrographs of the eukaryotic ribosomal RNA being transcribed from *Drosophilia melanogaster* ribosomal genes shown in Figure 4. The answer, then, to the first question is that the information contained in a protein assembly map, determined *in vitro*, cannot be correlated with the *in vivo* assembly without additional information on the RNA domains to which the proteins assemble.

The second question is: What does the assembly map tell us about the structure and topography of the ribosome? The answer is that if it is done very accurately it tells us a lot about cooperative interactions between the proteins either through direct contact or

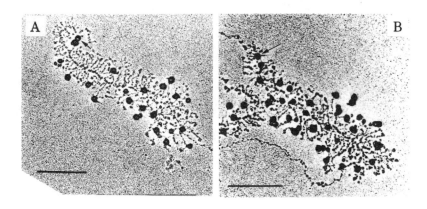

Fig. 4. Electron microscope visualization of *Drosophilia melanogaster* ribosomal RNA being transcribed by RNA polymerase. Antibodies raised against eukaryotic ribosomal protein S14 are visible as blobs on the nascent RNA. The first antibody visualized is indicated by arrows. The sample in B was treated with a fivefold higher concentration of antibody than that in A. Scale bars represent 0.5 μm. Reproduced from reference 53 in *Proceedings of the National Academy of Sciences* with the permission of W. Chooi.

via the RNA and, probably, it yields information about protein-protein neighbourhoods in the ribosome. Two general mechanisms of protein-RNA assembly have emerged, for example. First, for a pair of proteins (or protein complexes) binding to the RNA, mutual stimulation of binding has been detected, leading to a stable intermediate complex (eg. 24). Second, certain proteins which contain more than one structural domain attach to the RNA through one domain and another provides a "link" to other proteins (reviewed in 17). The main difficulty inherent in assembly mapping is that if the isolated ribosomal proteins are not completely pure (and many are very difficult to purify), stimulation effects from contaminants may be detected; if such effects occur early in the mapping they may distort a whole region of the map.

Below, we consider the assembly of the 50S subunit of $E.$ $coli$; the assembly of the 30S subunit has been discussed in detail else-where (54); also, to a large extent, this latter information has been superseded by binding and assembly studies of proteins to the three major RNA domains of 16S RNA (reviewed in 55). The present state of the map for the 50S subunit is presented in Figure 5. The studies were made on the 23S RNA in two steps, at low Mg^{++} (4 mM) and at high Mg^{++} (20 mM) and then correlated with assembly studies with the large RNA fragments 13S RNA, 8S RNA and 11/12S RNA (31,56). Binding and assembly results obtained with RNA fragments in high Mg^{++} (20 mM) are summarized in Figure 6. Several differences exist. For example, the weak RNA binding effects presented in the assembly map for L10, L12, L18, L22 and L29 and the stronger effect for L9 have not generally been detected in single protein binding experiments (re-viewed in 17,33) and although RNA binding properties have been

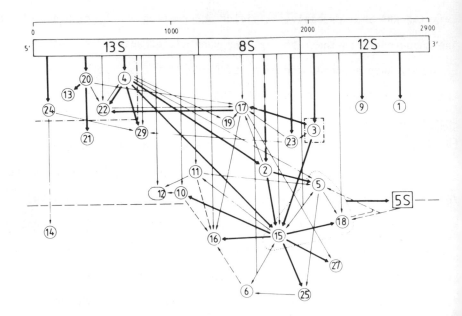

Fig. 5. Assembly map of the 50S subunit. The intensity of the arrows reflects the degree of interdependence of the proteins during assembly. --·--·· encloses proteins which are important for the first stage of assembly. ..... indicates proteins which are involved in 5S RNA - 23S RNA assembly. The major fragments, 13S, 8S and 12S RNA, derived from mild ribonuclease treatment of 23S RNA are indicated. Reproduced from reference 31 in *Proceedings of the National Academy of Sciences* with permission of K. Nierhaus.

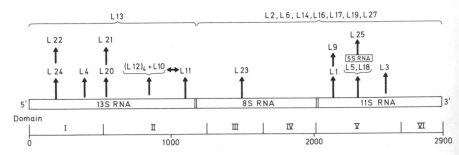

Fig. 6. Compilation of 23S RNA binding and assembly data summarized in refs. 17 and 33. Contact of an arrow with the RNA indicates that the primary RNA binding site of the protein has been localized at this position.

reported for L15, L16, L17 and L19, no reproducible evidence for site specificity has been reported (57-60). Other discrepancies also arise. For example, protein binding studies indicate that L25 forms a site-specific complex with 5S RNA (see Figure 2) whereas in the assembly map there is no interrelationship; L5 and L18 have also been shown to bind to, and couple, 5S RNA to 23S RNA (reviewed in 34). The data on isolated proteins are generally more rigorous because various criteria can be applied to establish the specificity of protein-RNA interactions.

The reconstitution procedure for the 50S subunit of *E. coli* is exceptional in that it requires two steps at the low and high magnesium concentrations, respectively (31,61; reviewed in 17,62). Other procedures for both *B. stearothermophilus* subunits and *E. coli* 30S subunits only require a high magnesium step. The reason for the low Mg^{++} step is unclear although it may be that the 23S RNA structure is too tight at 20 mM Mg^{++} and that opening is necessary, particularly for those proteins such as L20 which are critical for early 5'-end assembly (it is also one of the few proteins which tends to show non specific binding at 20 mM Mg^{++}). Possibly, *in vivo*, where the average cellular magnesium concentration is closer to the low value, protamines supplement the Mg^{++} in stabilizing the RNA structure. However, the studies in low Mg^{++} *in vitro* bring other disadvantages; some of the proteins bind partly unspecifically at low Mg^{++} (63) (at very low Mg^{++} (< 1 mM) all the proteins bind to the RNA). It may be, therefore, that some of the weak protein-RNA binding effects, indicated in Figure 5, are of a non-specific nature.

PROTEIN TOPOGRAPHY

The three dimensional distribution of the proteins and RNA in the ribosome has been investigated by many different and novel approaches. Early studies with ribonucleases revealed that large amounts of RNA could be digested without appreciably affecting the hydrodynamic properties of the subunits; similarly single protein-specific antibodies and various chemical reagents were employed to demonstrate that most, if not all, of the proteins were partially accessible on the surface of the ribosome (reviewed in 17). While studies on the RNA, using nucleotide-specific reagents, as well as ribonucleases, have served to identify accessible sequences both on 70S ribosomes and at the subunit interface (see chapter by Noller and Van Knippenberg), progress on protein accessibility has been less certain.

As mentioned earlier data from the immune electron microscopy has been confusing because there are major differences between the models of the Berlin group (5) and the Los Angeles group (64) who are using the same approach; a recent analysis revealed that up to 80% of the putative antigenic sites *differ* (65). It now appears that cross contamination of antibodies with those specific for other purified ribosomal proteins has been the major cause of errors in the Berlin model and their extensive measurements are currently being repeated and the data extensively revised (9). This reevaluation has been necessary largely as a result of inconsistencies with the data from a neutron triangulation technique which has been pioneered by Moore and Engelman at Yale University (reviewed in 17). The approach involves measuring the distances between the centres of mass of pairs

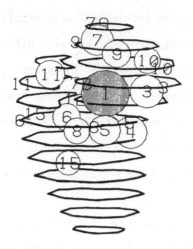

Fig. 7. Superposition of the neutron scattering model on the electron microscopic image of the 30S subunit. The neutron model is shown in a stereo view, superimposed on the electron microscopic model presented by Lake and colleagues (64,66) so as to minimize the distance between the antigenic determinant positions and corresponding protein centers of mass. Proteins are presented as spheres of the appropriate volumes drawn to scale. The distance between contour planes is 20 Å. Reproduced from reference 4 in *Journal of Molecular Biology* with permission of P. Moore.

of deuterated proteins in the ribosome and building up a three dimensional map of the proteins. These distances have been measured for many combinations of the 30S subunit proteins S1, S3, S4, S5, S6, S7, S8, S9, S10, S11, S12 and S15 (4) and they are presented in Figure 7 as encircled numbers. The positions have been coordinated with the surface antigen positions determined by the Los Angeles group (64,66) which are shown as numbers that are not encircled. The data show a good degree of agreement and suggest that most of the proteins are concentrated towards the central and upper parts of the model. The lower region consists mainly of RNA and Lake and co-workers have suggested that this may be the region of the ribosome attached to the cell membrane (67).

The data correlate fairly well with results on protein pairs which have been chemically cross-linked with bifunctional reagents and/or shown to interact in solution. For example, S5-S8 and S13-S19 where the centres of mass are very close have been cross-linked in high yield. Moreover, in general the protein pairs with widely separate centres of mass such as S3-S11 and S7-S15 have either not been chemically cross-linked, or the products have only been detected in very low yield.

At this stage we have very little insight into the relative spatial arrangement of the proteins and the RNA in the subunits, although such information is being gradually accrued both from identification of the RNA binding sites of the proteins (see preceding section on Protein-RNA Recognition) and from identification of the sequences of oligonucleotides which have been chemically cross-linked to the proteins. Whereas the former method has tended to identify the RNA domain or subdomain to which a protein, or a group of proteins, associates, the latter approach has served to localize, more precisely, adjacent regions of protein and RNA within the ribosome which are not necessarily the same as the binding site. Using the latter approach several close contact points have been identified between proteins S1, S4, S7 and S20, and 16S RNA and proteins L4, L6, L21, L23, L27 and L29, and 23S RNA (68-70). Clearly, a large catalogue of such information on the proximal protein and nucleic acid sequences will prove an important factor in elucidating the three dimensional organization of the ribosome.

The results on the 50S subunit are less complete. Only a few protein antigenic sites have been placed by the Los Angeles group (71)

and although many have been localized by the Berlin group these cannot
be considered reliable at present. Moreover, there are so far no comprehensive data on the neutron triangulation method, although such
studies have commenced (72). The chemical cross-linking approach provides the most complete data on the surface arrangement of the proteins (73,74). In particular, two protein-rich domains at the peptidyl
transferase centre and at the elongation factor binding site, have
been characterized and are considered further below; the former is
depicted in Figure 8. A model of the subunits is also presented
(Figure 9) in the ensuing sections which describe the main functional
domains on the *E. coli* ribosome.

mRNA BINDING

The interaction of mRNA with the ribosome is an essential
first step in protein biosynthesis. While comparing sequences of
viral mRNA's and 16S RNA, Shine and Dalgarno (75) noted a highly
conserved complementarity between a pyrimidine-rich sequence at the
3'-terminus of 16S RNA and a purine-rich sequence at the 5'-end of a
mRNA. Various lines of evidence indicate that this RNA-RNA interaction
does occur on the ribosome and is an essential part of mRNA-ribosome
recognition (eg. 76). However, this is not the only important interaction since polynucleotides which cannot recognize this rRNA
sequence are also processed.

Ribosomal proteins, in particular S1 and S21, have been implicated directly in the mRNA recognition process. Employing a deoxyoctanucleotide (5'-3') d(A-A-G-G-A-G-G-T) which is complementary to
the 3'-terminal sequence of 16S RNA, it was demonstrated that the

latter sequence was accessible in active 30S subunits but not when phage RNA was bound, nor in subunits which had been inactivated at low magnesium concentrations, nor in free 16S RNA (77). Supporting evidence for this change of accessibility came from chemical modification studies and immunochemical studies with antibodies raised against neighbouring modified bases (78). It was shown, further, that the active subunits would only bind the deoxyoctanucleotide if protein S21 was present; this result correlated with the observation that the ribosome can only initiate on viral or *E. coli* mRNA (as opposed to polynucleotides) if S21 is present; the mRNA binding capacity of the ribosome could also be eliminated by adding antibodies raised against protein S21 (79). These data, coupled with the evidence for a ribosomal binding site for protein S21 near the 3'-end of 16S RNA, strongly suggest that S21 modulates the 16S RNA in this region and opens up the "Shine and Dalgarno" sequence for the incoming mRNA.

Protein S1 is also essential for natural mRNA binding and it is a component of certain RNA phage replicases. This is the largest ribosomal protein (M.W. ~ 65,000) and is highly elongated (15). It assembles with the 30S subunits through its N-terminal domain (80) in the neighbourhood of the 3'-terminus of 16S RNA, probably mainly through protein-protein interactions (81). It has two nucleic acid binding sites and the primary binding site is located in the central sequence region (82). Most importantly, it has the capacity to disrupt the secondary structure of helical polynucleotides (eg. 83). Although its mode of action has been studied less rigorously than that of protein S21, it does not appear to influence the binding of the deoxyoctanucleotide to the terminal 16S RNA sequence; the current view is that it complements the action of S21 on the 16S RNA by associating with, and opening, the complementary sequence of the incoming mRNA, so that an RNA-RNA interaction can occur.

PEPTIDYL TRANSFERASE CENTRE

Facilitating the transfer of the peptide of the peptidyl tRNA in the P-site to the α-amino group of the incoming amino acyl tRNA in the adjacent A-site is a primary function of the ribosome; another possible function of this centre is the hydrolysis of the peptidyl-tRNA during protein chain termination. The peptidyl transferase constitutes a protein-rich domain that is independent from that of the elongation factor-dependent GTP hydrolysis domain (84; see next section). While Chladek and colleagues using various chemical derivatives and analogues of the A and P site-bound tRNA's to probe the peptidyl transferase centre conclude that it resembles that of a proteolytic enzyme (85,86), another possibility is that the ribosome acts as a rigid template and simply aligns the two interacting groups (87).

Several proteins have been implicated in the peptidyl transferase centre by partial reconstitution procedures (84,88,89) or by the affinity label method (reviewed in 90). In the latter approach an antibiotic such as chloramphenicol or puromycin, which binds to the centre, is substituted with a chemically active group and then bound specifically to the ribosome when the substituted group reacts with a ribosomal component. The proteins identified are listed in Table 1. Neither method can distinguish between proteins involved in a catalytic centre and those which simply help to structure, or are adjacent

TABLE 1.

Summary of proteins identified at the peptidyl transferase centre.

Reconstitution method		Affinity label procedures
B. stearothermophilus (84)	E. coli (88,89)	E. coli (90)
L2	L2	L2
L3 (or L6)	L3	L11
L4	(L4?)	(L14?)
L5	L15	L15
L14	L16	(L16?)
L16	L18	L18
L20	L25	L23
		L27

to, the centre. However, proteins involved in catalysis are likely to be detected by most methods. While some of the proteins occur in two of the columns, only L2 and, possibly, L16 are common to all three These two proteins are also the earliest proteins to have been implicated in the catalytic activity (91,92) and modification of one or more histidines in each protein leads to loss of peptidyl transfer activity, as do other chemical treatments (92-94).

There is always the possibility that the catalytic site is a composite one consisting of more than one protein and possibly RNA. There is increasing evidence for a functional role of the RNA, especially for those regions which are both conserved in sequence and accessible on the ribosome. Ribosome reconstitution experiments with (a) kethoxal-modified RNA and unmodified proteins and (b) unmodified RNA and kethoxal-modified proteins indicate that activity is lost only when the RNA is modified (95). This may reflect, however, that the tRNA cannot bind to the ribosome through RNA-RNA interactions (possibly at the conserved -C-C-A end (85,96)). On the other hand, further support for an RNA role comes from the finding that chloramphenicol-resistants mutants in mitochondrial ribosomes occur at sites in domain V of the large ribosomal RNA secondary structure; the sequence regions are so highly conserved that they can be correlated with nucleotides 2447, 2457, 2503 and 2504 of the *E. coli* 23S RNA (96).

A map showing the neighbourhoods of most of the proteins implicated in the peptidyl transferase centre is shown in Figure 8. The results are based on chemical cross-linking studies with a bifunctional reagent (73). The identified cross-linked pairs of proteins are connected by lines and are close neighbours on the ribosome surface; most of the proteins listed in Table 1 are present. The L12 domain, adjacent to L11, is on the left and the 5S RNA-L5-L18-L25 complex is on the right. The position of the peptidyl transferase centre is shown in Figure 9 on the Los Angeles model of the 50S subunit. It was localized by electron microscopic visualization of antibodies attached to pyromycin covalently bound to the ribosome (97, 98). The L12 domain is also indicated and the 5S RNA-protein complex lies in the central protuberance. The tRNA molecule is drawn to scale in the ribosomal A-site; tRNA bound in the P site would lie very close to the A site-bound tRNA (99).

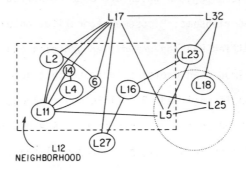

Fig. 8.   A map of the proteins which have been cross-linked with the
          reversible disulphide cross-linking reagent 2-iminothiolane.
          Cross-linked proteins are joined by lines. Proteins impli-
          cated in the peptidyl transferase centre are encircled.
          Those proteins which have been cross-linked to protein L12
          are enclosed in a box with dashed lines. The 5S RNA-binding
          proteins are enclosed in a dotted circle. Reproduced from
          reference 73 in *"Ribosomes"* (G. Chambliss *et al.* eds.) with
          permission of Rob Traut.

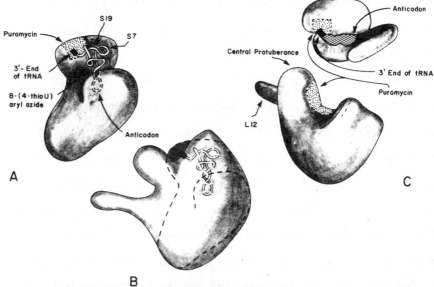

Fig. 9.   Models for tRNA binding to (A) 30S subunits and (B) A site of
          70S ribosomes. The 50S subunit is shown on top of the 30S sub-
          unit. In C, the ribosome is viewed from above the cleft of
          the 30S subunit (rotated 90° from the plane of the paper in
          (B)), and the subunits are separated for clarity. Reproduced
          from reference 97 in *Journal of Biological Chemistry* with
          permission of D. Glitz.

ELONGATION FACTOR BINDING SITE

Another important functional domain in the 50S subunit which is involved in the binding of elongation factors EF-Tu and EF-G and possibly also of initiation factor 2 and the release factor 2 is the $(L12)_4$-L10 protein complex and a small group of proteins including L11; it lies adjacent to the peptidyl transferase centre (see Figures 8 and 9). L12 was the first ribosomal protein to be purified, by Möller and co-workers, who showed subsequently that it was very exceptional in being both acidic and multicopied; their sequence determination also revealed a hydrophobic core region (reviewed in 100). They speculated, on the basis of these unusual properties, that it might constitute an important functional domain and were able to show subsequently that its presence was essential for EF-G catalysed GTP hydrolysis (100). The protein domain is represented as a stalk-like projection in the model in Figure 9, and it is considered that both EF-Tu and EF-G, which are required at different stages of the protein synthesis cycle, bind on or near this appendage.

L12 exists as a dimer in solution; probing of the structure, with bifunctional cross-linking reagents, has produced a model in which the two monomers are arranged in a shifted parallel alignment, with protruding N-terminal sequences, by means of which they interact with protein L10 (101). The $(L12)_4$-L10 complex binds to 23S RNA and, whereas one dimer of L12 dissociates readily from the complex, the other remains strongly bound (24). This observation correlates with further studies of Möller *et al.* who have demonstrated that whereas the two dimers are anchored on L10 they may not necessarily both lie in the stalk-like structure of the ribosome (102).

Circumstantial evidence from electron microscopy studies suggests that only one dimer may be located in this projection and that the other may turn into the interface region of the subunit (6,102). Moreover, only one dimer is required for EF-G dependent GTPase activity; the reactivity is retained when the other site is vacant (102).

Although the L12 dimer is required for factor dependent GTP hydrolysis and a mechanism for the interaction of the EF-Tu-bound GTP with the C-terminal domain of the protein has been proposed (see section on Protein Structure), the γ-phosphate of the GTP has been chemically cross-linked primarily to the adjacent protein L11 and also to the neighbouring 5S RNA-binding proteins L5 and L18 (103). There is also no direct evidence for EF-G associating with the dimer. Immune electron microscopy studies place the factor at the base of the stalk-like structure (104) and photochemical cross-linking experiments implicated a group of proteins in the factor binding site; these include L1, L3, L11 on the 50S subunit and S3 and S4 on the 30S subunit compatible with the occurrence of the factor in the subunit interface region (105). L11 is, almost certainly, a neighbour to the $(L12)_4$-L10 complex on the 23S RNA but the other proteins are more widely separated in the subunit interface region perhaps reflecting the high molecular weight of the factor (M. Wt. 77,321). The other factor binding sites have been less well characterised but EF-Tu has been assigned a site overlapping with that of the EF-G (29) and release factor 2 also has a partially overlapping site (106).

CONCLUSION

All the indications are that a primitive and inefficient translation system consisted of RNA (1,2) possibly working together with small organic molecules (107). Such a system would have required the RNA to make some kind of distinction between amino acids. Subsequently, proteins would have been incorporated into the translation machine to make it more efficient and more accurate. Given such a concept the primary role of the proteins becomes to modulate the RNA and to complement and supplement its function. They may do this in various ways. For example, in selecting one of a few equally stable conformers or in effecting switches between active/inactive conformers such as opening/closing the Shine and Dalgarno sequences for 16S RNA-mRNA recognition; other possibilities include facilitating folding and assembly of the large and complex RNA molecules, stabilizing/destabilizing RNA-RNA interactions at the subunit interface, helping in the recognition of mRNA, tRNA and ribosomal factors and in stabilizing their ribosomal interactions. In certain roles where a protein may be able to do a job better, such as in catalyzing peptidyl transfer, the proteins may have replaced a simpler and less effective templating role of the RNA. Also, proteins are better at forming hydrophobic surfaces or pockets for binding other proteins such as ribosomal factors; this may be one role of the $(L12)_4$-L10 complex but even there we cannot eliminate the involvement of the RNA.

This general view of the evolution of ribosomal proteins receives some support from the experiments of Dabbs and colleagues who have isolated several mutants of *E. coli* which lack different ribosomal proteins and yet have normal growth rates (108,109). It then becomes difficult to determine their functions without investigating, in a detailed way, their interaction, direct or indirect, with the RNA structure. However, the secondary structural models of the ribosomal RNA's which are now being refined provide a basis, for the first time, for well conceived experiments of this type to be undertaken with confidence. Nevertheless, recent evidence has indicated the importance of keeping an open mind about ribosomal protein function; their capacity to inhibit their own synthesis when produced in excess by binding to their own mRNA, and the capacity of a 5S RNA-binding protein in eukaryotic oocytes to bind its own DNA gene and to control transcription of 5S RNA were major surprises; there may be many more to come.

ACKNOWLEDGEMENTS

I appreciate the help of Birgitte Buus, Jonna Christensen and Arne Lindahl in preparing this Chapter, and thank the Danish Science Research Council for financial support.

REFERENCES

1. Crick, F.H.C. *J. Mol. Biol. 38*, 367-379 (1968).

2. Woese, C.R. in *"Ribosomes"* (G. Chambliss *et al.*, eds.) pp. 357-376, University Park Press, Baltimore.

3. Wittmann, H.G., Littlechild, J.A. and Wittmann-Liebold, B., in *"Ribosomes"* (G. Chambliss *et al.*, eds.) pp. 51-88, University Park Press, Baltimore (1980).

4. Ramakrishnan, V.R., Yabuki, S., Sillers, I.Y., Schindler, D.G., Engelman, D.M., and Moore, P.B. *J. Mol. Biol. 153*, 739-760 (1981).

5. Stöffler, G., Bald, R., Kastner, B., Lührmann, R., Stöffler-Meilicke, M. and Tischendorf, G. In *"Ribosomes"* (G. Chambliss *et al.*, eds.) pp. 171-205, University Park Press, Baltimore (1980).

6. Tischendorf, G., Zeichhardt, H. and Stöffler, G. *Proc. Natl. Acad. Sci. USA 72*, 4820-4824 (1975).

7. Lake, J.A., Pendergast, M., Kahan, L. and Nomura, M. *Proc. Natl. Acad. Sci. USA 71*, 4688-4692 (1974).

8. Winkelmann, D. and Kahan, L. in *"Ribosomes"* (G. Chambliss *et al.*, eds.) pp. 255-265, University Park Press, Baltimore (1980).

9. Noah, M., Kastner, B., Stöffler-Meilicke, M. and Stöffler, G. Abstract, *7th Annual EMBO Symp. on Ribosomes*. Heidelberg (1981).

10. Morrison, C.A., Garrett, R.A. and Bradbury, E.M. *FEBS Lett. 82*, 435-439 (1977).

11. Morrison, C.A., Bradbury, E.M., Littlechild, J. and Dijk, J. *FEBS Lett. 83*, 348-352 (1977).

12. Kime, M.J., Ratcliffe, R.G., Moore, P.B. and Williams, R.J.P. *Eur. J. Biochem. 110*, 493-498 (1980).

13. Kime, M.J., Ratcliffe, R.G. Moore, P.B. and Williams, R.J.P. *Eur. J. Biochem. 116*, 269-276 (1981).

14. Serdyuk, I.N., Gogia, Z.V., Venyaminov, S.Yu., Khechinashvili, N.N., Bushuev, V.N. and Spirin, A.S. *J. Mol. Biol. 137*, 93-107 (1980).

15. Laughrea, M. and Moore, P.B. *J. Mol. Biol. 112*, 399-421 (1977).

16. Österberg, R., Sjöberg, B., Garrett, R.A. and Littlechild, J. *FEBS Lett. 73*, 25-28 (1977).

17. Garrett, R.A. *Int. Rev. Biochem.* (R.E. Offord, ed.) Vol. 25, pp. 127-177, University Park Press, Baltimore (1979).

18. Leijonmarck, M., Eriksson, S. and Liljas, A. *Nature 286*, 824-826 (1980).

19. Liljas, A. and Newcomer, M.E. *J. Mol. Biol. 153*, 393-398 (1981).

20. Appelt, K., Dijk, J., Reinhardt, R., Sanhuesa, S., White, S.W., Wilson, K.S. and Yonath, A. *J. Biol. Chem. 256*, 11787-11796 (1981).

21. Pettersson, I., Hardy, S.J.S. and Liljas, A. *FEBS Lett. 64*, 135-138 (1976).

22. Marquis, D.M. and Fahnestock, S.R. *J. Mol. Biol. 142*, 161-180 (1980).

23. Dijk, J., Littlechild, J. and Garrett, R.A. *FEBS Lett. 77*, 295-300 (1977).

24. Dijk, J., Garrett, R.A. and Müller, R. *Nucleic Acids Res. 6*, 2717-2730 (1979).

25. Dijk, J. and Littlechild, J. *Methods Enzymol. 59*, 481-502 (1979).

26. Rohde, M.F., O'Brien, S., Cooper, S. and Aune, K.C. *Biochem. 14*, 1079-1085 (1975).

27. Tindall, S.H. and Aune, K.C. *Biochem. 20*, 4863-4866 (1981).

28. Expert-Bezançon, A., Barritault, D., Milet, M., Guérin, M.-F. and Hayes, D.H. *J. Mol. Biol. 112*, 603-120 (1977).

29. Kurland, C.G. *Ann. Rev. Biochem. 46*, 173-200 (1977).

30. Held, W.A. and Nomura, M. *Biochem. 12*, 3273-3281 (1973).

31. Röhl, R. and Nierhaus, K.H. *Proc. Natl. Acad. Sci. USA 79*, 729-733 (1982).

32. Newberry, V.N. and Garrett, R.A. *Nucleic Acids Res. 8*, 4131-4142 (1980).

33. Zimmermann, R.A. in *"Ribosomes"* (G. Chambliss *et al.*, eds.) pp. 135-170, University Park Press, Baltimore (1980).

34. Garrett, R.A., Douthwaite, S. and Noller, H.F. *Trends Biochem. Sci. 6*, 137-139 (1981).

35. Douthwaite, S., Garrett, R.A., Wagner, R. and Feunteun, J. *Nucleic Acids. Res. 6*, 2453-2470 (1979).

36. Garrett, R.A., Ungewickell, E., Newberry, V., Hunter, J. and Wagner, R. *Cell Biol. Internat. Reps. 1*, 487-502 (1977).

37. Douthwaite, S., Christensen, A. and Garrett, R.A. *Biochem. 21*, 2313-2320 (1982).

38. Schulte, C., Schiltz, E. and Garrett, R.A. *Nucleic Acids. Res. 2*, 931-942 (1975).

39. Newberry, V.N. Doctorate Thesis, University of Birmingham, U.K. (1978).

40. Noller, H.F. in *"Ribosomes"* (G. Chambliss *et al.*, eds.) pp. 3-22, University Park Press, Baltimore (1980).

41. Peattie, D.A., Douthwaite, S., Garrett, R.A. and Noller, H.F. *Proc. Natl. Acad. Sci. USA 78*, 7331-7335 (1981).

42. Gausing, K. *Trends Biochem. Sci. 7*, 65-67 (1982).

43. Nomura, M., Dean, D. and Yates, J.L. *Trends Biochem. Sci. 7*, 92-95 (1982).

44. Branlant, C., Krol, A., Machatt, A. and Ebel, J.P. *Nucleic Acids Res. 9,* 293-307 (1981).

45. Sakonju, S., Brown, D.D., Engelke, D., Ng, S-Yu, Shastry, B.S. and Roeder, R.G. *Cell 23,* 665-669 (1981).

46. Pelham, H.R.B. and Brown, D.D. *Proc. Natl. Acad. Sci. USA 77,* 4170-4174 (1980).

47. Honda, B.M. and Roeder, R.G. *Cell 22,* 119-126 (1980).

48. Pelham, H.R.B., Wormington, W.M. and Brown, D.D. *Proc. Natl. Acad. Sci. USA 78,* 1760-1764 (1981).

49. Nazar, R.N., Yaguchi, M. and Willick, G.E. *Can. J. Biochem. 60,* 490-496 (1982).

50. Nomura, M., Mizushima, S., Ozaki, M., Traub, P. and Lowry, C.V. *Cold Spring Harb. Symp. Quant. Biol. 34,* 49-61 (1969).

51. Vasiliev, V.D. and Zalite, O.M. *FEBS Lett. 121,* 101-104 (1980).

52. Cabezón, T., Herzog, A., Petre, J., Yaguchi, M. and Bollen, A. *J. Mol. Biol. 116,* 361-374 (1977).

53. Chooi, W.Y. and Leiby, K.R. *Proc. Natl. Acad. Sci. USA 78,* 4823-4827 (1981).

54. Held, W.A., Ballou, B., Mizushima, S. and Nomura, M. *J. Biol. Chem. 249,* 3103-3111 (1974).

55. Noller, H.F. and Woese, C.R. *Science 212,* 403-411 (1981).

56. Chen-Schmeisser, U. and Garrett, R.A. *Eur. J. Biochem. 69,* 401-410 (1976).

57. Stöffler, G., Daya, L., Rak, K.H. and Garrett, R.A. *J. Mol. Biol. 62,* 411-414 (1971).

58. Garrett, R.A., Müller, S., Spierer, P. and Zimmermann, R.A. *J. Mol. Biol. 88,* 553-557 (1976).

59. Littlechild, J., Dijk, J. and Garrett, R.A. *FEBS Lett. 74,* 292-294 (1977).

60. Marquardt, O., Roth, H.E., Wystrup, G. and Nierhaus, K.H. *Nucleic Acids Res. 6,* 3641-3650 (1979).

61. Amils, R., Matthews, E.A. and Cantor, C.R. *Nucleic Acids Res. 5,* 2455-2470 (1978).

62. Nierhaus, K.H. *Curr. Topics Microbiol. Immunol. 97,* 81-155 (1982).

63. Schulte, C., Morrison, C.A. and Garrett, R.A. *Biochem. 13*, 1032-1037 (1974).

64. Lake, J.A. in *"Ribosomes"* (G. Chambliss *et al.*, eds.) pp. 207-236, University Park Press, Baltimore (1980).

65. Gaffney, P.T. and Craven, G.R. in *"Ribosomes"* (G. Chambliss *et al.* eds.) pp. 237-253, University Park Press, Baltimore (1980).

66. Kahan, L., Winkelmann, D.A. and Lake, J.A. *J. Mol. Biol. 145*, 193-214 (1981).

67. Lake, J.A. Abstract, *7th Annual EMBO Symp. on Ribosomes*, Heidelberg (1981).

68. Wower, I., Wower, J. and Brimacombe, R. *Nucleic Acids Res. 9*, 4285-4301 (1981).

69. Ehresmann, B., Backendorf, C., Ehresmann, C. and Ebel, J.P. *FEBS Lett. 78*, 261-266 (1977).

70. Golinska, B., Millon, R., Backendorf, C., Olomucki, M., Ebel, J.P. and Ehresmann, B. *Eur. J. Biochem. 115*, 479-484 (1981).

71. Lake, J.A. and Strycharz, W.A. *J. Mol. Biol. 153*, 979-992 (1981).

72. Stuhrmann, H.B. Abstract, *7th Annual EMBO Symp. on Ribosomes*, Heidelberg (1981).

73. Traut, R.R., Lambert, J.M., Boileau, G. and Kenny, J.W. in *Ribosomes* (G. Chambliss *et al.*, eds.) pp. 89-110, University Park Press, Baltimore (1980).

74. Cover, J.A., Lambert, J.M., Norman, C.M. and Traut, R.R., *Biochem. 20*, 2843-2852 (1981).

75. Shine, J. and Dalgarno, L. *Eur. J. Biochem. 57*, 221-230 (1975).

76. Steitz, J.A. and Jakes, K. *Proc. Natl. Acad. Sci. USA 72*, 4734-4738 (1975).

77. Backendorf, C., Overbeek, G.P., Van Boom, J.H., van der Marel, G. Veeneman, G. and van Duin, J. *Eur. J. Biochem. 110*, 599-604 (1980)

78. van Duin, J. and Wijnands, R. *Eur. J. Biochem. 118*, 615-619 (1981)

79. Backendorf, C., Ravensbergen, C.J.C., van der Plas, J., van Boom, J.H., Veeneman, G. and van Duin, J. *Nucleic Acids Res. 9*, 1425-1444 (1981).

80. Giorginis, S. and Subramanian, A.R. *J. Mol. Biol. 141*, 393-408 (1980).

81. Boni, I.V., Zlatkin, I.V. and Budowski, E.I. *Eur. J. Biochem. 121*, 371-376 (1982).

82. Subramanian, A.R. Abstract, *7th Annual EMBO Symp. on Ribosomes*, Heidelberg (1981).

83. Bear, D.G., Ny, R., Derveer, D.V., Johnson, N.P., Thomas, G., Schleich, T., and Noller, H.F. *Proc. Natl. Acad. Sci. USA 73*, 1824-1828 (1976).

84. Auron, P.E. and Fahnestock, S.R. *J. Biol. Chem. 256*, 10105-10110 (1981).

85. Quiggle, K., Kumar, G., Ott, T.W., Kyu, E.K. and Chládek, S. *Biochem. 20*, 3480-3485 (1981).

86. Bhuta, A., Quiggle, K., Ott, T., Ringer, D. and Chládek, S. *Biochem. 20*, 8-15 (1981).

87. Nierhaus, K.H., Schulze, H. and Cooperman, B.S. *Biochem. Int. 1*, 185-192 (1980).

88. Hampl, H., Schulze, H. and Nierhaus, K.H. *J. Biol. Chem. 256*, 2284-2288 (1981).

89. Hayes, D.H., Guerin, M.F. and Nierhaus, K.H. Abstract, *7th Annual EMBO Symposium on Ribosomes*, Heidelberg (1981).

90. Cooperman, B.S. in *"Ribosomes"* (G. Chambliss *et al.*, eds.) pp. 531-554, University Park Press, Baltimore (1980).

91. Moore, V.G., Atchison, R.E., Thomas, G., Morgan, M. and Noller, H.F. *Proc. Natl. Acad. Sci. USA 72*, 844-848 (1975).

92. Fahnestock, S.R. *Biochem. 14*, 5321-5327 (1975).

93. Bernabeu, C., Vazquez, D. and Ballesta, J.P.G. *Eur. J. Biochem. 79*, 469-472 (1977).

94. Baxter, R.M., White, V.T. and Zahid, N.D. *Eur. J. Biochem. 110*, 161-166 (1980).

95. Amils, R., Ballesta, J.P.G., Juan, F., Tejedor, F., Teixidó, J. and Léon, G. Abstract, *7th Annual EMBO Symp. on Ribosomes*, Heidelberg (1981).

96. Noller, H.F., Kop, J., Wheaton, V., Brosius, R., Gutell, R.R., Kopylov, A.M., Dohme, F., Herr, W., Stahl, D.A., Gupta, R., and Woese, C.R. *Nucleic Acids Res. 9*, 6167-6189 (1981).

97. Olson, H.M., Grant, P.G., Cooperman, B.S. and Glitz, D.G. *J. Biol. Chem. 257*, 2649-2656 (1982).

138

98. Lührmann, R., Bald, R., Stöffler-Meilicke, M. and Stöffler, G. *Proc. Natl. Acad. Sci. USA 78*, 7276-7280 (1981).

99. Johnson, A.E., Adkins, H.J., Matthews, E.A. and Cantor, C.R. *J. Mol. Biol. 156*, 113-140 (1982).

100. Möller, W. in *"Ribosomes"* (M. Nomura *et al.*, eds.) pp. 711-731, Cold Spring Harb. Lab. (1974).

101. Maassen, J.A., Schop, E.N. and Möller, W. *Biochem. 20*, 1020-1025 (1981).

102. Möller, W., Amons, R., Cremers, A.F.M., Maassen, J.A., Mellema, J.E., Roobol, K., Schop, E.N., Schrier, P.I. and Zantema, A. Abstract *7th Annual EMBO Symp. on Ribosomes*, Heidelberg (1981).

103. Maassen, J.A. and Möller, W. *J. Biol. Chem. 253*, 2777-2783 (1978).

104. Girshovich, A.S., Kurtskhalia, T.V., Ovchinnikov, Yu, A. and Vasiliev, V.D. *FEBS Lett. 130*, 54-59 (1981).

105. Maassen, J. and Möller, W. *Eur. J. Biochem. 115*, 279-285 (1981).

106. Stöffler, G., Tate, W.P. and Caskey, C.T. *J. Mol. Chem. 257*, 4203-4206 (1982).

107. Crick, F.H.C. in *"Life Itself: Its Origin and Nature"*, pp. 82, Simon and Schuster, New York (1981).

108. Stöffler, G., Cundliffe, E., Stöffler-Meilicke, M. and Dabbs, E.R *J. Biol. Chem. 255*, 10517-10522 (1980).

109. Dabbs, E.R. *J. Bacteriol. 140*, 736-737 (1979).

Genes: *Structure and Expression*
Edited by A. M. Kroon
© 1983 John Wiley & Sons Ltd.

REGULATORY STEPS IN THE INITIATION OF PROTEIN SYNTHESIS

Harry O. Voorma

INTRODUCTION

The sequence of events leading to a 80S initiation complex is a multistep process, starting with the dissociation of 80S ribosomes into 40S and 60S ribosomal subunits. Dissociation is a prerequisite for initiation since the initial binding of Met-tRNA occurs exclusively on a 40S subunit. Spontaneous dissociation of 80S ribosomes leads to free 40S ribosomal subunits, which are prevented to re-associate with 60S by the binding of the eukaryotic initiation factor eIF-3 and presumably eIF-4C to 40S (see Fig. 1 and Table 1). Both factors seem to act as anti-association factor (1-3).

TABLE 1. Properties of eukaryotic initiation factors.

eIF	Disso-ciation of 80S	Met-tRNA binding to 40S	mRNA binding to 40S	Joining 40S with 60S	Met-puro reaction	Required in pH 5 system	$M_r \times 10^{-3}$
eIF-1	-	+	+	+	-	+	12
eIF-2	-	+	+	+	+	+	125
eIF-3	+	+	+	+	+	+	600
eIF-4A	-	-	+	+	-	+	50
eIF-4B	-	-	+	+	-	+	80
eIF-4C	+	+	+	+	+	+	17.5
eIF-4D	-	-	-	-	+	-	16
eIF-4E	-	-	+	+	-	+	24
eIF-5	-	-	-	+	+	+	125-160

The complex of 40S.eIF-3.eIF-4C participates in the next step which is the binding of Met-tRNA giving rise to a 40S preinitiation complex. The factor involved in this binding reaction is initiation factor eIF-2, which is capable to form a ternary complex with Met-tRNA and GTP (4-7). eIF-2 consists of three subunits (α β and γ).

Abbreviations: eIF: eukaryotic initiation factor, Met-tRNA: methionyl-tRNA, Met-puro: methionyl-puromycin, mRNA: messenger RNA, HRI: hemin-regulated inhibitor, SFV: Semliki Forest Virus, EMC: encephalomyocarditis, CBP: capbinding protein = eIF-4E.

The α-subunit is involved in the binding of GTP, the γ-subunit binds Met-tRNA, whereas the β-subunit appears to play a role in the recycling of the factor and complex formation with eRF (see next section). It is generally accepted that the components in the ternary complex are present in a 1:1:1 ratio. The ternary complex binds to the 40S

EUKARYOTES

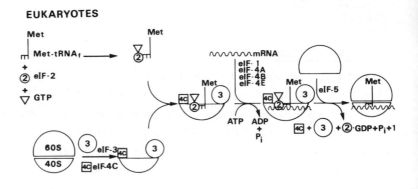

Fig. 1. Initiation of protein synthesis.
②, eIF-2; ③, eIF-3; 4C .eIF-4C: ∇ GTP: ⊓ Met-tRNA. For details see text of introduction, from (42).

subunit, facilitated by the presence of eIF-3 and eIF-4C on the 40S subunit. eIF-4C stabilizes the binding resulting in a two-fold stimulation. It is reasonable to assume that the three factors mentioned occur in stoichiometric amounts on these 40S-preinitiation complexes (2,3,8). Whereas the conditions for the three preceding steps: dissociation of 80S, ternary complex formation and 40S pre-initiation complex formation are fairly understood, the next step binding of messenger RNA to the 40S in a complicated one, in which four factors play a role. These factors are eIF-1, eIF-4A, eIF-4B and the cap binding protein eIF-4E.

Most of the eukaryotic mRNA's do have some structural peculiarities which are absent in prokaryotic messengers. The 3'-end generally contains a large stretch of poly(A) from 50-200 nucleotides (9-11). The function of the poly(A) tail is not entirely clear; it affects the stability of the messenger (12). At present binding to the cytoskeleton via poly(A) binding protein cannot be excluded (13). The 5'-ends of most eukaryotic messengers have a so-called cap structure, $m^7G(5')ppp(5')X(m)pY.......3'$, which promotes mRNA binding during

initiation of translation (14-16). Binding of mRNA by the 40S preinitiation complex can be divided into a number of steps involving
recognition of the capstructure of the mRNA by the capbinding protein

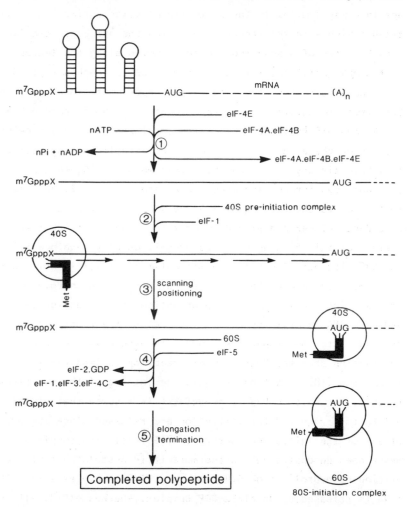

Fig. 2. Functional characterization of eukaryotic initiation
factors involved in cap recognition, melting of secondary
structure, mRNA binding and Met-tRNA positioning.
① melting, ② mRNA binding, ③ scanning and positioning,
④ joining of 60S, release of factors, ⑤ translation of
message.

eIF-4E, followed by binding of eIF-4B and eIF-4A. These proteins, for
which binding ATP hydrolysis is a prerequisite, play a role in making

the leader sequence accessible for the 40S preinitiation complex. A
requirement for mRNA binding is the removal of the secondary structure prior to the binding of the 40S. It is assumed that the factors
act in a way similar to the unwinding proteins during DNA-replication,
establishing a melted structure on which the 40S preinitiation complex by means of a scanning mechanism moves along the leader sequence
of the mRNA until the initiation codon AUG is reached (17). Reason
for such a multistep model stems from experiments in which it has
been tried to determine the stoichiometry of the factors eIF-4A,
eIF-4B and eIF-4E in 40S initiation complexes. The existence of these
factors on these complexes could not be established suggesting that
they have already fulfilled their role when the 40S initiation complex has been formed. The model, presented in Fig. 2, postulates
that prior to the binding of 40S the secondary structure is melted
by eIF-4E, eIF-4B and eIF-4A, whereafter binding of 40S takes place.
Whether at this binding step caprecognition plays again a role cannot be ruled out. The scanning mechanism by itself is not understood,
since movement along the messenger may also be an energy-requiring
process. The final step, the codon-anticodon interaction between AUG
and Met-tRNA is mediated by eIF-1 by means of a repositioning of
Met-tRNA into the proper position. This hypothesis is based on the
observation that the Met-puromycin reaction is impaired when eIF-1
is omitted (18). The final step is the joining of the 60S ribosomal
subunit, for which eIF-5 is needed. It is assumed that at this point
all factors required for initiation are released since the presence
of any of the initiation factors on 80S initiation complexes has
never been demonstrated. Furthermore it is postulated that the GDP
arising by hydrolysis of GTP from the ternary complex eIF-2.GTP.Met-
tRNA is released as an eIF-2.GDP complex. Whereas eIF-3, eIF-4C and
eIF-1 are released and can be used in another round of initiation,
the eIF-2.GDP complex cannot serve directly in a new round. Recycling
of this factor becomes possible after complex formation with eRF, eukaryotic recycling factor (see following section).

Summarizing the initiation of eukaryotic protein synthesis is a
process in which distinct steps can be distinguished. The Met-tRNA
binding precedes mRNA binding and results in the formation of a 40S

preinitiation complex (19). The availability of eIF-2 for ternary complex formation shall exert a direct effect on the number of 40S initiation complexes assembled. Regulation at this stage should have an overall effect on the rate of initiation. It is known that during the cell cycle the initiation rate changes considerably, which makes it feasible to assume that this rate is regulated by the availability of eIF-2 molecules.

A similar mechanism has been demonstrated for the inhibition of protein synthesis by hemin-deficiency.

Another level at which more specifically protein synthesis is regulated is at the messenger selection point. Since different messengers display a varying degree of secondary structure in the leader sequence, it has been postulated that the availability of eIF-4B, eIF-4E and perhaps eIF-4A decides to a very great extent which messengers are being translated. A low concentration of these factors shall allow the translation of low-requirement messengers, i.e. a low degree of secondary structure in the leader sequence, whereas messengers for which much factor is required are being excluded from translation. It appears that animal viruses infecting eukaryotic cells are using such a strategy.

In the following sections these two examples shall be dealt with more extensively.

REGULATION OF THE RATE OF INITIATION

The most extensively studied model system in which the rate of initiation drops considerably when hemin-deficiency is established is the rabbit reticulocyte lysate (for reviews, see 20,21). In this lysate incorporation of amino acids into globin chains proceeds for over 60 min in the presence of hemin, whereas in case of hemin-deficiency the original rate levels off sharply and a low rate is achieved within 5 min as is shown in Fig. 3, panel a. The arrest in protein synthesis is due to a block in initiation (22). Hemin-deficiency triggers the activation of a protein kinase, the hemin-regulated inhibitor HRI, which phosphorylates the α-subunit of eIF-2 (23-25). It is hard to comprehend which mechanism is the basis of this event since by HRI phosphorylated eIF-2 proved to be as active

144

in model-assay system as ternary complex formation and 40S initiation complex formation as non-phosphorylated eIF-2 (8,26-28).

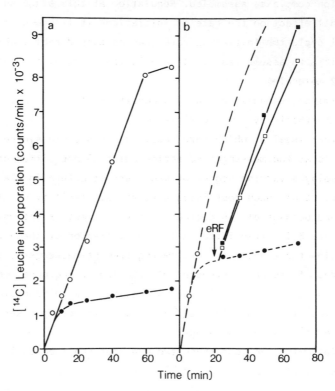

Fig. 3. Time-course of amino acid incorporation in a control and hemin-deficient reticulocyte lysate (o) plus hemin, (•) minus hemin (panel a).

Fig. 4. Relief of inhibition in a hemin-deficient lysate by eRF. (o) control, (•) hemin-deficient lysate, broken lines, (□,■) eRF added, 1 and 2 µg respectively (panel b).

However, very recently a number of data became available which may be the clue for the understanding of the mechanism of the hemin-regulation. First of all a new factor was discovered and characterized, which was capable to relieve the inhibition of protein synthesis in a hemin-deficient lysate (27). It was known already that eIF-2 could restore protein synthesis in such a lysate, but this factor could achieve the same thing and pointed to a relationship between this factor and eIF-2 (see Fig. 4, panel b). Moreover, it has been found that the factor, which has been designated eRF, eukaryotic recycling factor, occurs in a complexed form together with eIF-2. Recently the factor, formerly called anti-HRI (27), was characterized in several laboratories as eIF-2 stimulating protein, ESP, and restoring factor, RF (29,30).

The activity of eRF was first found to be stimulatory on the formation of 40S-preinitiation complex to such a level that a complete transfer of Met-tRNA from the ternary complex to 40S seems to occur. A second observation was that GDP exchange from an eIF-2.GDP complex is greatly impaired when the α-subunit is phosphorylated (31), whereas the following observations demonstrated both that complex formation between eIF-2 and eRF was impaired in case of phosphorylated eIF-2 and that ternary complex formation was greatly inhibited when mixtures of GTP and GDP (25:1 ratio were employed, which inhibition is completely overcome by eRF (32). Putting these facts together makes the following model very likely, which is depicted in Fig. 5. Upon joining of the 40S initiation complex with the 60S subunit GTP is hydrolyzed and the initiation factors are released in a free form except for eIF-2 which is released as an eIF-2.GDP complex. Further utilization is only possible when an exchange reaction occurs, which bears a very good resemblance with the EF-Tu.GDP/EF-Tu.EF-TS

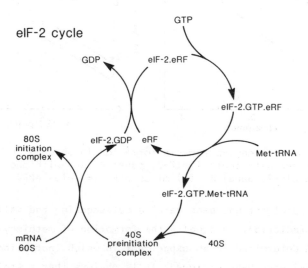

Fig. 5. Proposed mechanism for the eIF-2 cycle.

exchange reaction. GDP is removed from the complex and a new eIF-2.
eRF complex is generated, which can bind GTP and Met-tRNA in subsequent steps. Dissociation of the eRF-moiety appears to occur after
Met-tRNA binding, since the complex has been detected when incuba-
tions with GTP have been carried out. The predictions from this model
are the following:

1. by omitting eRF in the reaction mixture eIF-2 can only carry out
   one round of initiation, which implies that eIF-2 is used stoichiometrically.
2. no catalytic use of eIF-2 should occur when complex formation between eIF-2 and eRF is impaired, for instance by phosphorylation
   of eIF-2 by HRI.

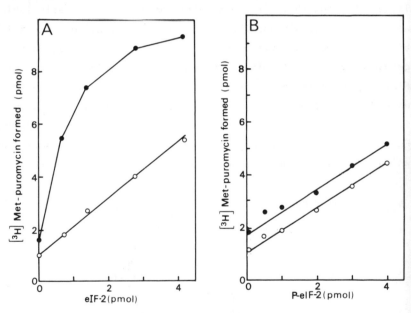

Fig. 6. Methionyl-puromycin formation in absence and presence of eRF with control eIF-2, panel A, and with phosphorylated eIF-2, panel B. (o) minus eRF, (•) plus eRF.

In Fig. 6 an experiment has been carried out checking the validity
of these two prediction. In panel A the formation of methionyl-puromycin has been determined in an experiment in which increasing
amounts of eIF-2 have been employed. It is obvious that a stoichiometric relationship between eIF-2 and the product, methionyl-puromy-

cin, exists. Addition of eRF stimulates at low input of eIF-2 the
methionyl-puromycin formation very markedly, indicating a catalytic
use of eIF-2 (a 4 to 5-fold stimulation). To stress this point, by
addition of eRF a cell-free system has been designed in which recycling of eIF-2 occurs; a condition not often met in cell-free systems
reconstituted with purified factors. In panel B a similar experiment
has been carried out, the only difference being that phosphorylated
eIF-2 has been used. Both in presence and absence of eRF eIF-2 is
used stoichiometrically. Although still a number of problems has to
be solved it is believed that the overall rate of initiation is regulated by this exchange reaction between eIF-2 and eRF. Careful experiments are needed to establish the precise mode of action of eRF
to explain all events that occur when the cell shifts down from a
high rate of initiation to a low rate.

REGULATION OF MESSENGER-RNA SELECTION

As was mentioned in the introduction four initiation factors are
involved in mRNA binding: eIF-1 and eIF-4B which are bound exclusively to ribosomes, eIF-4A which occurs mainly in the postribosomal
supernatant, whereas the cap-binding protein eIF-4E has been isolated
from both eIF-4B and eIF-3 preparations (33,34). Some of these factors are bound to play a role in mRNA selection since it has been
shown that mRNA's with a low degree of secondary structure possess a
diminished dependence for factors in the mRNA binding reaction,
whereas a high degree of secondary structure requires much more factor for binding. It seems that the mechanism of host shutoff in
virus-infected cells is using this discrepancy in factor dependence
to shutoff host cell and early viral protein synthesis in favour of
late viral protein synthesis. Some of the evidence recently accumulated points towards altered or inactivated initiation factors in infected cells. In extracts of poliovirus-infected HeLa cells (34) and
reovirus-infected L-cells (35) capped mRNA's are translated at reduced efficiencies as compared with control extracts. The translation of capped mRNA's in poliovirus-infected HeLa cell extracts can
be stimulated by the addition of eIF-4B (36), although it is not
clear whether eIF-4B or a contaminating eIF-4E fraction was respon-

sible for the restoration of protein synthesis (see 34). Semliki Forest Virus is another virus that causes the shutoff of host protein synthesis (37). Both the 42S viral mRNA which is utilized as an early mRNA for the synthesis of non-structural proteins and the subgenomic 26S mRNA, which encodes for the structural proteins late in infection, are capped (38,39, see Fig. 7). Late in infection only virus-encoded structural proteins are synthesized, which suggested that the host

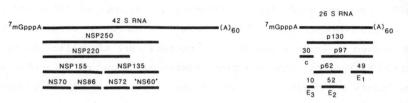

Fig. 7. Structural and non-structural polypeptides encoded by 42S and 26S Semliki Forest Virus RNA.

mRNA's are outcompeted by the late viral mRNA's because of a higher affinity of the latter mRNA's for the protein synthesizing machinery. In this system inactivation of one or more initiation factors would not be an absolute prerequisite for shutoff. However, we found inactivation of factors implying that the factor dependence on the 5' cap structure of late viral mRNA is less stringent.

Table 2 shows the results of these experiments in which the ability of partly purified ribosomal washes, containing the mRNA binding

TABLE 2. Effect of ribosomal washes from uninfected and infected cells on translation of different mRNA's.

| mRNA added | $[^{35}S]$Methionine incorporation (counts/min) | | |
	washes from control cells	washes from infected cells	% inhibition
early SFV (42S)	138.714	40.215	71
late SFV (26S)	150.583	135.985	10
EMC	196.675	164.695	16
host neuroblastoma	84.855	9.429	89

For details see (40)

factors, from uninfected and infected neuroblastoma cells were tested in order to determine the dependence of these washes in a standard pH 5 cell-free system for protein synthesis programmed with a number of mRNA's. It is clear that the transition from control to infection-exposed washes results in a severe inhibition of protein synthesis with host neuroblastoma mRNA and 42S Semliki Forest Virus RNA, whereas translation of no cap-carrying encephalomyocarditis virus RNA and late 26S SFV-RNA is almost not affected. This finding indicates that the alteration in the protein synthesizing machinery occurred at the level of the initiaton factors. It was confirmed in a further characterization that only the fraction containing eIF-3, eIF-4B and cap binding protein eIF-4E was affected (40). The observation that the translation of late SFV-RNA and EMC-RNA was not affected by the transition from control to viral-exposed initiation factors evoked the important question whether the utilization of these mRNA's as template is eIF-4B and eIF-4E independent. Therefore, a comparison has been made with the requirement for these factors for optimal translation of early SVF-RNA (42S) late SFV-RNA (26S), EMC-RNA and host cell mRNA, which is given in Fig. 8. It is striking that opti-

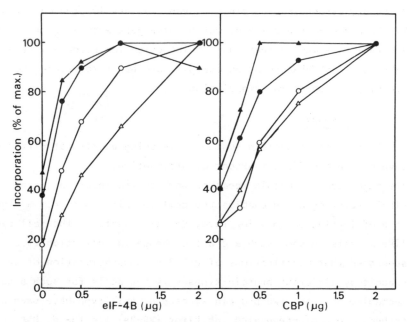

Fig. 8. Determination of eIF-4B and eIF-4E requirement for optimal translation of different mRNA's.
(▲) EMC-RNA, (●) 26S SFV-RNA, (Δ) 42S SFV-RNA, and (o) host mRNA, for details see (41).

mal protein synthesis with 42S SFV-RNA and host cell mRNA requires 2 to 4 times more eIF-4B than translation of late 26S RNA. On the other hand it was shown that chemical decapping affected the translation of 26S SFV-RNA to a much greater extent than 42S SFV-RNA, suggesting that extraordinary structural features are present in either cap or leader sequence (41).

So in contrast to the strategy used in poliovirus infection in which the viral RNA does not possess a cap and its translation is cap-independent, host cell shutoff can easily be achieved by inactivation of eIF-4B and eIF-4E. However, in Semliki Forest Virus infection the host-cell shutoff also takes place, wiping out early viral mRNA and host mRNA translation, whereas late viral mRNA translation continues. In this case we have to bear in mind that all mRNA's do contain a cap, translation is even highly cap-dependent. Nevertheless the late viral RNA is translated and early viral RNA and host mRNA are excluded from translation.

The hypothesis being put forward here is that the degree of secondary structure determines, given the very low concentration of eIF-4B and eIF-4E available, which messengers are being translated. The rationale may be that late in infection low secondary structure mRNA's, 26S SFV-RNA being an example, are built to escape the restraint put upon normal translation.

SUMMARY

Regulation of protein synthesis is being exerted at different levels with a different extent of attenuation.
The major control module seems to work by the inactivation of the eIF-2 recycling which enables the cell to shift down from a high rate of initiation to a low rate. Certain events in the cell cycle like mitosis do show such a drastic change in initiation rate. It is suggested that modifications of eIF-2 by phosphorylation of the α-subunit by different protein-kinases is the basis for such a control mechanism. Already two protein kinases of this type have been described, the hemin-regulated inhibitor (20-24) and the ds-RNA activated inhibitor from interferon-treated cells (21). On the other hand modifications of the β-subunit by other metabolic events, for in-

stance low NADH/NAD$^+$ ratio, can as yet not be excluded (20).

Other conditions like amino acid starvation, serum deprivation, heat-shock and virus-infection seem to evoke quite different strategies. In some cases it has been demonstrated that inactivation of mRNA binding factors as eIF-4B and eIF-4E, favour the translation of low-dependence, i.e. low secondary structure, messengers. It shall be worthwhile to establish whether the mRNA's with such low degree of secondary structure encoded proteins that are aimed at the survival of the cell under extreme metabolic or stress conditions. Much more work on the structure and nucleotide sequences of the leader sequence is needed to prove these hypothetical points.

ACKNOWLEDGEMENT

Part of the hypotheses presented in this paper have been discussed at the International Conference on Translation/Transcriptional Regulation of Gene Expression, Fogarty International Center - National Institute of Health, Bethesda, April 7-9, 1982, in which discussion B. Safer, R. Jagus, N. Gupta, J. Siekierka, D. Levin, E. Henshaw, M. Kozak, H. Trachsel, N. Sonenberg and W. Merrick participated.

REFERENCES

1. Davis, B.D., Nature *231*, 153-157 (1971)
2. Benne, R. and Hershey, J.W.B., Proc. Natl. Acad. Sci. USA *73*, 3005-3009 (1976)
3. Thomas, A., Goumans, H., Voorma, H.O. and Benne, R., Eur. J. Biochem. *107*, 39-45 (1980)
4. Safer, B., Anderson, W.F. and Merrick, W.C., J. Biol. Chem. *250*, 9067-9095 (1975)
5. Lloyd, M., Osborne, J.B., Safer, B., Powell, G.M. and Merrick, W.C., J. Biol. Chem. *255*, 1189-1193 (1980)
6. Barrieux, A. and Rosenfeld, M.G., J. Biol. Chem. *252*, 392-398 (1977)
7. Barrieux, A. and Rosenfeld, M.G., J. Biol. Chem. *252*, 3843-3847 (1977)
8. Trachsel, H. and Staehelin, T., Proc. Natl. Sci. USA *75*, 204-208 (1978)
9. Brawerman, G., Progr. Nucl. Acid Res. and Mol. Biol. *17*, 117-148 (1976)

152

10. Lewin, B., Cell *4*, 11-20 (1975)

11. Greenberg, J.R., J. Cell Biol. *64*, 269-288 (1975)

12. Huez, G., Marbaix, G., Burny, A., Hubert, E., Leclerq, M., Cleuter, Y., Chantrenne, H., Soreq, H. and Littauer, U.Z., Nature *266*, 473-474 (1977)

13. Setyono, B. and Greenberg, J.R., Cell *24*, 775-783 (1981)

14. Shatkin, A.J., Cell *9*, 645-653 (1976)

15. Rottman, F., Shatkin, A.J. and Perry, R.P., Cell *3*, 197-199 (1974)

16. Filipowicz, W., FEBS Lett. *96*, 1-11 (1978)

17. Kozak, M., Cell   , 1109-1123 (1978)

18. Thomas, A., Spaan, W., van Steeg, H., Voorma, H.O. and Benne, R., FEBS Lett. *116*, 67-71 (1980)

19. Schreier, M.H. and Staehelin, T., Nature New Biol. *242*, 35-38 (1973)

20. Jagus, R., Anderson, W.F. and Safer, B., Progr. Nucl. Acid Res. and Mol. Biol. *25*, 127-185 (1980)

21. Austin, S.A. and Clemens, M.J., FEBS Lett. *110*, 1-7 (1980)

22. Legon, S., Jackson, R.J. and Hunt, T., Nature New Biol. *241*, 150-152 (1973)

23. Gross, M. and Rabinowitz, M., Proc. Natl. Acad. Sci. USA *69*, 1565-1568 (1972)

24. Gross, M., Biochim. Biophys. Acta *366*, 319-332 (1974)

25. Farrell, P.J., Balkow, K., Hunt, T., Jackson, R.J. and Trachsel, H., Cell *11*, 187-200 (1977)

26. Merrick, W.C., Peterson, D.T., Safer, B., Lloyd, M. and Kemper, W.C., Proc. 11th FEBS Meet., Vol. 43, pp 17-26 (1977) Pergamon Oxford, New York

27. Amesz, H., Goumans, H., Haubrich-Morree, T., Voorma, H.O. and Benne, R., Eur. J. Biochem. *98*, 513-520 (1979)

28. Benne, R., Salimans, M., Goumans, H., Amesz, H. and Voorma, H.O., Eur. J. Biochem. *104*, 501-509 (1980)

29. Siekierka, J., Mitsui, K. and Ochoa, S., Proc. Natl. Acad. Sci. USA *78*, 220-223 (1981)

30. Ralston, R.O., Das, A., Grace, M., Das, H.K. and Gupta, N.K., Proc. Natl. Acad. Sci. USA *76*, 5490-5494 (1979)

31. Clemens, M.J., Pain, V.M., Wong, S-T. and Henshaw, E.C., Nature *296*, 93-94 (1982)

32. Safer, B., International Conference on translational/transcriptional regulation of gene expression, Fogarty Internation Center, N.I.H. Bethesda, April 7-9, 1982

33. Thomas, A., Goumans, H., Amesz, H., Benne, R. and Voorma, H.O., Eur. J. Biochem. *98*, 329-337 (1979)

34. Trachsel, H., Sonenberg, N., Shatkin, A.J., Rose, J.K., Leong, K., Bergmann, J.E., Gordon, J. and Baltimore, D., Proc. Natl. Acad. Sci. USA *77*, 770-774 (1980)

35. Skup, D. and Milward, S., Proc. Natl. Acad. Sci. USA *77*, 152-156 (1980)

36. Rose, J.K., Trachsel, H., Leong, K. and Baltimore, D., Proc. Natl. Acad. Sci. USA *75*, 2732-2736 (1978)

37. Keränen, S. and Kääriäinen, L., J. Virol *16*, 388-396 (1975)

38. Kääriäinen, L. and Söderlund, H., Curr. Top. Microbiol. Immunol. *82*, 15-69 (1978)

39. Pettersson, R.F., Söderlund, H. and Kääriäinen, L., Eur. J. Biochem. *105*, 435-443 (1980)

40. Van Steeg, H., Thomas, A., Verbeek, S., Kasperaitis, M., Voorma, H.O. and Benne, R., J. Virol. *38*, 728-736 (1981)

41. Van Steeg, H., Van Grinsven, M., Van Mansfeld, F., Voorma, H.O. and Benne, R., FEBS Lett. *129*, 62-66 (1981)

42. Thomas, A.A.M., Benne, R. and Voorma, H.O., FEBS Lett. *128*, 177-185 (1981)

Genes: *Structure and Expression*
Edited by A. M. Kroon
© 1983 John Wiley & Sons Ltd.

TRANSPORT OF PROTEINS FROM THE SITES OF GENETIC EXPRESSION TO THEIR
SITES OF FUNCTIONAL EXPRESSION: PROTEIN CONFORMATION AND
THERMODYNAMIC ASPECTS.[a]

Christoph Kempf[+], Richard D. Klausner[+], Robert Blumenthal[*],
and Jos van Renswoude[+]

[+]Laboratory of Biochemistry and Metabolism, NIADDK, and [*]Laboratory
of Theoretical Biology, NCI, National Institutes of Health, Bethesda,
MD 20205, U.S.A.

[a]Dedicated to the memory of Mones Berman

Introduction

During the past decades our knowledge of protein biosynthesis has
steadily grown. In particular, a great deal of the molecular details
of the translation process itself and of the regulation thereof have
been elucidated. At present, we more or less know how and where in
the cell proteins are made. Only during the past ten years questions
as to how proteins travel from the site(s) of biosynthesis to their
final destination within or outside the cell have gained attention.
Proteins are biosynthesized at three major locations within the cell:
the cytosol (eukaryotic and prokaryotic cells) mitochondria and
chloroplasts (eukaryotic cells). Some proteins remain "soluble" after
synthesis, i.e., they fulfill their role in the fundamentally aqueous
environment of cytosol, nucleosol, mitochondrial matrix and the
stroma of chloroplasts. Others become membrane proteins or are
destined to be exported from the cell ("secretory" proteins). Both
membrane- and secretory proteins must interact intimately with
membranes at some point during their intracellular journey. This
interaction occurs either during (cotranslationally) or after
(posttranslationally) the actual synthesis of the peptide backbone.
Many examples are known, to date, of cotranslational secretion across
or insertion into cellular membranes. Virtually all membrane- and
secretory proteins emerge from the ribosome as preproteins; they
contain a very hydrophobic N-terminal prepiece, consisting of 15-30
amino acid residues. This prepiece is called "signal sequence" or
"leader sequence"[1]. Upon passage of the membrane- or secretory
protein into/across the appropriate biological membrane, the signal
sequence is removed through action of a membrane-associated enzyme:
signal peptidase. It is now well documented (Reviews: Refs. 1-4) that
the signal sequence plays a key role in the initiation of the
interaction between membrane- and secretory proteins on one hand and

[1]the terms "signal sequence" and "leader sequence" will be used
interchangeably

the "acceptor" membrane[1] on the other. Currently, two theories are available to assist in explaining the role of the signal sequence. In one theory, most often referred to as "signal hypothesis" (Blobel and coworkers, Refs. 1,5), the signal sequence is believed to be recognized specifically by an acceptor-membrane bound receptor protein. Via a still undefined guidance system the nascent peptide chain would then make its way into and across the acceptor membrane. The other theory, known as the "trigger hypothesis", was developed by Wickner and colleagues (for reviews see Refs. 1,6). According to this hypothesis, the leader sequence as soon as it emerges from the ribosome would act as a "conditioner" of the amino-terminus of the nascent chain, i.e., it would force upon the growing peptide chain a conformation compatible with penetration into the acceptor membrane. The first-penetrating part of the peptide chain may well, by virtue of its overall hydrophobicity, cause a perturbation of the lipid bilayer of the acceptor membrane. Such a perturbation may be large enough to significantly reduce the energy needed to allow passage of the rest (often very hydrophilic and even charged) of the protein molecule into and/or across the membrane. The trigger hypothesis predicts, in practice, that following a large enough initial perturbation of the acceptor membrane, insertion and/or translocation proceed(s) more or less automatically, i.e., according to a "self-assembly" mechanism. It has been suggested that the hydrophilic and charged amino acid residues of the nascent peptide chain would be lined by a relatively hydrophilic (proteinaceous) channel, on their way across the hydrophobic core of the bilayer. If a protein can adapt its conformation (see below), however, to the highly hydrophobic environment in the center part of membrane, there may be no need for such "channels". Whatever the precise function of leader sequences may be, it is now well established that this part of a membrane- or secretory protein in-statu-nascendi plays a key role in the interaction between the protein and the acceptor membrane. Leader sequences apparently interact specifically with appropriate acceptor membranes, and not with just any intracellular membrane. This means that an element of recognition is involved in their interaction with biological membranes. We do not know whether leader sequences are recognized through their primary structure, their overall hydrophobicity, their inherent secondary and higher structure, or any combination of the foregoing features.

As indicated above, many examples are available today of proteins that interact with an appropriate acceptor membrane in a cotranslational fashion. If elongation of the peptide chain keeps pace with the feeding of the chain into and across the membrane, we can envisage the protein being "threaded" through the membrane and - upon emerging from it on the far side - acquiring a final conformation compatible with an aqueous environment. A tight coupling of chain elongation and feed-in on the proximal side will probably restrict the peptide chain in attaining a great deal of secondary and higher structure before actually entering the membrane. The folding

---

[1] i.e., the endoplasmic reticulum in eukaryotic cells and the cell membrane in prokaryotic cells

process on the other (extracytoplasmic) side of the membrane may
yield energy usable in "pulling" the peptide chain completely
(secretory proteins) or partly (membrane proteins) across the
membrane. That part of the N-terminal leader sequence that initiates
the interaction with the acceptor membrane may have a very particular
conformation allowing it to be "recognized" by the acceptor
membrane. Also, that part of the peptide chain that resides – at any
given time – inside the acceptor membrane, will have a specific
folding state compatible with the hydrophobic environment of the
membrane (see below). It is far more difficult to understand how
certain proteins can insert into and even translocate across
biomembranes in a posttranslational fashion, i.e., after having been
synthesized on free cytoplasmic ribosomes and fully folded in the
aqueous environment of the cytosol. Examples of these are nuclear-DNA
coded proteins destined for import into mitochondria and chloroplasts
as well as proteins that become plasma membrane-associated
posttranslationally. Evidently, such proteins – while in a folding
state compatible with an aqueous environment – retain the ability to
specifically bind to an appropriate acceptor membrane. If they are to
completely or partially traverse a membrane, they will have to adapt
their conformation to the new environment. It seems that for such
proteins, e.g. the large subunit of Na^+,K^+ - ATPase (plasma
membrane), cytochrome C (mitochondria), ribulose 1,5-biphosphate
carboxylase (chloroplast), and β-lactamase (bacterial cell membrane),
the problems of folding are quite different than for those proteins
that negotiate the bilayer in a cotranslational way. It is obvious
that in posttranslational protein-membrane interaction, dramatic
changes in protein conformation, as an adaptation to a new
environment, will have to take place. An exciting, very recent
discovery may open new lines of thought on how membrane- and
secretory proteins in general initiate cotranslational contact with
their acceptor membrane. Meyer and colleagues (7) have found that the
rough endoplasmic reticulum contains a 72 kilodalton protein which
functions as a universal docking protein for signal sequences
complexed to "signal recognition protein" (SRP) first described by
Walter et al. (8). Meyer and colleagues report that SRP exists in two
forms, a "soluble", cytoplasmic one, and one bound to the rough
endoplasmic reticulum. Binding of SRP to a newly synthesized
amino-terminal leader sequence brings chain elongation to a halt.
This translation-block is relieved upon binding of the complex of
SRP, nascent peptide chain, ribosome and mRNA to the docking protein.
This march of events provides a very secure link between
translation and insertion of the peptide chain into the acceptor
membrane. It also explains why some proteins, whose synthesis is
initiated on free cytoplasmic ribosomes, still interact
cotranslationally with the rough endoplasmic reticulum. One could
say, teleologically, that through the combined action of SRP and
docking protein it is avoided that the nascent chain acquires a
folding state which is incompatible with membrane-crossing. In this
light, the question as to how certain proteins are able to negotiate
an acceptor membrane in a posttranslational way remains the more
puzzling. For some of these proteins precursor forms have been found.
There is, however, no evidence that the extra, N-terminal amino acid
sequence ("transit peptide") in these precursors has a structure

and/or function comparable to that of a "classical" leader sequence, as outlined above. It is, however, likely that through its precursor form the protein is recognized by a particular cell organelle.

There is one other aspect of protein biosynthesis that needs to be brought up briefly; after synthesis of their peptide backbone many proteins acquire some biochemical modification. This is commonly referred to as "posttranslational modification" and comprises, e.g., glycosylation, acetylation, acylation, phosphorylation, sulfatation, and proteolytic processing such as seen with proproteins like proinsulin and proalbumin, as well as in certain activation steps of the complement system. The function of many posttranslational modifications is still unknown. It is not inconceivable, however, that they play an important role in "fine-tuning" the secondary and higher structure of proteins, and in serving as "traffic labels" to guide their intracellular disposition (see below).

Much of our limited knowledge of in vivo conformation of proteins has arisen from the study of in vitro systems, in particular for integral membrane proteins such as bacteriorhodopsin, and the erythrocyte membrane proteins glycophorin and Band III. Since with the existing physical-chemical techniques it is virtually impossible to directly measure the folding state of a protein in the intact cell, conclusions with respect to such folding state(s) will have to come from good theoretical considerations, direct in vitro experiments, and indirect in vivo evaluation.

In the following sections we will discuss some general physical-chemical aspects of protein folding in aqueous and non-aqueous environments and address some specific questions concerning biosynthetic protein folding and -transport.

Aspects of in vivo protein-folding: the mixing of oil and water

The three-dimensional structure of proteins has clearly been demonstrated to be the result of the sequence of amino acids and the environment in which the protein folds. Anfinsen and his coworkers (9) provided the fundamental observations that led to the concept that the native structure of a protein is the one representing the thermodynamic free energy minimum for the sum of all of the interactions within the protein and between the protein and the solvent. This basic concept is well founded upon experiments in which native proteins are denatured by exposure to chemical or physical denaturants, and subsequently recovered as native proteins upon return of the denatured proteins to the original environmental conditions. Most denaturation studies involve two types of alterations of the protein: (i) reduction and breakage of disulfide bonds, and (ii) destruction of the secondary and tertiary structure by denaturants, such as urea or guanidine hydrochloride, that alter the non-covalent interactions within the protein. The lesson from all of these studies is, stated simply, that many denatured proteins are able to regain their native structure and -function in a spontaneous way, simply by being returned to their original environment, which lacks the denaturant or denaturing conditions.

Early studies showed that denaturation and renaturation could be well described by a two-state model (i.e., native and denatured) in which a first-order transition would connect the two states. More

detailed studies have revealed a more complicated pathway, involving
multiple metastable intermediate structures during the folding
process. The observation of folding intermediates, both in a
theoretical and an experimental sense, has led to a refinement of our
concepts of the folding process. One of the striking features of
protein folding is that unfolded proteins can regain their native
state relatively rapidly. If a protein had to randomly sample all
possible conformations in order to "find" the native one, the folding
process would be extremely slow. This notion led to the idea that
proteins fold via specific pathways leading through defined and
reproducible intermediate states.

The principle behind these preferred pathways is that certain
segments of the amino acid sequence serve as nucleation sites for the
folding process. Thus, specific regions of the protein fold
independently and acquire the correct secondary structure. The
interplay of these nucleation regions then results in the final
conformation of the protein. The nucleation theory is a first step in
efforts to explain the rapid time course of correct folding in the
intact cell. The observed rapidity of folding has also led to the
question whether the cell possesses a mechanism for catalyzing
folding, specifically with respect to the formation of correct
disulfide linkages. Several studies have reported the existence of a
non-specific (with regard to a particular protein) cellular system
for catalyzing the rapid formation and breakage of protein
intramolecular disulfide bonds. Although such a system does not per
se lead to the formation of correct crosslinks, it would greatly
speed up the search for the correct, i.e., energetically most
favorable, protein structure.

As mentioned above, the native state of a protein is believed to
be the one that represents its minimum free energy structure in a
given environment. In virtually all studies on protein folding, the
folding process has been examined in an aqueous environment. Although
this applies readily to cytosolic and secreted proteins, protein
folding in conjunction to the biosynthesis of membrane proteins is
governed by different rules. The lipid bilayer, which is generally
accepted to be the basic structural unit of membranes, provides a
very specific environment for proteins, quite different from an
aqueous phase. The most prominent feature of this environment is that
the acyl chain interior of the bilayer is inert, with respect to
intermolecular interactions. It is a region marked by a relatively
low dielectric constant, it provides no environmental source of
hydrogen bonding. This has two major consequences for protein
folding: (i) it is energetically very expensive to bury a charged
amino acid in the bilayer, and (ii) bilayer-spanning regions of a
membrane protein will tend to adopt a regular secondary structure
which satisfies the hydrogen bonding requirements within the protein
rather than being compatible with an aqueous environment. As a
result, an intramembraneous region (usually poor in amino acids with
charged side groups) of a membrane protein, folds into an α-helix. In
the α-helix all backbone hydrogen bonds are accomplished within the
helix. This structure will probably prove to be a relatively constant
one for trans-membrane proteins.

Amphipotential proteins

Because they contain long stretches of uncharged amino acids, most integral membrane proteins are not stable in water. It is as if they are "built" to fold into a hydrophobic environment. How then can the cell synthesize a water-insoluble protein? It appears that the cell has solved this problem by carrying out the synthesis of these proteins on ribosomes that are bound to membranes. The protein is inserted into the membrane as it is synthesized, cotranslationally. This process has been most clearly demonstrated for the G protein of vesicular stomatitis virus. The folding problem, i.e., first finding a conformation compatible with the aqueous cytosol and then refolding into the membrane environment, is avoided by inserting the protein into the membrane as it is being synthesized. Many, but not all, integral membrane proteins appear to employ this route. The coat protein of filamentous bacteriophage as well as those membrane proteins of the mitochondria and chloroplasts that are synthesized in the cytosol are examples of membrane proteins that exist as water-soluble proteins and only insert into the membrane posttranslationally. Such proteins have the conformational flexibility to exist either in a water-soluble or in a lipid-soluble state. Such proteins have been called amphipotential or amphipatic, because of this conformational flexibility.

Some non-membrane proteins are also amphipotential; apolipoproteins generally coexist with lipids in lipoproteins. Several of these lipoproteins, including A1, A2, and E, have been freed of all lipid and studied as water-soluble proteins. Despite the fact that these proteins tend to aggregate in aqueous solution, they can exist as water-soluble monomers. Addition of lipids to such proteins results in reformation of lipoprotein particles. For these proteins the change of environment caused by the addition of lipid invokes profound changes in their folding state, most markedly reflected by an increase of α-helix content.

We have studied tubulin as an example of an amphipotential protein. Tubulin, the basic structural unit of microtubules, is a heterodimer made up of two non-covalently linked subunits. For several years it has been known that tubulin can exist both as an integral membrane protein and as a soluble, cytosolic protein. More recently, several groups have shown that one single mRNA can code both physical forms of the protein. According to the rules of folding discussed above, we would predict that tubulin would undergo a structural transition if its soluble form could be induced to enter a lipid bilayer. We found that purified, water-soluble tubulin could indeed become integrally associated with a pure lipid membrane (10). This process required mixing of the protein with vesicles made up of lipids with a defined melting temperature, at which the packing of the lipid acyl chains changes from a solid crystalline array to a liquid crystal. When the temperature of the tubulin-vesicle mixture was raised across this melting point, the protein inserted into the membrane. In this state, the tubulin would mediate vesicle-vesicle fusion in the presence of calcium. Circular-dichroism measurements clearly indicated that the environmental transition of this protein was accompanied by a structural alteration, to more α-helical, consistent with folding in a hydrophobic environment.

Interaction of proteins with biological membranes; insertion and translocation

All of the above-mentioned examples point out the crucial relationship between protein folding and environment. This relationship has to be kept in mind when trying to understand how (in vivo) protein biosynthesis is linked to intracellular protein transport. All proteins that are secreted from the cell or transported into cellular organelles, such as lysosomes, mitochondria, etc., must cross one or more biological membranes. One can safely make the assumption that a protein which has fully folded in an aqueous environment cannot - as such - translocate across a membrane. Therefore, in order to enter a membrane, a secretory- or membrane protein must either not be allowed to acquire any significant degree of "aqueous environment"-folding, or profoundly alter its conformation to one compatible with a hydrophobic (membrane-) environment. Most membrane- and secretory proteins are synthesized with a leader sequence, as discussed above. The ultimate function of this sequence is to initiate the interaction of the nascent peptide chain with the acceptor membrane in a cotranslational fashion, ensuring that "aqueous environment"-folding of the nascent chain cannot occur to any significant extent. The leader sequence must be "exposed" and presumably correctly folded itself, in order to mediate the binding of the nascent chain to the acceptor membrane. An example which illustrates this issue is provided by ovalbumin. Ovalbumin (OA) is the major secreted protein of the hen oviduct. In contrast with other secreted proteins, OA is not first synthesized as a preprotein; it contains no "classical" (i.e., cleaved) leader sequence. Nevertheless, OA is secreted in a manner identical to other secreted preproteins; it is translocated across the endoplasmic reticulum cotranslationally, and apparently interacts with the same sites on the membrane as other secretory preproteins do. This led to the question as to what part of the OA molecule replaces the function of the classical leader sequence. Lingappa et al. (11) proposed that the leader sequence was located in the center of the polypeptide chain. Such a location would invalidate our assumption that the peptide must be translocated before it folds in the aqueous environment of the cytosol. Recently, we and other groups (12,12a,13) have shown that the amino-terminal region of the protein functions as the leader sequence. We have compared this unusual secretory protein with one that has a classical, cleaved leader sequence: pre-kappa immunoglobulin light chain. We have measured the specific binding of in vitro synthesized, leader sequence containing peptides of each of these proteins to sites on the rough endoplasmic reticulum (Fig. 1). The amount of pre-kappa bound to the microsomes increases with time of synthesis. Even at late times after the initiation of synthesis, where most of the protein is present as completed pre-kappa (as judged by polyacrylamide gel electrophoresis; data not shown), significant binding remains. This is consistent with a persistent exposure of the leader sequence, even in the fully synthesized preprotein. OA, on the other hand, displays a dramatically different binding curve. The amount of OA bound to these membranes rises steeply during the early minutes after initiation of synthesis, and then - as the average OA-chain length increases - steadily declines;

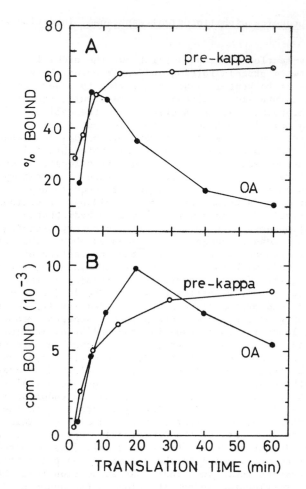

Fig. 1. <u>Posttranslational</u> <u>binding</u> <u>of</u> <u>ovalbumin-</u> <u>and</u> <u>pre-kappa</u> <u>light</u>
<u>chain</u> <u>nascent</u> <u>peptides</u> <u>to</u> <u>rough</u> <u>microsomes</u>
Purified pre-kappa light chain mRNA (mouse myeloma cell line) and
ovalbumin mRNA were used to direct protein synthesis in a
reticulocyte lysate system. At various times after initiation of
synthesis, samples were withdrawn from the translation system and
treated with puromycin, to release nascent chains from polysomes.
Subsequently, stripped rough microsomes were added and the mixture
was incubated for 15 min at 27°C. The microsomes were then spun down
in an Airfuge and assayed for TCA-precipitable radioactivity derived
from ^{35}S-methionine in the cell free translation system.
    Results are expressed as the % of total TCA-precipitable radio-
activity, that became microsome associated (A), or as absolute
microsome associated, TCA-precipitable counts (B).

complete OA binds very poorly to the membranes. The binding of the
early translation products can be blocked by pre-kappa but not by
mature kappa, which lacks the leader sequence. Polyacrylamide gel
electrophoresis revealed that OA peptides of molecular weight less
than about 15,000 daltons were capable of binding to the microsomes.
Our interpretation of this finding was that the early peptides of OA
contain an amino-terminal, functional leader sequence, through which
the peptide chain binds to the membranes. Upon elongation of the
chain the amino-terminal region apparently becomes inaccessible to
the membrane binding site. A surface active, hydrophobic leader
sequence that is normally proteolytically removed from a preprotein
is in this case functionally removed, not by excision but by
sequestration through folding. This explains why mature OA possesses
no characteristics of preproteins.

The folding of proteins influences many aspects of protein
biosynthesis. The dual environment (aqueous and membrane-) present in
the cell places several constraints on the folding of proteins.
Proteins that exist in both environments must be able to adopt
conformations compatible with both. This implies that such proteins
probably possess more than one stable conformation. This condition is
fulfilled by the type of protein which we termed amphipotential.
Proteins destined for the membrane that cannot fold into a stable
"water-soluble" conformation most probably avoid this state by being
inserted into the membrane during synthesis so that they can fold
into their proper environment. Finally, proteins that are secreted
across a membrane must be able to, at least transiently, adopt a
conformation compatible with translocation across a lipid bilayer. As
we discussed earlier, the absence of a source of hydrogen bonding
within the bilayer may be one of the strongest forces in determining
acceptable conformational transients. The α-helix seems to be of
physiologic importance; bacterial mutant preproteins, carrying
"helix-breaking" amino acid substitutions in the leader sequence,
cannot be secreted. This fact has recently been incorporated by
Engelman and Steitz into a model for protein secretion, the helical
hairpin hypothesis (14). According to this model, the membrane is
traversed by two antiparallel helices. The role of the leader
sequence is to provide one (helical) arm of this hairpin and also to
provide the hydrophobicity to drive the initial partitioning into the
membrane. In order to process the initial pair of helices, the
carboxy-terminal region of the leader sequence must be able to adopt
a chain reversal configuration. As the chain elongates during
synthesis, the second helix receives continuous input of new amino
acids entering the membrane-spanning helix, while residues on the
opposite side are excluded into the aqueous region. As more and more
residues are pushed out of the membrane they begin folding in the
transmembrane aqueous space. This folding may enhance the
continuation of the threading process. Thus, at least three folding
processes combine to allow translocation. (i) The hydrophobic leader
sequence adopts an α-helical conformation followed by a chain
reversal, (ii) the transiently membrane-spanning segment of the
growing chain must enter the membrane as an α-helix, and (iii) the
lengthening chain folds into a water-soluble, globular form on the
opposite side of the membrane. This entire process must take place
cotranslationally, to succesfully compete with folding of the chain

before the membrane is spanned by the protein.

If the cotranslational requirement for protein insertion into or
-secretion across membranes reflects the kinetic competition between
aqueous-phase folding and bilayer interaction, then the interaction
with the membrane must occur soon after peptide synthesis has been
initiated. Alternatively, if the interaction with the membrane does
not take place immediately, it must be prevented in some way that the
protein is fully synthesized in the cytosol where it would attain a
conformation incompatible with insertion into and/or secretion across
the membrane. This dilemma seems to be resolved by the discovery of a
dual form of SRP (a membrane-bound as well as a soluble one) together
with a universal docking protein, as mentioned in the Introduction.

One aspect of any folding model in conjunction with protein
movement into or across membranes involves the energetics of
translocation. Engelman and Steitz (14) presented estimates of
energies of transfer which suggested that the cost of transferring
polar and even charged amino acids across the membrane could be paid
for by other aspects of the folding and partition events. This work
led to the helical hairpin hypothesis, as discussed above. Recently,
we have developed a new algorithm for predicting the three-
dimensional structure of membrane proteins in general (15). In
applying this algorithm to predict the three-dimensional structure of
a leader sequence we used the conformational energy minimization
calculations developed in the laboratory of Harold Scheraga. The
result was a long α-helix followed by a tight chain reversal at the
carboxy-terminal region of the leader sequence, exactly as predicted
by the helical hairpin hypothesis.

Although hydrophobic forces probably dominate in determining
whether or not a protein will insert into or translocate across a
membrane, recent studies in our laboratory suggest that electrostatic
forces may play a significant role. A negatively charged hepatic
receptor protein was found to respond to a trans-positive membrane
potential by "electrophoresing" into a bilayer (16). In the presence
of ligand, the protein appeared to cross the membrane and expose
binding sites on the opposite side. Similarly, a positively charged
portion of the peptide mellitin crosses a lipid membrane in response
to a trans-negative potential (17). A role for the membrane potential
in assembly and translocation of proteins has been described for a
number of other systems: (i) Date et al. (18) have shown that
dissipation of the electrochemical gradient by uncouplers prevents
proper insertion and processing of M13 viral coat protein in E. coli
membranes; (ii) Oxender an coworkers (19) presented evidence that
membrane depolarization prevents secretion of leucine binding protein
and β-lactamase into the periplasm of E. coli; (iii) Randall et al.
(20) have shown that an electrochemical gradient is essential for
secretion of several E. coli periplasmic- and outer membrane
proteins; (iv) Schleyer et al. (21) have recently demonstrated that a
transmembrane potential is required for posttranslational transfer of
proteins into mitochondria; (v) a membrane potential is required for
channel formation by diphteria toxin in planar lipid bilayers (22).

Although it is not clear which role the membrane potential might
play in protein secretion, the above-cited findings have led us to
postulate a role for the membrane potential in the orientation of
membrane proteins (23). Since most cells are inside-negative, we

postulated that transmembrane proteins would have hydrophobic transmembrane segments bracketed by positively charged residues on the cytoplasmic side and negatively charged residues on the extra-cytoplasmic side. Thermodynamically, these asymmetrically placed charge clusters would create a preference for correct orientation of the protein. This prediction is borne out by examination of the few transmembrane proteins (glycophorin, M13 coat protein, H-2k^b, HLA-A$_2$, HLA-B$_7$, and mouse Ig μ heavy chain) for which sufficient information on both sequence and orientation exists. In addition to the macroscopic (i.e., measurable with electrodes) membrane potential, the microscopic membrane potential (e.g. surface potential and dipole potential) could play a role in protein orientation and -translocation (23).

Intracellular protein traffic: the problem of sorting newly synthesized proteins

In the preceding section we focused on some of the physicochemical issues involved in the biosynthesis of proteins. We examined the relationship between the folding requirements of aqueous vs membrane environments and biosynthesis. The task to explain protein traffic movement into intracellular organelles is more formidable. From the thermodynamic point of view we are not only faced with explaining movement across membranes (and e.g. in the case of mitochondria two membranes), we also have to explain the specificity of the process. Why do mitochondrial proteins only go to mitochondria, peroxisomal proteins to peroxisomes, lysosomal proteins to lysosomes? Here we are mainly concerned with proteins , but it should be noted, that lipids also exhibit exquisite selectivity for different organelles (24). For instance, whereas the plasma membrane usually contains 50% cholesterol, the mitochondrial membrane contains no cholesterol at all. This observation is particularly interesting considering the fact that cholesterol can spontaneously exchange between the membranes of liposomes. The maintenance of a specific lipid composition of the membranes of intracellular organelles poses a challenging problem for the thermodynamicist.

The ultimate thermodynamic description of the biogenesis of organelles is self-assembly. Just as virus particles self-assemble spontaneously (25), without the help of a complex guidance system of "ropes and pulleys", organelles may self-assemble in the same way. We can envisage for instance, that a particular mitochondrial protein will find its way inside the mitochondrion, simply because it fits there, like a missing piece in a jigsaw puzzle. We can then picture the cell as a multidimensional jigsaw puzzle or perhaps a multifacetted Rubik's cube with particular colors not only facing the outside, but also with all kinds of specific inner surfaces. The "correct" solution for the Rubik's cube paradigm is that all the blue pieces (mitochondrial proteins) end up on a blue surface, and the red pieces (lysosomal proteins) on a red surface, and so forth. Is the "correct" solution to the Rubik's cube guided by the fingers and brain of a skillful Rubik's cube operator ("ropes and pulleys") or will the Rubik's cube spontaneously find the "correct" solution (thermodynamics)? In the latter case a scrambled Rubik's cube will unscramble because of the inherent energetics built in the system.

For instance red-red and blue-blue interactions are much more favorable than red-blue interactions. In either case we are faced with solving the specific algorithm for the correct solution of our multifacetted Rubik's cube model for the cell. The ultimate test for the thermodynamic self-assembly model is to take two organelles (for instance, peroxisomes and mitochondria), scramble them (e.g. by sonication), and observe whether they will spontaneously reassemble.

It is clear that many proteins are not imported into intracellular organelles (mitochondria, chloroplasts, peroxisomes) by "vectorial translation". Pulse-labelled intraorganellar proteins are initially found outside; they are chased into the organelle only after a lag (which is different for different proteins). Translocation into the organelle is not inhibited by inhibitors of protein synthesis. Moreover, studies with cell-free systems showed posttranslational movement. The proteins are initially made as precursors, which usually are of higher molecular weight (some precursors are not, see below), than the mature proteins. The precursors are initially cytosolic, which raises the questions how precursors to insoluble membrane proteins can exist in the cytosol. It seems that the precursors do not exist as monomers, but rather as larger aggregates (protein micelles). We do not know, whether those aggregates are homogeneous (all containing the same monomer), or whether they complex with other proteins. Many of those proteins are amphipatic (see previous section) and therefore have the potential to form micelles with their hydrophobic moiety buried inside.

How are the proteins imported into the intracellular organelles? The first step is obviously precursor synthesis, followed by discharge into a cytosolic pool. In the case of mitochondria there is evidence for a "receptor" for the precursor on the mitochondrial surface. Finally, there is a translocation step, the mechanism of which is totally unknown. The field of protein transport across membranes is very much in its infancy. We are now at the stage where the field of transport of small molecules across membranes was 40 years ago. For instance, the controversy whether the transport of a protein across a membrane is mediated by specific carriers or channels, or whether the protein just moves through the lipid bilayer is very much alive. A few studies (26, 27) indicate that the precursor portion of the protein is totally unnecessary for tranlocation. It has, for instance, been shown that the mitochondrial matrix protein aspartate aminotransferase is transported into the mitochondrion. According to those studies the role of the precursor piece is to keep the protein monomeric. After processing inside the mitochondrion the monomers can then self-assemble into oligomers which are no longer translocation-competent.

Protein crystallization might be brought up as an interesting, possible mechanism to explain the unidirectional movement of matrix proteins of mitochondria, peroxisomes or chloroplasts into the intraorganellar space. Crystallization is most certainly a thermodynamic process. Rat liver mitochondrial matrix contains approximately 56% w/w of closely packed protein molecules. If there are spherical molecules packed in a cubical array (six protein protein contact points per molecule) then the protein concentration is about 60%. This is the protein concentration in a crystal. According to this paradigm the matrix proteins would find their lowest energy situation inside the matrix and therefore spontaneously find their place there. A crystallization process might even compensate for the formidable energetic barriers of crossing cell membranes.

The question of organellar specificity can then be taken care of by hypothesizing a receptor for the precursor form on the surface of a given organelle: each protein carries in itself a specific structure that recognizes a complementary structure on the membrane across or into which it has to move. In a thermodynamic sense this receptor will provide an intermediary lower energy stage (due to complementary structures) in the pathway of the transported molecule to its final "lowest energy" destination. A more complex guidance system is then not required.

There seems to be a number of variations on the theme of movement of proteins into organelles. The precursor forms of secretory proteins, which are cotranslationally secreted into the lumen of the endoplasmic reticulum, are usually about 3,000 daltons larger than the mature protein. For nuclear-DNA coded proteins that are imported into mitochondria and chloroplasts the molecular weight of the N-terminal prepiece may measure up to 6,000 daltons. On the other hand certain heme-proteins do not seem to require a larger precursor form. The apoprotein (without heme) presumably has the correct disposition for proper processing (see below).

One possible role for a precursor in the transport of an intraorganellar protein could be that the precursor piece forms its own channel for the rest of the molecule to pass through. For instance, in chloroplasts the small subunit of ribulose-1,5-biphosphate carboxylase is synthesized on free cytoplasmic polysomes as a precursor which is 4,000 daltons larger than the mature polypeptide. After translation the precursor is transported into intact chloroplasts, processed to the mature form and assembled with the endogenous large subunits in the chloroplast stroma to form the holoenzyme. The 4,000 dalton precursor piece may by itself form a channel for the rest of the molecule to pass through. An analogy with diphteria toxin comes to mind. The toxin has an A and a B portion. The A portion is the physiologically important one, since it kills the cell by interacting with elongation factor 2. Experiments with planar lipid bilayer membranes have shown (22) that at low pH the B-portion forms channels. Presumably the toxin is endocytosed and in the low pH environment of the lysosome the B fragment forms a channel for the A portion to enter the cytoplasm and exert its toxic effect.

As mentioned above, in a few of the cases the precursor form does not have a larger molecular weight than the mature protein. Peroxisomal biosynthesis (28) represents an interesting case in this context. For instance, catalase is synthesized in a precursor form, which has the same subunit molcular weight as the mature form. the precursor is recognized by anticatalase antibody, but it lacks the heme group. Thus, the precursor is an apomonomer. It is converted into monomeric catalase by taking up a heme and subsequently the functionally active tetramer is formed. Addition of the heme is mediated by an enzyme which is located in the intraperoxisomal space. The translocation step is carried out by the apomonomer (precursor). The transported species is apparently trapped inside by the addition of the heme group. How the protein is specifically directed into the peroxisomal membrane seems to be an unresolved question. Whether there is a receptor for the apomonomer or whether the inherent structure of the membrane will permit crossing is unclear. We also do not know whether the apomonomer can potentially cross any membrane, but will only remain at its final destination (the peroxisome) because it is trapped there by the addition of heme.

Concluding remarks

We have attempted to look at several levels of the problem of protein biosynthesis from a physicochemical point of view. First of all, we have emphasized that constraints in protein folding are an important aspect of synthesis. In particular, the insertion of membrane proteins into the hydrophobic lipid bilayer and the movement of secretory or intraorganellar proteins across this structure involve complex folding problems. How proteins are moved to and accumulate at their ultimate sites of residence within the cell, is unknown. This "sorting" may ultimately be accomplished by self-assembly in which the specific composition of organelles is maintained by attractive physical forces just as in crystallization-processes or in the self-assembly of viruses. Biophysical, biochemical and genetic models will prove to be important tools in finding the ultimate answers to these central biologic questions.

J.v.R. gratefully acknowledges the financial support from the Niels-Stensen Stichting, The Netherlands, and the Netherlands Organization for the Advancement of Pure Scientific Research (Z.W.O.), The Netherlands.

References

1. Sabatini, D.D., Kreibich, G., Morimoto, T. and Adesnik, M. (1982). "Mechanisms for the incorporation of proteins in membranes and organelles", J. Cell Biol. 92, 1-22
2. Kreil, G. (1981). "Transfer of proteins across membranes", Ann. Rev. Biochem. 50, 317-348
3. Emr, S.D., Hall, M.N. and Silhavy, T.J. (1980). "A mechanism of protein localization: The signal hypothesis and bacteria", J. Cell Biol. 86, 701-711
4. Davis, B.D. and Tai, Ph.-C. (1980). "The mechanism of protein secretion across membranes", Nature 283, 433-438
5. Blobel, G., Walter, P., Chang, C.N., Goldman, B.M., Erickson, A.H. and Lingappa, R. (1979). "Translocation of proteins across membranes: The signal hypothesis and beyond", in: "Secretory Mechanisms" C.R. Hopkins and C.J. Duncan, eds., Cambridge University Press, Cambridge, pp. 9-36
6. Wickner, W. (1980). "Assembly of proteins into membranes", Science 210, 861-868
7. Meyer, D.I., Krause, E. and Dobberstein, B. (1982). "Secretory protein translocation across membranes - the role of the 'docking' protein", Nature 297, 647-650

8. Walter, P., Ibrahimi,I. and Blobel, G. (1981). "Translocation of proteins across the endoplasmic reticulum. I.Signal recognition protein (SRP) binds to in vitro assembled polysomes synthesizing secretory protein", J. Cell Biol. 91, 545-550

9. Anfinsen, C.B. (1967). "The formation of the tertiary structure of proteins", Harvey Lect. 61, 95-116

10. Klausner, R.D., Kumar, N., Weinstein, J.N., Blumenthal, R. and Flavin, M. (1981). " Interaction of tubulin with phospholipid vesicles. I.Association with vesicles at the phase transition", J. Biol. Chem. 256, 5879-5885

11. Lingappa, V., Lingappa, J. and Blobel, G. (1979). "Chicken ovalbumin contains an internal signal sequence", Nature 281, 117-121

12. Meek, R.L., Walsh, K.A. and Palmiter, R. (1980). "Ovalbumin contains an NH_2-terminal functional equivalent of a signal sequence", Fed. Proc. 39, 1867

12a. Kempf, C., van Renswoude, J., Weinstein, J.N., Blumenthal, R. and Klausner, R.D. (1982). "Unique folding pathway and "hidden" leader sequence of ovalbumin", Fed. Proc. 41, 1191

13. Braell, W.A. and Lodish, H.F. (1982). "Ovalbumin utilizes an NH_2-terminal signal sequence", J. Biol. Chem. 257, 4578-4582

14. Engelman, D.M. and Steitz, T.A. (1981). "The spontaneous insertion of proteins into and across membranes: The helical hairpin hypothesis", Cell 23, 411-422

15. Pincus, M.R. and Klausner, R.D. (1982). "Prediction of the three-dimensional structure of the leader sequence of pre-kappa light chain, a hexadecapeptide", Proc. Natl. Acad. Sci. USA 79, 3413-3417

16. Blumenthal, R., Klausner, R.D. and Weinstein, J.N. (1980). "Voltage-dependent translocation of the asialoglycoprotein receptor across lipid membranes", Nature 288, 333-338

17. Kempf, C., Klausner, R.D., Weinstein, J.N., Van Renswoude, J., Pincus, M. and Blumenthal, R. (1982). "Voltage-dependent trans-bilayer orientation of melittin", J. Biol. Chem. 257, 2469-2476

18. Date, T., Goodman, J.M. and Wickner, W.T. (1980). "Procoat, the precursor of M 13 coat protein requires an electrochemical potential for membrane insertion", Proc. Natl. Acad. Sci. USA 77, 4669-4673

19. Daniels, C.J., Bole, D.G., Quay, S.C. and Oxender, D.L. (1981). "Role of membrane potential in the secretion of protein into periplasm of Escherichia coli", Proc. Natl. Acad. Sci. USA 78, 5396-5400

20. Enequist, H.G., Hirst, T.R., Harayama, S., Hardy, S.J.-S. and Randall, L.L. (1981). "Energy is required for maturation of exported proteins in Escherichia coli", Eur. J. Biochem. 116, 227-233

21. Schleyer, M., Schmidt, B. and Neupert, W. (1982). "Requirement of a membrane potential for the posttranslational transfer of proteins into mitochondria", Eur. J. Biochem. 125, 109-116

22. Donovan, J.J., Simon, M.I., Draper, R.K. and Montal, M. (1981). "Diphteria toxin forms transmembrane channels in planar lipid bilayers", Proc. Natl. Acad. Sci. USA 78,172-176

23. Weinstein, J.N., Blumenthal, R., Van Renswoude, J., Kempf, C. and Klausner, R.D. (1982). "Charge clusters and the orientation of membrane proteins", J. Membrane Biol. 66, 203-212

24. Quinn, P.J. (1976). "The molecular biology of cell membranes", University Park Press, Baltimore, pp. 26-38

25. Fraenkel-Conrat, H. (1969). "The Chemistry and Biology of Viruses", Academic Press, New York, N.Y., Chapter X, pp. 173-181

26. Waksman, A., Hubert, P., Cremel, G., Rendon, A. and Burgun, C. (1980). "Translocation of proteins through biological membranes. A critical view", Biochim. Biophys. Acta 604, 249-296

27. Passarella, S., Marra, E., Doonan, S. and Quagliariello, E. (1980). "Selective permeability of rat liver mitochondria to purified malate dehydrogenase isoenzymes in vitro", Biochem. 192, 649-658

28. Lazarow, P.B., Shio, H. and Robbi, M. (1980). "Biogenesis of peroxisomes and the peroxisome reticulum hypothesis", In: Biological Chemistry of Organelle Formation. T. Buecher, W. Sebald and H. Weiss, eds. Springer Verlag, Heidelberg, pp.187-206

Genes: *Structure and Expression*
Edited by A. M. Kroon
© 1983 John Wiley & Sons Ltd.

APPROACHES TO THE STUDY OF HORMONAL
REGULATION OF GENE EXPRESSION

A.J. Wynshaw-Boris and R.W. Hanson

Department of Biochemistry, Case Western Reserve
University, School of Medicine,
Cleveland, Ohio, U.S.A.

INTRODUCTION

An understanding of the control of gene expression in eukaryotes
has advanced rapidly since the recent development of recombinant
DNA technology. Evidence has accumulated supporting a variety of
levels of control of gene expression (Brown, 1981 and Darnell,
1982), of which transcriptional control appears to be the most
frequent type of regulation (Darnell, 1982). The investigation of
the mechanisms by which gene expression is transcriptionally
regulated has focused on regions of the gene responsible for
transcriptional control and for factors which might interact with
these regions in order to modulate gene expression.

Hormonally regulated genes offer an ideal system in which to study
the genomic regions and factors responsible for transcriptional
control of gene expression. Many hormones acutely regulate gene
expression by altering transcription rates of responsive genes
(McKnight and Palmiter, 1979; Guyette et al., 1979; Lamers et al.,
1982). In this chapter we will examine the hormonal regulation of
transcription in eukaryotic cells by reviewing techniques which
have been used to study sequences and factors responsible for
transcriptional regulation in general and how they have been
applied to the study of hormonally regulated genes. The utility
of DNA mediated gene transfer and in vitro transcription systems
to investigate these questions will be highlighted. We will then
use the hormonally regulated gene phosphoenolpyruvate carboxykinase (GTP) (EC 4.1.1.32) to illustrate our own approach to the
investigation of mechanisms of hormonal regulation of gene
transcription. It is, of course, impossible to review all of the
literature on the hormonal regulation of gene expression or to
comprehensively cover all of the techniques used to study gene
expression. Instead, we have selected specific approaches
currently used to study the general mechanisms controlling gene
transcription and have focused on their use in the study of
hormonal regulation of gene expression.

I. APPROACHES USED TO STUDY SEQUENCE REQUIREMENTS OF GENE
    TRANSCRIPTION.

Our discussion of the various approaches used to study sequence
requirements for gene transcription will follow the general
outline shown in Figure 1. We will not cover the techniques used
to clone DNA sequences of interest into phage or specific
plasmids, since these procedures have been discussed in detail in
excellent reviews, such as the one published recently by Cold
Spring Harbor (1982). Increasingly, cDNA or genomic clones to
hormonally regulated genes are becoming generally available, so
that the cloning step itself may not always be necessary for an
individual investigator interested in studying gene expression.

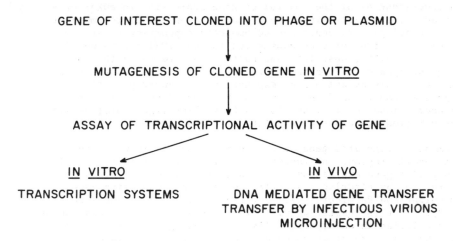

Fig. 1. A general approach to the study of the
transcriptional control regions.

A. Assay of Gene Transcription.

The transcriptional activity of a gene under investigation can be
determined by a variety of procedures which can be broadly
classified as in vitro or in vivo assays. In vitro gene transcription assays are completely cell free, where the gene of
interest is transcribed in a cellular extract. In vivo assays
involve the transfer of genes into whole cells, oocytes, or
fertilized eggs via a variety of techniques which will be
discussed below. Each of the currently available procedures for
the measurement of gene transcription have advantages and limitations, especially when used to study the hormonal regulation of
gene expression. For example, the chicken chromosomal ovalbumin
gene can be expressed when injected into frog oocyte nuclei

(Wickens et al., 1980). However, since frog oocyte nuclei in vitro are not sensitive to hormonal stimulation, this system is not suitable for the study of hormonal regulation of this gene.

1. In Vitro Assays of Gene Transcription. Currently, there are two cell-free transcription systems that are in wide use. The first of these is the system of Weil et al. (1979), which employs a cytosolic extract of KB cells to accurately transcribe eukaryotic RNA polymerase II genes when RNA polymerase II is added. A second procedure developed by Manley et al. (1980), uses an extract of HeLa cells which can transcribe genes accurately in the absence of added RNA polymerase II. In both of these assays, a truncated gene is incubated with radio-labeled nucleotides and the transcripts separated by size on high resolution gels to analyze the accuracy of initiation and the extent of transcription. A truncated gene is used because these systems generally transcribe to the end of a DNA fragment (run-off transcription), and the fidelity of initiation can be easily determined by sizing the transcripts on sequencing gels.

Recently, Zarucki-Schulz et al. (1982) have used the in vitro transcription system of Manley et al. (1980) to study the sequence requirements in the 5'-control regions of the chicken ovalbumin gene. A site specific mutation in the 5' flanking region of the gene was constructed in which the sequence 5'-GACTATATTCCCC-3' was altered to 5'-GGCTGTATTCCCC3' via synthetic oligonucleotide primers. This region contains the TATA box, a highly conserved region in eukaryotic genes which is located approximately thirty nucleotides 5' from the initiation site of transcription. While correct initiation of ovalbumin RNA transcripts was not altered by this single-base substitution in the TATA box, the efficiency of transcription of the template containing the mutant sequences was greatly reduced as compared to the normal wild-type gene. These findings demonstrate the usefulness of the in vitro transcription system in determining the sequence requirements within a transcriptional control area.

However useful these cell-free systems may be in determining sequence requirements for transcription, it is important to remember that they have major limitations for studies of hormonal regulation of gene expression. In vitro transcription systems have not as yet been found to respond directly to hormonal stimulation or to cellular intermediates of hormone action. Also, when eukaryotic genes have been analyzed different sequences have been found to be important in vitro and in vivo (for review see Shenk, 1981). In vitro, the TATA homology is indispensable for proper initiation and efficiency of transcription. In vivo the TATA box is needed to position correct transcriptional initiation, but the only indispensable control sequences are more than fifty nucleotides 5' from the transcriptional initiation site. Two possible explanations for this lack of correspondence between the in vivo and in vitro transcription studies are: 1) factors may be lost from the

in vitro system during purification; and 2) the chromatin structure of templates in vitro might not correspond with that of
genomic chromatin (Mathis and Chambon, 1981). Clearly, more work
needs to be done to improve in vitro transcription systems so that
in vivo type transcriptional control can be obtained. It is
possible that future refinements of these cell-free systems will
render them responsive to regulation, at which time they will be
ideal candidates for detailed analysis of hormonal control of gene
expression.

2. In Vivo Assays of Gene Transcription. The transcriptional
activity of a number of genes has been measured using in vivo
techniques. These procedures involve the introduction of the DNA
into cells, oocytes, or fertilized eggs. DNA can be introduced
passively by layering a calcium phosphate precipitate of DNA onto
the cells (Graham and vander Eb, 1973), by the use of viral
packaged DNA, (Mulligan and Berg, 1980) or by microinjection of
the DNA directly into cells (Graessman and Graessman, 1976). DNA
has been microinjected into oocytes (Kressman et al., 1977), the
male pronucleus of fertilized mouse eggs (Brinster et al., 1981)
or teratocarcinoma cells (Wagner et al., 1981). In some cases,
the microinjection has resulted in phenotypic expression of a
specific gene in several tissues of the adult animal, including
the germ line (Palmiter et al., 1982). We will outline the usefulness of each of these techniques and indicate their applicably
to studies of hormonal regulation of gene expression. But first we
will consider the vectors commonly used to introduce genes into
mammalian cells. In this discussion, we will follow the conven-
tions of Hanahan et al. (1980) and refer to the process of DNA-
mediated gene transfer as convection while the term transformation
will refer to the phenotype associated with altered patterns of
cellular growth regulation.

a. Choice of Vectors for Introduction of DNA Sequences. Many
types of vectors may be used to introduce eukaryotic genes into
mammalian cells, and the choice of vector depends primarily upon
which cell system is to be used as the recipient of the gene and
whether stable integration or rapid transient expression of genes
is desired. Genes can basically be introduced into cells as naked
DNA, as metaphase chromosomes or as packaged viral peptides. We
will confine our discussion to naked DNA since this is most
applicable to the study of gene regulation.

Figure 2 outlines the basic vectors used to introduce naked DNA
into mammalian cells. The advantages of using naked DNA are that
there are no limitations on the size of DNA introduced, and, as we
will discuss below, various modifications can be made to the gene
and to the DNA used to introduce the gene. The major disadvantages are that it is relatively inefficient as a means of stably
introducing genes into cells, and it is time consuming to select
cells which have stably incorporated a convected gene. However,
if one is interested only in expression of the convected gene,

transient gene expression from appropriate vectors allows 15–20%
efficiency of expression (Chu and Sharp, 1981) within 2–4 days of
transformation.

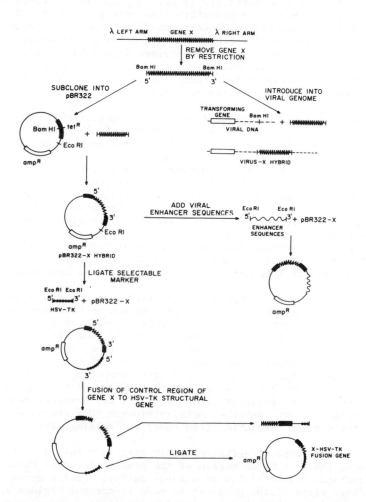

Fig. 2. Several vectors used to convect naked DNA into
mammalian cells. See text for details.

Genes are isolated from genomic libraries which are packaged in
engineered bacteriophage λ vectors or plasmids (Fig. 2). The gene
contained in phage vectors can be used to convect cells directly.
If there is a method of selecting the gene of interest, then the
gene can be introduced into cells alone. Co-transformation with a

selectable marker, such as the herpes simplex virus thymidine
kinase gene (HSV-TK) into thymidine kinase deficient mouse L cells
(Ltk⁻ cells), can be used if no method exists for selection of
the gene of interest. Kurtz (1981) used a phage vector containing
an α_{2u} globulin gene to co-transform Ltk-cells with an HSV-TK
gene. Bacteriophage clones contain considerable vector sequence
(40 kilobases, kb, for λ vectors) and are difficult to grow. For
this reason, the gene is usually subcloned into a bacterial
plasmid, such as pBR322, which has the advantages of ease of
purification and manipulation. Gene-containing plasmids can be
used for convection or co-transformation.

Plasmids may also be manipulated and further engineered. There
are three basic types of engineered vectors. In the first type, a
selectable marker is inserted into the plasmid containing the gene
of interest, or vice versa. For example, globin genes have been
ligated to HSV-TK containing plasmids for transformation (Mantei
et al., 1979). The advantage of ligating the gene of interest
with a selectable gene is the certainty of having incorporated the
gene of interest when cells are selected by the transformed
phenotype. Vectors utilizing viral sequences can also be used,
and the transformed phenotype can be selected (Mulligan et all,
1979). SV40 is commonly used to provide transforming sequences,
although bovine papilloma virus (Sarrer et al., 1981) and
retroviral vectors (Doehmer et al., 1982) have also been used.

A second, commonly used type of vector involves a plasmid containing the gene of interest to which viral sequences have been added,
enhancing transcription of convected genes (Moreau et al., 1981).
The best studied enhancer sequences are the 72 base pair (bp)
repeats found near the SV40 origin of replication, which are part
of the viral promoter. These repeats enhance the transcription of
convected genes when they are part of the convecting vector
(Moreau et al., 1981) without an apparent effect on the regulation
of the gene. These enhancer sequences will, for example, enhance
the transcription of convected metallothionein (MT) genes, while
permitting cadmium regulation (see Hamer and Walling, 1982).

The third basic type of vector commonly used is a fusion gene,
containing the 5' control region of the gene of interest fused to
the structural region of the selectable marker. Fusion genes
using other parts of the gene can be made if control regions are
found in regions other than 5' flanking sequences. The gene of
interest is ligated into a plasmid containing a suitable selectable marker. Restriction enzymes which cut selectively in the 5'
untranslated region of each gene are used to remove sequences
between the 5' end of the gene to be studied and the structural
gene of the selectable marker and these two genes are ligated to
form a selectable, fusion gene. It is then possible to monitor
regulation of the selectable marker by the new control region.
These types of fusion genes are very useful in the study of
sequences responsible for the hormonal regulation of gene trans-

cription since it is possible to confer hormonal sensitivity to genes normally unresponsive to a specific hormone. Fusion genes have been constructed between the 5' control region of the MT gene and the HSV-TK structural gene (Mayo et al., 1982), and the 5' control region of human growth hormone (GH) and the HSV-TK structural gene (Robins et al., 1982). These studies are discussed in more detail below.

Recently, it has been possible to use vector-host systems in which the convected gene is rapidly expressed in a transient manner. These transient expression systems are useful because mRNA production can be assayed 2-4 days after convection. Normally, it takes 1 to 2 months to assay mRNA from selected cells containing stably incorporated convection genes. The vectors used for transient expression are constructed with viral sequences as outlined above. We will now discuss the basic components used to construct transient expression systems for SV40-containing vectors. Other viral vectors, such as bovine papilloma virus (Zinn et al., 1982) also can provide transient expression of convected genes.

Two components seem to be necessary for SV40 recombinants to replicate efficiently as extrachromosomal elements in permissive cells: a functional origin of replication and the presence of T antigen encoded from the vector or the host genome (see Elder et al., 1981 for review). A T antigen producing cell line, COS-7, has been constructed by transforming permissive CV-1 monkey kidney cells with an origin defective mutant of SV40 (Gluzman, 1981). It is then possible to convect a vector into these cells containing the SV40 origin of replication, and the vector will be replicated as an extrachromosomal element. In non-permissive cells, such as HeLa cells, extrachromosomal maintenance of a convected vector appears to depend on the presence of an intact copy of a tandemly repeated 72 bp sequence next to an SV40 origin of replication in that vector (Banerji et al., 1981). Transient expression systems have been used to study the expression of globin genes (Banerji et al., 1981; Humphries et al., 1982), as well as the mouse MT gene (Hamer and Walling, 1982).

An example of a recent vector which should prove very useful for studies of the hormonal regulation of gene expression is a plasmid which contains the pBR322 origin of replication and the β-lactamase gene (for bacterial growth), coupled to an SV40 early promoter region into which chloramphenicol acetyltransferase coding sequences have been inserted (Gorman et al., 1982). When this recombinant was introduced into African green monkey kidney cells, chloramphenicol acetyltransferase (a protein not present in mammalian cells) could be assayed after 48 hours. In order to assay regulatory sequences at the 5'-end of a gene being studied, the entire SV40 promoter region was removed and a unique Hind III site was substituted. The 5'-regulatory sequences from HSV-TK, α-2 type I collagen, and the long terminal repeat of the Rous sarcoma virus were inserted at this site, and chloramphenicol acetyltransferase was produced. Using this approach, it might be possible to assay 5'-regulatory sequences of hormonally regulated genes within a relatively short period of time (48 hours).

We have not discussed all vectors which can be used for the convection of genes into mammalian cells but have emphasized general types of vectors and examples of their use in the study of the regulation of gene transcription. We will now discuss methods of convection of these vectors into cells.

b. Cell Convection. Cells are generally convected directly with the types of vectors outlined in the previous section. The inclusion of a suitable marker in the vector simplifies the selection of convected cells. Most in vivo assays of gene transcription reported to date involve some type of DNA mediated gene transfer. The calcium phosphate precipitation technique of Graham and vander Eb (1973) is generally used, but a high efficiency convection technique has recently been described (Sheu et al., 1982) where as many as 70% of cells are capable of expressing the convected gene in transient expression systems.

A common assay system used in DNA mediated gene transfer involves co-transformation of the gene of interest with HSV-TK into mouse Ltk⁻ cells. The use of this system is limited to cells deficient in cellular thymidine kinase. However, since L cells are responsive to glucocorticoid hormones (Lipman and Thompson, 1974), they can be used to study glucocorticoid responsive genes. NIH/3T3 cells are easily convected to a transformed phenotype and can be used if the selectable marker is a transforming virus. Dihydrofolate reductase deficient CHO cells can be used if the selectable marker is dihydrofolate reductase. NIH/3T3 cells (Huang et al., 1981) and CHO cells (Lee et al., 1981) are also glucocorticoid responsive.

However useful these systems are in studying the expression of convected genes, the host range of most of the commonly used vectors is narrow. Also, some cell lines for which vectors are available may not be able to express certain convected genes. For this reason the availability of a system with a wide host range is desirable. Colbere-Garapin et al. (1981) developed a vector which expresses a bacterial kanamycin resistance gene linked to an HSV-TK control region for expression. Convection of this vector into mammalian cells imparts resistance to the antibotic G-418, which is lethal to mammalian cells. Southern and Berg (1982) have constructed a similar vector currently in wide use. These vectors extend the range of cell lines which can be used for studies of this type to include any cell which has a reasonable convection frequency.

The factors which are important in selecting a system in which to convect genes into cells can be summarized as follows. First, one must decide whether stable integration or transient expression is desirable. This will be determined mainly by the cell lines and vectors which are available. Cell lines must be chosen based on the type of regulation required for the convected gene. For example, a glucocorticoid responsive cell line is needed if a glucocorticoid sensitive gene is studied. A suitable vector can then be selected, as well as a suitable method of convecting that vector.

c. <u>Microinjection</u>. It is also possible to study the regulation of gene transcription <u>in vivo</u> by the microinjection of specific sequences into cells, oocytes, or fertilized eggs. Microinjection is the most efficient means of convecting genetic material. However, it is technically difficult, and relatively few cells or eggs can be injected in a given period of time. It is the only means of introducing genes into fertilized eggs to study developmentally regulated gene expression. An excellent example of the potential of this approach is the recent work of Brinster <u>et al</u>. (1981) in which a mouse metallothionein-HSV-TK fusion gene was introduced into the male pronucleus of a fertilized mouse egg by microinjection. The fusion gene was inherited as though it was on a single chromosome and, as we shall discuss in the next section of this review, the expression of the gene in the livers of several of the off-spring could be markedly enhanced by the administration of cadmium. The potential of this approach is that it allows the introduction of a heritable, functional gene of heterologous origin into an animal in a manner which results in the phenotypic expression of that gene in a target tissue. Since the tissue of the host maintains its normal hormonal sensitivity, it is possible to study the regulation of gene expression <u>in vivo</u>. Of course, this technique has wider implications for gene therapy via direct gene replacement.

However, there are limitations to this approach. Only 10% of microinjected eggs retained the injected sequences in their DNA after development and only 20% of these animals expressed the gene. Furthermore, the inheritance of expression has so far been unpredictable (Palmiter <u>et al</u>., 1982). In other studies, the β-globin gene was expressed in tissues which normally do not contain β globin, yet no expression was noted in hematopoeitic tissues (Costantini and Lacy, unpublished observations). Although this approach is promising, there are still many difficulties in trying to introduce genes into fertilized eggs and to study their developmental regulation.

d. <u>Conclusions</u>. There are a number of vectors which can be used to introduce genes into a variety of cell types, oocytes and eggs. At the present time, <u>in vivo</u> transcription systems are the only suitable assays for the regulation of gene transcription. However, there are still general limitations with <u>in vivo</u> transcription assays which may present problems. For example, genes are often recombined and rearranged when introduced into cells (Scangos and Ruddle, 1981) which could disrupt regulation. Also, multiple copies of genes are inserted into cells. Copy number does not seem proportional to expression of genes (Scangos and Ruddle, 1981), and is also unpredictable. In spite of these problems, in vivo transcriptional assays are important for studying the sequence requirements for gene expression (see below).

B. Mutagenesis of Cloned Genes In Vitro

1. Deletion mutants. The ability to study sequence requirements for transcription of a particular gene depends upon the modification of the gene and the effects of such modification on its expression. Ideally, the best approach would be to specifically modify regions of the gene in situ or to mutagenize cells and isolate mutants specific for the gene of interest. However, it has not yet been possible to target mutations in situ to specific genomic sites, and the time required to isolate and characterize eukaryotic mutants has limited the usefulness of the latter approach. Consequently, research has focused on in vitro deletion and mutagenesis of cloned genomic fragments. Deletions are constructed by digesting these cloned sequences with restriction enzymes that cleave within the control region of a gene (Fig. 3). The restricted DNA is then digested with exonucleases, the single stranded overhang is digested with S1 nuclease, and the blunt ends are ligated to form the deletion. A double-stranded exonuclease such as Bal 31 can also be used to create deletions. In this case, no S1 nuclease treatment is needed. By varying the time of exonuclease treatment, the extent of the deletion can be varied. Deletions can be made throughout the control region by using various restriction sites in that region as start sites for exonuclease digestion.

2. Site Directed Mutagenesis. The disadvantage in using deletion mutants is that it is impossible to determine whether differences in transcriptional activity between deleted and undeleted control regions are due to the loss of specific and necessary sequences or simply changes in spatial arrangements. For this reason, site-directed mutagenesis is preferred for determining transcriptional control sequences. Recently, McKnight and Kingsbury (1982) have published a general procedure for in vitro mutagenesis in which a small cluster of nucleotide residues (usually 10) are substituted within the regulatory area of a specific gene in a highly directed manner without causing either the addition or deletion of other sequences. In this approach, shown in Figure 4, a portion of the cloned gene to be investigated is digested with a restriction enzyme which cuts within the control region of the gene. This is followed by exonuclease and S1 nuclease treatment to form blunt ends. A synthetic Bam H1 linker, for example, is then ligated to the blunt ends and the fragments sequenced to determine the extent of the deletion. Fragments are then matched by sequence so that after digestion with Bam H1 and religation, the added Bam H1 linker substitutes for 10 nucleotides. By choosing proper restriction sites, it is

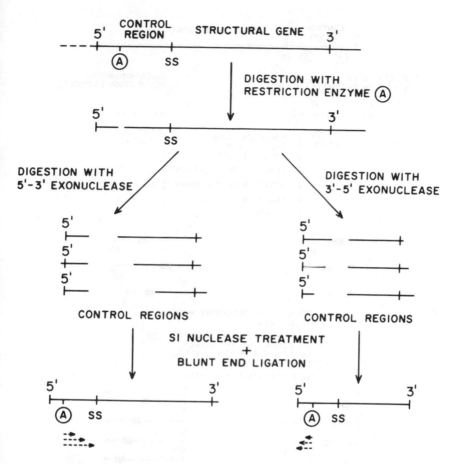

Fig. 3. The in vitro deletion of specific sequences from
cloned genomic fragments. The gene is schematically
illustrated as consisting of two regions: the structural
gene, which provides the linear template for mRNA synthesis (from SS to 3' in the Figure); and the control region,
which contains the information regulating the transcription of the structural gene (from 5' to SS). SS refers to
the start site of transcription. A is a restriction site
in the control region for restriction enzyme A. The
control region is digested with 5'-3' or 3'-5' exonucleases, treated with S1 nuclease and the fragments blunt-end
ligated causing specific deletions. These deletions are
indicated by the dashed lines below the intact control
regions at the bottom of the Figure. By varying the time
of exonuclease treatment, deletions of varying sizes can
be formed. Also, by using other restriction enzymes which
cleave in the control region, different groups of
deletions can be constructed.

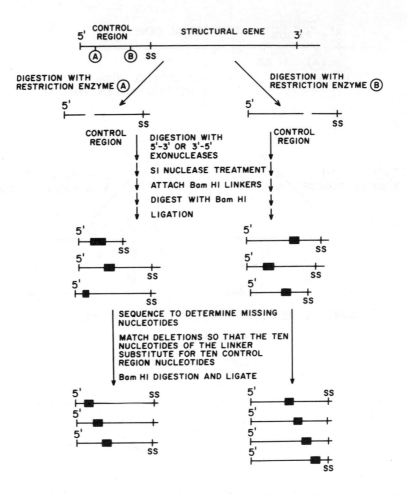

Fig. 4. Analysis of transcriptional control regions using linker-scanning mutants (McKnight and Kingsbury, 1982). See the legend of Figure 2 and the text for details. ━━━■━━━ refers to synthetic Bam Hl linkers. A and B are sites for restriction enzymes A and B, respectively.

possible to scan the entire 5'-regulatory region of a gene, creating mutants in which all of the nucleotides in the region are substituted in groups of 10. Since nucleotides are replaced, instead of deleted, it is possible to more confidently assign importance to certain sequences with a putative control region. Of course, the substitution of bases might also introduce "poison"

sequences into control regions of genes. These sequences could
then turn off transcription of a gene, even though the sequence
from which it is derived is not important for transcriptional
regulation.

Using this "linker-scanning" method of mutagenesis McKnight and
Kingbury (1980) surveyed the 120 base control region of the HSV-TK
gene by microinjecting intact HSV-TK genes and linker scanning
mutants into Xenopus oocytes. They noted that mutations in the
region 16 to 32 bases 5' to the start of transcription (-16 to
-32, where +1 is the start site of transcription), an area which
contains the TATA box, greatly decreased HSV-TK transcription.
Also, mutations in a guanine-rich area between -47 and -61 and a
cytosine-rich segment between -80 and -105 resulted in a 10-fold
reduction in transcription. However, a mutation at -74 to -84,
which includes the sequence 5'-GGCGAATT-3', (a highly conserved
sequence thought to be an important control region in eukaryotic
genes) had no effect on either the efficiency or accuracy of transcription of the HSV-TK gene. This experimental approach should
provide a direct, in vivo method for analysis of the control of
transcription of hormonally regulated genes once control regions
within these genes have been identified.

A wide variety of other techniques are available for directed
mutagenesis. Gilliam and Smith (1979a,1979b) have used site-
specific mutagenesis with synthetic oligonucleotide primers. There
are also chemical techniques available for "region" directed
mutagenesis of genes (Weiher and Schaller, 1982). The choice of
technique for mutagenesis or deletion depends upon the gene being
studied. A more detailed treatment of this subject is contained
in a review by Shortle et al. (1981).

C. Assay of Transcriptional Activity of Modified Genes

Once a transcriptionally active gene has been modified, its
activity is assayed and compared with the expression of other,
unmodified genes, using one of the assay techniques discussed
earlier (see section I.A). We will now discuss studies in which
the sequence requirements for the hormonal regulation of the
expression of specific genes have been determined.

D. Studies of Sequence Requirements for Hormonal Regulation of
   Gene Expression.

We have selected for discussion four hormonally regulated genes
which have been studied using the techniques outlined in the
previous sections of this review (Table 1). There are, of course,
other hormonally regulated genes that have been studied, but to
date only the four reviewed here have been investigated at the
structural level using in vivo cell transfection techniques. The
assumption has been made that the 5' flanking sequences contain
transcriptional control signals, since this appears to be the case

TABLE 1. Summary of Hormonal Regulation of Gene Expression of Four Transfected Genes

Gene	Regulators of Transformed Genes	Source of Gene	Cloning Vector	Method of Determining Sequence Requirements	Recipient of Gene	Method of Introduction	References
Mouse Mammary Tumor Virus (MMTV)	Glucocorticoids	unintegrated virus from infected rat hepatoma cells	bacteriophage λ	not done	Ltk⁻ cells	DNA mediated gene transfer (HSV-TK cotransformation)	Buetti and Diggelman, 1981
		Proviral DNA from mouse liver	bacteriophage λ	not done	Ltk⁻ cells	DNA mediated gene transfer (HSV-TK cotransformation)	Hynes, et al. 1981
		Viral DNA	SV40-pBR322 LTR-dhfr expression plasmid	gene fusion	CHO dhfr⁻ cells	DNA mediated gene transfer	Lee, et al. 1981
		Viral DNA	pBR322-HaMu SV p21-LTR plasmid	gene fusion	NIH/3T3 cells	DNA mediated gene transfer	Huang, et al. 1981
α_{2u} globulin	Glucocorticoids	rat chromosomal library	bacteriophage λ	not done	Ltk⁻ cells	DNA mediated gene transfer (HSV-TK cotransformation)	Kurtz, 1981

Gene product	Induction	Library	Construct	Functional analysis	Cell type	Method	Reference
Metallothionein (MT)	Cadmium but not glucocorticoids	mouse chromosomal library	MT-HSV-TK fusion	not done	fertilized mouse eggs-adults	microinjection	Brinster, et al. 1981
			MT-HSV-TK fusion	not done	oocytes	microinjection	Brinster, et al. 1982
			MT-HSV-TK fusion, MT-gpt plasmid	deletion mapping	Ltk⁻ cells HeLa cells	DNA mediated gene transfer	Mayo, et al. 1982
			MT-HSV-TK fusion	not done	Fertilized mouse eggs	microinjection	Palmiter et al. 1982
			MT-SV40 plasmid	not done	Monkey kidney cells	DNA mediated gene transfer	Hamer and Walling, 1982
Growth Hormone (GH)	glucocorticoids	rat chromosomal library	bacteriophage λ and pBR322	not done	NIH/3T3 cells	DNA mediated gene transfer (integration into Mo-MSV DNA)	Doehmer, et al. 1982
		human chromosomal library	bacteriophage λ	5' deletion of control region; GH-HSV-TK fusion	Ltk⁻ cells	DNA mediated gene transfer (HSV-TK cotransformation)	Robins, et al. 1982

Abbreviations: Ltk⁻, thymidine kinase deficient mouse L cells; CHO-dhfr⁻, dihydrofolate reductase deficient Chinese hamster ovary cells; pBR322, an E. coli plasmid; gpt, bacterial xanthine-guanosine phosphoribosyl transferase fused to a eukaryotic promoter; HSV-TK, herpes simplex virus thymidine kinase; Mo-MSV, Moloney murine sarcoma virus; SV40, simian virus 40; dhfr, dihydrofolate reductase; HaMuSV-p21, Harvey murine sarcoma virus p21 protein.

for those genes which have been studied. It is entirely possible, however, that transcriptional control regions are located in 3' flanking sequences, introns, and/or exons of structural genes.

These genes were all isolated from genomic libraries packaged in modified bacteriophage λ. The recombinant λ vectors were used for introduction of the genes into cells directly, or were used to subclone the genes into other modified vectors in order to enhance expression or to construct fusion genes. Two other regulated genes have also been studied, although hormones do not control their expression. A cloned Drosophila heat shock gene introduced into Ltk⁻ cells by co-transformation with HSV-TK (Corces et al., 1981) appears to be induced by raising the temperature of the convected cells. Human β interferon cDNA (Pithi et al., 1981) and a β interferon gene cloned into bovine papilloma virus (Zinn et al., 1982) have been introduced into mouse cells. Poly (rI-rC), which induces β interferon in human cell culture, also induces either the cDNA or the gene which have been convected into mouse cells. Since these genes are not hormonally regulated, we will not discuss them further. These examples underscore the useful-ness of the convection of defined DNA sequences as a method for studying the regulation of eukaryotic gene expression.

1. Mouse mammary tumor virus (MMTV). There are numerous strains of MMTV, an RNA tumor virus, that are transmitted in the milk of inbred strains of mice or as a DNA provirus in the germ line (for review, see Bentvelzen and Hilgers, 1980). Integrated proviral DNA is transcribed by RNA polymerase II and the produc-tion of viral RNA is stimulated by glucocorticoid hormones (see Varmus et al., 1979). The sequences responsible for glucocorti-coid regulation of MMTV have been roughly mapped. When uninte-grated (Buetti and Diggelman, 1981) or proviral (Hynes et al., 1981) DNA was co-transformed into Ltk⁻ cells with the HSV-TK gene, viral gene expression was greatly stimulated by glucocorti-coid hormones. This demonstrated that the hormonally responsive transcriptional signals reside in the virus. These sequences were localized to the proviral long terminal repeat by fusing this sequence to a dihydrofolate reductase cDNA (Lee et al., 1981) or to the Harvey murine sarcoma virus p21 gene (Huang et al., 1981) to form a chimeric gene which was regulated by glucocorticoids. A summary of the regulation of this gene is presented in Table 1.

2. α-$_{2u}$ Globulin. α$_{2u}$ globulin is a protein produced in the liver of male rats under the complex control of a variety of hormones. Androgens, glucocorticoids, thyroid hormone and growth hormone stimulate α$_{2u}$ globulin synthesis, while estrogen inhibits synthesis of the protein (Kurtz and Feigelson, 1977; Kurtz, et al., 1978). Kurtz (1981) has co-transformed Ltk-cells with the HSV-TK gene and an α$_{2u}$ globulin genomic fragment packaged in λ Charon 4A. Eight of twelve colonies which contained α$_{2u}$ globulin sequences produced glucocorticoid regulated α$_{2u}$ globulin mRNA and protein. Ltk⁻ cells do not contain androgen and estrogen

receptors, so that the sequences responsive to these hormones could not be studied. The glucocorticoid regulatory sequences of α_{2u} globulin have not been mapped beyond their inclusion in the cloned genomic fragment, which contained 8 kb of 5' flanking and 2 kb of 3' flanking sequences.

3. Metallothionein (MT). The most thoroughly studied hormonally regulated gene is mouse MT. Metallothioneins are important in zinc homeostasis and in resistance to heavy metal toxicity. Mouse MT is induced transcriptionally by both heavy metals and glucocorticoids (Beach and Palmiter, 1981; Durnam and Palmiter, 1981; Hager and Palmiter, 1981; Mayo and Palmiter, 1981). The gene has been cloned (Durnam et al., 1980) and found to be 1.1 kb in size. The small size of this gene facilitates cloning and manipulative procedures such as constructing deletions and gene fusions. A fusion gene was made between the 5' flanking sequences of the MT gene and the HSV-TK structural gene. The transcriptional activity of this fusion gene has been determined in vivo by injection into fertilized mouse ova (Brinster et al., 1981). The embryos were then implanted into pseudopregnant females and allowed to develop. The fusion gene was found to be integrated in all mouse tissues of 15% of surviving offspring, and one mouse expressed high levels of HSV-TK protein. The tissue distribution of this protein paralleled that of endogenous MT, suggesting that there were enough regulatory sequences present to ensure proper tissue distribution and regulated expression of the fusion gene.

The regulation of this fusion gene was tested by microinjection into mouse oocytes (Brinster et al., 1982), or by DNA mediated gene transfer into cells. The intact MT gene was assayed for activity in Ltk- cells (Mayo et al., 1982), human HeLa cells (Mayo et al., 1982) and cultured monkey kidney cells (Hamer and Walling, 1982). In all cases the introduced gene was expressed and regulated by heavy metals such as cadmium. The regulatory sequences for cadmium are located within 90 nucleotides 5' to the start site of RNA transcription by deletion mapping (Brinster et al., 1982). However, glucocorticoid regulation of the intact gene or the fusion gene has never been observed. Two explanations for this have been proposed (Mayo et al., 1982). First, the sequences necessary for glucocorticoid regulation are not located within the cloned gene used for transformation; or 2) the chromatin structure surrounding transfected MT genes does not allow glucocorticoid regulation. Palmiter's group has also found that the MT gene is amplified when cells are exposed to high levels of cadmium (Beach and Palmiter, 1981). The amplified genes are likewise induced by heavy metals but not by glucocorticoids (Mayo and Palmiter, 1982), even though at least 25 kb of DNA is amplified 5' to the structural gene. This suggests that either the transfected gene is in a non-permissive region of chromatin with respect to glucocorticoid inducibility, or that glucocorticoid regulatory sequences are modified or lost during amplification and cloning. Recent data suggest that DNA methylation is correlated with fusion gene expression in transgenic mice and their offspring (Palmiter et al., 1982). It is possible that methylation of regulatory sequences in a convected gene could account for the lack of glucocorticoid regulation observed for the MT gene. A causal relationship between any of the possibilities and the lack of glucocorticoid responsiveness of the convected MT gene awaits further investigation.

4. <u>Growth Hormone (GH)</u>. Studies with cultured pituitary cells have shown that the levels of GH mRNA can be markedly induced by either glucocorticoids or thyroid hormone (Martial et al., 1977). Rat GH has been introduced into mouse cells via a Moloney mouse sarcoma virus (Mo-MSV) vector (Doehmer et al., 1982). Viral DNA was then purified and the GH gene was inserted into the virus. Calcium phosphate precipitation was used to introduce this recombinant into NIH/3T3 cells, and viral transformed cells were selected. A clonal line was selected that synthesized GH mRNA which is slightly larger than mature rat GH mRNA from pituitary cells, but it also produced GH protein which was induced by glucocorticoids. A human GH gene was introduced into Ltk- cells by co-transformation with HSV-TK (Robins et al., 1982) and the GH mRNA as well as the protein itself were increased by glucocorticoids in many of the co-transformed cells. When 500 bases of the 5' flanking sequences of the GH gene were fused to the coding region of HSV-TK, thymidine kinase is also hormonally regulated. Unlike the metallothionein gene, the glucocorticoid regulatory sequences could be localized by an <u>in vivo</u> transcription assay to within 500 bases of the 5' flanking sequences of the GH gene.

5. <u>Conclusions</u>. The tools used to study transcriptional regulatory sequences in general have just begun to be applied to hormonally regulated genes, and from the studies outlined above, it is apparent that these approaches are promising. At this time, only glucocorticoid regulated genes have been studied, but as different hormonally regulated cell lines are found which are useful for DNA mediated gene transfer, other hormonally regulated genes will be studied. It would obviously be useful if regulatable <u>in vitro</u> transcription assays were developed so that mechanisms by which regulatory sequences alter gene expression could be studied directly. There are also many unanswered questions even for the genes which have been studied to date. Is the effect of the hormones on gene expression transcriptional or post transcriptional? This question has been answered only for the cadmium regulation of metallothionein. It was found that part of the effect of cadmium on the expression of metallothionein was via an increased rate of transcription, but part of the effect could not be accounted for based on the increased transcription rate alone (Mayo et al., 1982). Why is glucocorticoid regulation of metallothionein not observed in transfected cells <u>in vivo</u>? Is there a consensus sequence for glucocorticoid regulatory signals

and are these signals localized in similar places in all genes
regulated by these hormones? How do the regulatory signals for
glucocorticoids and cadmium compare? Clearly the approaches
described in this review can address these questions, and shortly
some of the answers may be forthcoming.

II. INTERMEDIATES IN THE HORMONAL REGULATION OF GENE TRANSCRIPTION.

In addition to the study of sequences required for the hormonal
regulation of gene expression, it is possible to investigate the
mechanism of this regulation. There are three apparent mechanisms
in which specific regulatory sequences can control gene expression. The hormone can bind directly to the regulatory sequences
to change gene expression. At the present time there is no
evidence to support direct binding of a hormone to DNA. The
second mechanism is that the hormone may bind to an intermediate
receptor and the hormone-receptor complex can then bind to DNA to
alter the transcription of a specific gene. Examples of this type
of control are steroid hormones such as glucocorticoids, estrogens
and progesterone as well as thyroid hormone. These hormones bind
to a cytoplasmic receptor, which is then translocated into the
nucleus. The hormone-receptor complex then binds to DNA, changing
the transcription rate of the hormonally regulated gene (for
reviews see, Gorski and Gannon, 1976; Yamamoto and Alberts, 1976).
The third type of mechanism is the binding of a hormone to surface
or cytoplasmic receptors, followed by the liberation of an intermediate. This intermediate is then transported to the nucleus,
changing specific gene transcription. Recent data on the
mechanism of action of prolactin on the expression of the β-casein
gene (Teyssot et al., 1981; Kelley et al., 1982) provides evidence
for this type of control. These mechanisms are presented as
simplest case examples. As we shall mention later in this review,
multiple levels of control may be necessary for hormonal stimulation of gene expression.

There are two recent reports in which these mechanisms outlined
above have been investigated. Both involve the direct addition of
estrogen-receptor complexes (Taylor and Smith, 1982) or supernatants from prolactin stimulated membranes containing prolactin
receptors (Teyssot et al., 1981), to isolated nuclei. We will
first discuss the isolated nuclei system used for measurements of
the rates of gene transcription, then give details of their use
for studying the possible effects of intermediates of hormone
action on the gene.

A. Isolated Nuclei as an Assay of Transcriptional Regulation.

To study the effects of macromolecules including hormones or their
intermediates, on the rate of transcription, an assay sensitive to
relative changes in the levels of transcription must be used which
can respond to added macromolecules. A completely in vitro trans-

cription system would be most useful if the rate of gene transcription and its regulation were the same as the rates of those processes for the gene in vivo. As was described above, these conditions have not been met for any eukaryotic genes studied to date, with the exception of T antigen regulation of the early SV40 genes (Myers et al., 1981). At the present time the most useful system for the study of the effects of macromolecules on gene transcription appears to be isolated nuclei (McKnight and Palmiter, 1979).

Nuclei are isolated from cells or from a tissue after incubation with or without hormones. These nuclei are then incubated in a suitably fortified medium to determine the transcriptional activity of a specific gene. Measurements of transcription rates require that newly synthesized RNA be distinguished from endogenous sequences present in all nuclei. Two methods have been developed for this purpose. Dale and Ward (1975) have incubated cells with mercurated-UTP and then isolated the newly synthesized RNA using a sulfhydryl-Sepharose column. Others have refined this technique to allow measurement of relative rates of transcription in isolated nuclei (O'Malley et al., 1977). Another procedure, introduced by McKnight and Palmiter (1979) involves incubation of isolated nuclei with [^{32}P] UTP, followed by the isolation of specific RNA by hybridizing total RNA from the nuclei with a filter-bound, cDNA probe. The amount of specific RNA synthesized can be calculated from the specific activity of the UTP, the length of the filter-bound cDNA, the efficiency of hybridization and the filter bound radioactivity. The rates of synthesis of specific RNA species under different hormonal conditions can easily be compared using this technique. An example of the use of this technique is given for the transcription of the gene for P-enolpyruvate carboxykinase by isolated rat liver nuclei (Table 2). However, it is important to note that only 5-10% of newly incorporated nucleotide in isolated nuclei is due to de novo initiation. This obviously places limits on this assay for studying factors which might alter initiation. However, the following studies indicate that this technique is useful for studying the effect of hormonal intermediates of the regulation of transcription.

B. The Use of Isolated Nuclei in Identifying Intermediates in the Hormonal Regulation of Gene Expression.

Isolated nuclei have been used to study macromolecular intermediates involved in the hormonal regulation of gene expression of ovalbumin by chicken oviduct (Taylor and Smith, 1982) and gene expression for the milk protein β-casein (Teyssot et al., 1981). Isolated nuclei have also been used to investigate intermediates in the regulation of the Drosophila heat shock genes by high temperatures (Craine and Kornberg, 1981). Since the latter gene is not hormonally regulated, we will not discuss it further.

1. <u>Regulation of Ovalbumin by Estrogen-Estrogen Receptor</u> <u>Complexes</u>. Taylor and Smith (1982) have purified a high affinity nuclear estrogen receptor which they have added to nuclei isolated from chicken oviduct which had been first stimulated with synthetic estrogens followed by the removal of the hormones. This procedure causes a rapid deinduction of ovalbumin gene expression and a subsequent decrease in concentration of ovalbumin RNA in the nucleus. These nuclei were then incubated with $[^{32}P]$ UTP and specific ovalbumin sequences were hybridized to ovalbumin cDNA bound to nitrocellulose. The addition of estrogen-receptor complexes (for only 30 min) to nuclei which had been purified from the oviduct of chronically stimulated chickens or from oviduct in which the estrogen stimulus was withdrawn from the animals for 24 hours increased ovalbumin transcription 1.6-1.9 fold over the same nuclei incubated without receptor complexes. However, if estrogens were withdrawn from the oviduct for 36 hours or longer before the nuclei were isolated, the addition of receptor-complexes was ineffective in stimulating ovalbumin transcription.

The degree of stimulation of ovalbumin transcription in isolated nuclei by receptor complexes was low but reproducible. The authors used only one highly purified estrogen receptor while there are at least two receptors and three estrogen binding proteins <u>in vivo</u>, any of which may be needed to give a maximum stimulation. The lack of stimulation of ovalbumin transcription by receptor complexes in oviduct nuclei isolated from chickens in which estrogen stimulation was withdrawn for 36 hours or longer meant that there was a high basal level of ovalbumin transcription in nuclei which were not treated with receptor complexes. It is likely that a greater stimulation could be observed if nuclei which are competent for receptor complex stimulation could be isolated from oviduct which has a lower basal level of ovalbumin transcription. However, since estrogen withdrawal of 36 hours or longer causes loss of sensitivity of nuclei to receptor complexes, it is logical to assume that there was a loss of nuclear factors necessary for ovalbumin transcription. This argues for the existence of more complex models for the hormonal regulation of gene transcription than a simple binding of a single receptor complex or intermediates to a gene. Despite its weaknesses, this approach to the estrogen regulation of ovalbumin gene transcription has considerable potential for the further study of macromolecules which regulate gene transcription.

2. <u>Regulation of β-Casein Gene Expression by a Low Molecular</u> <u>Weight Protein Factor Released from Prolactin Stimulated Mammary</u> <u>Membranes</u>. Houdebine and co-workers (Teyssot <u>et al</u>., 1981; Kelley <u>et al</u>., 1982) have isolated a polypeptide factor which is synthesized by plasma membranes which had been incubated with prolactin. This factor stimulates the transcription of the β-casein gene when incubated with nuclei isolated from the mammary gland of lactating rabbits which had been treated for four days with the prolactin antagonist bromocryptin to decrease β-casein gene transcription.

The nuclei were then incubated with mercurated–CTP to allow isolation of only the newly synthesized RNA on a sulfhydryl-Sepharose column. RNA was eluted from this column with 2–mercaptoethanol and β–casein mRNA molecules present were assayed by their ability to protect β–casein cDNA from S1 nuclease digestion.

Microsomes isolated from the mammary gland of lactating rabbits when incubated with prolactin produced a soluble compound which stimulated nuclear transcription of the β–casein gene 5-10 fold. Other lactogenic hormones such as human growth hormone and bovine placental lactogen also stimulated β–casein gene expression, but bovine growth hormone, insulin, parathyroid hormone, luteotropic hormone and epidermal growth factor did not. Supernatant fractions from microsomes prepared from other prolactin receptor containing tissues, which were incubated with prolactin also increased β–casein gene transcription by isolated nuclei. However, microsomes from tissues which do not contain prolactin receptors did not respond to added prolactin. The effect could be observed with nuclei from mammary glands isolated from pseudopregnant or bromocryptin treated lactating rabbits, but not with nuclei from the mammary glands of untreated, lactating rabbits nor from rabbit liver or reticultocyte nuclei.

The factor produced by microsomes after prolactin stimulation appears to be a polypeptide with a molecular weight 1000-1500 daltons (Kelley et al., 1982). This protein factor could be inactivated by trypsin but not by heating for ten minutes at 100°C and its effect was blocked by the addition of α–amanitin to the nuclei. An antibody to the prolactin receptor mimics the action of prolactin in intact mammary gland and also stimulates the release of this polypeptide from microsomes. This factor does not simulate globin gene expression, nor the expression of 28S rRNA, which is stimulated by prolactin but via a different mechanism than the milk protein.

This study establishes the likely existence of a second messenger for prolactin. The authors were able to combine the use of isolated nuclei as an assay of gene transcription with a sensitive method for purifying newly synthesized RNA from nuclei containing low basal levels of specific β–casein RNA. This significantly increased the proportion of β–casein mRNA relative to other RNA species present in nuclei. These reports indicate the potential usefulness of isolated nuclei for studies of the mechanism by which hormones and their intermediates regulate gene transcription.

C. Future Directions in the Study of Hormonal Intermediates in the Regulation of Gene Transcription.

   1. Purification of Intermediates with DNA Cellulose Affinity Columns. Mulvihill et al. (1982) have devised a competitive binding assay to determine the DNA sequences which bind to progester-

one receptors. In this procedure, progesterone receptors were bound to a DNA cellulose column, and small subcloned regions of progesterone sensitive genes were used to elute the receptors from the column. Since the DNA sequences of the subcloned fragments were known, a consensus sequence could be devised. This technique may be modified in order to isolate intermediates for a variety of other hormones. Putative regulatory sequences for a specific gene could be bound to cellulose, and an affinity column constructed for molecules which bind to this region of the DNA. The fractions from the column could then be assayed with an appropriate type of isolated nuclei or, perhaps eventually, with a completely in vitro transcription system since factors isolated by the techniques outlined above may not be able to directly enter nuclei.

2. The Use of In Vitro Transcription to Assay a Regulatory Protein of Eukaryotic Gene Expression. Myers et al. (1981) have studied the mechanisms of T antigen regulation of the SV40 early region genes. T antigen is produced early in the course of SV40 infection and is known to decrease the expression of early region genes and to stimulate viral DNA replication. They were able to show that purified T antigen selectively represses transcription of viral early RNA synthesis in the cell free transcription system of Manley et al. (1980). They were also able to define the spatial organization of T antigen binding sites relative to the early region promoter. This is the only instance so far studied where there is in vivo type regulation using an in vitro transcription system and suggests that improvements in responsiveness of these systems is possible.

III HORMONAL REGULATION OF GENE EXPRESSION FOR PHOSPHOENOL-PYRUVATE CARBOXYKINASE (GTP) FROM RAT CYTOSOL.

A. Introduction

In the following section we will review studies from our own laboratory on the regulation of gene expression for the cytosolic form of P-enolpyruvate carboxykinase (PEPCK) from the rat. The gene for this enzyme is very actively transcribed under the control of a variety of hormones, including cAMP and glucocorticoids (Lamers et al., 1982). Although our studies have not as yet employed many of the techniques outlined in this review, the responsiveness of the PEPCK gene suggests that it will make an excellent model for determining the mechanism of action of peptide hormones, acting via cAMP, on gene transcription. PEPCK is also one of several metabolically important enzymes whose levels are coordinately regulated by hormones, which act by directly altering gene transcription.

PEPCK is a gluconeogenic enzyme present in the liver and kidney of all mammalian species. The cytosolic form of this enzyme is inducible by glucocorticoids and by peptide hormones acting via cAMP, while its synthesis is deinduced by insulin (see Tilghman et al., 1976 for review a of the earlier literature on the hormonal induction of this enzyme). These hormones act by rapidly altering the synthesis rate of the enzyme. The administration of Bt_2cAMP to a glucose-fed rat results in an 8-fold induction in the synthesis rate of hepatic PEPCK within 90 min (Iynedjian and Hanson, 1977). Conversely, the injection of insulin to a diabetic rat causes a rapid decrease in the synthesis rate of the enzyme, with a half- time of synthesis of 40 min (Tilghman et al., 1974). Subsequent studies from our laboratory (Iynedjian and Hanson, 1977; Kioussis et al., 1978; Yoo-Warren et al., 1981; Cimbala et al., 1982) have demonstrated that these rapid changes in PEPCK synthesis rate caused by cAMP and by insulin are accompanied by equally rapid alterations in the levels of enzyme mRNA.

There has been considerable conjecture concerning the mechanisms by which cAMP and insulin alter the induction of enzymes such as PEPCK. Most of the theories proposed to date have relied on indirect experiments using inhibitors. For example, the stimulatory effect of cAMP on PEPCK activity can be blocked by the administration of actinomycin D (Yeung and Oliver, 1968a and 1968b) suggesting that the cyclic nucleotide acts at the level of gene transcription. In contrast, there are numerous other studies which indicate an effect of cAMP on the translation of PEPCK mRNA (Wicks and McKibbin, 1972, Tilghman et al., 1975). Recently, cloned cDNA to PEPCK mRNA has been isolated (Yoo-Warren et al., 1981) which has allowed a direct determination of the hormonally induced changes in the sequence abundance of PEPCK mRNA in the liver. It has also been possible to quantitate the rates of transcription of the PEPCK gene by nuclei isolated from rat liver using the techniques outlined in previous sections of this review (see Section IIIC and Table 2). While we have not as yet carried out cellular convection studies with the PEPCK gene, we do have evidence that cAMP acts directly on the gene to stimulate its transcription (Lamers et al., 1982).

B. Regulation of PEPCK mRNA levels in the nucleus and cytosol by cAMP. Liver nuclei contain several RNA species which hybridize to a cDNA probe for PEPCK (Fig. 5). The nuclear RNAs range in size from approximately 1.0 kb to 6.5 kb and as many as 5 of these nuclear RNA species are larger than the 2.8 kb size of the mature cytosolic PEPCK mRNA. There are also 3 smaller nuclear RNA species, possibly derived from the larger RNAs which hybridize with our probe. Both the putative PEPCK mRNA precursor and the smaller related RNA species are polyadenylated (Cimbala et al., 1982) and are not observed in cytosolic RNA (see lane C in Fig. 5). There is a rapid alteration in the sequence abundance of these nuclear RNAs. For example, refeeding glucose to a starved rat causes a rapid decrease in the abundance of nuclear precursors for the enzyme, so that within 2 hr after glucose refeeding, these RNAs are barely detectable. However, Bt_2cAMP administration to these animals causes a 5 to 8-fold increase in all of the nuclear precursors within 20 min, after which their sequence abundance decreases markedly.

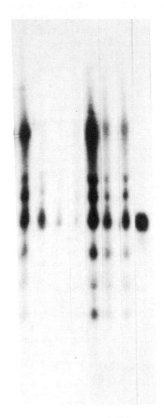

Figure 5. Changes in PEPCK RNA in Rat Liver nuclei. Adrenalectomized rats, starved for 24 hrs, were refed glucose at zero time and then injected with Bt_2cAMP at 120, 140, and 160 nm. RNA was extracted from liver nuclei, separated by agarose gel electrophoresis, transferred to nitrocellulose and hybridized to ^{32}P-labeled PEPCK cAND. For comparison, cytosolic PEPCK mRNA (size 2.8 kb) from the liver of a rat starved for 24 hrs is also shown (lane C). For experimental details see Lamers, Hanson and Meisner (1980). Figure taken from Meisner et al., 1983.

Although the previous sections of this review have focused on the regulation of gene transcription, it is also possible that hormones control other processes in the pathway leading to the synthesis of enzyme proteins. Therefore the ability to detect nuclear precursors for PEPCK provides a unique opportunity to study the processing of RNA transcripts to mature mRNA. The largest of these precursors is 6.5 kb and may be a complete transcript of the PEPCK gene, which is about 6.5 kb in length (Yoo-Warren et al., unpublished observations). The gene for the enzyme contains 8 intervening sequences, which agrees with the number of nuclear precursors shown in Figure 5. It is probable that the large 6.5 kb nuclear precursor is processed to the mature, 2.8 kb cytosolic form by removal of intervening sequences. This is supported by the change in the ratio of the large precursor RNA species to the smaller nuclear RNA which occurs after the addition of Bt_2cAMP. However, definitive evidence in support of the kinetic changes responsible for PEPCK mRNA maturation in the nucleus must await the availability of specific intron probes.

The sequence abundance of the cytosolic mRNA for PEPCK is also altered rapidly. Glucose refeeding of a starved rat caused a marked reduction in enzyme mRNA which paralleled the changes in the rate of PEPCK gene transcription (Lamers et al., 1982). The administration of Bt_2cAMP to glucose-fed rats rapidly induced cytosolic PEPCK mRNA so that within 3 hrs its sequence abundance was induced 8-fold. This increase in enzyme mRNA corresponds directly to the rapid increase in PEPCK synthesis observed after the administration of Bt_2cAMP (Iynedjian and Hanson, 1977).

C. Cyclic AMP and the Regulation of Transcription of the PEPCK Gene. The rate of PEPCK RNA synthesis by isolated nuclei from the livers of rats fed glucose or fed glucose and injected with Bt_2cAMP is shown in Table 2. The transcription rate was measured essentially as described by Palmiter and McKnight (1981). The rate of PEPCK RNA synthesis in starved rats is 0.37%, which is reduced over 90% by α-amanitin. The basal rate of PEPCK gene transcription in glucose fed rats is about 0.06%. However, within 20 min after the injection of Bt_2cAMP the rate of PEPCK gene transcription increases about 6-fold. This induced rate of transcription is among the highest reported for any single copy eukaryotic gene. A detailed discussion of the regulation of PEPCK by cAMP is contained in a review by Meisner et al. (1983).

The mechanism(s) by which cAMP stimulates PEPCK gene transcription are not understood. The responsiveness of the gene to cyclic nucleotide injected into animals prior to the isolation of hepatic nuclei suggests that the regulation of this gene may be amenable to direct stimulation of these nuclei in vitro. The putative 5'-regulatory region of the gene for PEPCK have also been isolated and the hormonal regulation of the gene is currently being tested using fusion genes in transfected cells by procedures similar to those described in detail in previous sections of this review.

Table 2. Effect of Glucose Refeeding and cAMP on PEPCK
Gene Transcription in Isolated Rat Liver Nuclei

Condition	Total RNA cpm x 10^6	pPCK2 cpm	pBR322 cpm	Transcription rate %
A. Starved	7.25	3120	350	0.37
B. Starved + α-amanitin	3.80	320	190	0.03
C. Glucose Refed (2 hr)	8.35	930	210	0.06
D. Glucose refed + cAMP (20 min)	7.35	3500	130	0.35

$$\text{Transcription rate} = \frac{\text{cpm (pPCK2–pBR322)}}{\text{cpm in total RNA}} \times \frac{100}{\text{efficiency}} \times \frac{2800}{600}$$

Where 2800 nucleotides is the length of the mature PEPCK mRNA and
600 bp is the length of the enzyme cDNA (pPCK2). The efficiency
was determined with ^{3}H–cRNA and ranged from 50 to 65%. Values
are the means of duplicate experiments. For details see Lamers et
al. (1982). Table taken from Meisner et al. (1983).

D. Conclusion

The purpose of this review has been to bring together many of the
techniques used to study the hormonal regulation of gene expression. As we have noted, there are currently very few examples of
hormonal control in which a direct effect on the gene of interest
has been demonstrated. However, this field is rapidly expanding
and perhaps, even before the publication of this chapter, other
genes will be added to the four which we have outlined. We can
also anticipate the rapid refinement of new methods for the study
of gene regulation, including the reconstitution of an in vitro
transcription system which is responsive to added regulatory
factors. Hopefully, this chapter will serve as a useful introduction to this exciting area of research.

IV. ACKNOWLEDGEMENTS

We would like to thank Drs. Herman Meisner, David Samols, David
Loose and Rick Woychik for critically reviewing this manuscript,
Jay Short for the photography, and Yvonne Coleman for preparation
of the manuscript. Supported in part by grant AM 25541 from the

198

National Institutes of Health and by a grant from the Kroc
Foundation. A.J.W-B. is a fellow supported by the Metabolic
Training Program, AM 07319.

REFERENCES

Banerji, J., Russoni, S., and Schaffner, W., 1981. Expression of
  a β-globin gene is enhanced by remote SV40 DNA sequences.
  Cell, 27, 299-308
Beach, L.R., and Palmiter, R.D., 1981. Amplification of the
  metallothionein-I gene in cadmium resistant mouse cells.
  Proc. Natl. Acad. Sci. USA, 78, 2110-2114.
Bentvelzen, P., and Hilgers, J., 1980. Murine mammary tumor
  virus, in Viral Oncology (Ed. Klein) pp. 311-355. Raven Press,
  New York.
Brinster, R.L., Chen, H.Y., Trumbauer, M., Senear, A.W., Warren,
  R., and Palmiter, R., 1981. Somatic expression of herpes
  thymidine kinase in mice following injection of a fusion gene
  into eggs. Cell, 27, 223-231.
Brinster, R.L., Chen, H.Y., Warren, R., Sarthy, A., and Palmiter,
  R.D., 1982. Regulation of metallothionein-thymidine kinase
  fusion plasmids injected into mouse eggs. Nature, 296, 39-42.
Brown, D.D., 1981. Gene expression in eukaryotes. Science, 211,
  667-674.
Buetti, E., and Diggelman, H., 1981. Cloned mouse mammary tumor
  virus DNA is biologically active in transfected mouse cells and
  its expression is stimulated by glucocorticoid hormones. Cell,
  23, 335-345.
Cimbala, M.A., Lamers, W.H., Nelson, K., Monahan, J.E.,
  Yoo-Warren, H., and Hanson, R.W., 1982. Rapid changes in the
  concentration of phosphoenolpyruvate carboxykinase mRNA in rat
  liver and kidney J. Biol. Chem., 257, 7629-7636.
Corces, V., Pellicer, A., Axel, R., and Meselson, M., 1981.
  Integration, transcription and control of a Drosophila heat
  shock gene in mouse cell. Proc. Natl. Acad. Sci. USA, 78,
  7038-7042.
Chu, G., and Sharp, P., 1981. SV40 DNA transfection of cells in
  suspension: Analysis of the efficiency of transcription and
  translation of T-antigen. Gene, 13, 197-202.
Colbere-Garapin, F., Horodniceanu, F., Kourilsky, P., and Garapin,
  A.C., 1981. A new dominant hybrid selective marker for higher
  eukaryotic cells. J. Mol. Biol., 150, 1-14.
Craine, B.L., and Kornberg, T., 1981. Activation of the major
  Drosophila heat-shock gene in vitro. Cell, 25, 671-681.
Dale, R.M.K., and Ward, D.C., 1975. Mercurated polynucleotides:
  new probes for hybridization and selective polymer
  fractionation. Biochemistry, 14, 2458-2469.
Darnell, J.E., 1982. Variety in the level of gene control in
  eukaryotic cells. Nature, 297, 365-371.
Doehmer, J., Barinaga, M., Vale, W., Rosenfeld, M.G., Verma, I.M.,
  and Evans, R.M., 1982. Introduction of rat growth hormone gene
  into mouse fibroblasts via a retroviral DNA vector: Expression
  and regulation. Proc. Natl. Acad. Sci. USA, 79, 2269-2272.

Durnam, D.M., Perriu, F., Gannon, F., Palmiter, R.D., 1980. Isolation and characterization of the mouse metallothionein-I gene. <u>Proc. Natl. Acad. Sci. USA</u>, 77, 6511-6515.

Durnam, D.M., and Palmiter, R.D., 1981. Transcriptional regulation of the mouse metallothionein-I gene by heavy metals. <u>J. Biol. Chem.</u>, 256, 5712-5716.

Elder, J.T., Spritz, R.A., and Weissman, S.M., 1981. Simian virus 40 as a eukaryotic cloning vesicle. <u>Ann. Rev. Genet.</u>, 15, 295-340.

Gillam, S., and Smith, M., 1979a. Site specific mutagenesis using synthetic oligonucleotide primers. I. Optimum conditions and minimum oligodeoxyribonucleotide length. <u>Gene</u>, 8, 81-97.

Gilliam, S., and Smith, M., 1979b. Site specific mutagenesis using synthetic oligodeoxyribonucleotide primers II. <u>In vitro</u> selection of mutant DNA. <u>Gene</u>, 8, 99-106.

Gluzman, Y., 1981. SV40-transformed simian cells support the replication of early SV40 mutants. <u>Cell</u>, 23, 175-182.

Gorman, C.M., Moffat, C.M., and Howard, B.J., 1982. Recombinant genomes which express chloramphenicol acetyltransferase in mammalian cells. <u>Mol. Cell. Biochem.</u>, 2, 1044-1051.

Gorski, J., and Gannon, F., 1976. Current models of steroid hormone action: A critique. <u>Ann. Rev. Physiol.</u>, 38, 425-450.

Graessman, M., and Graessman, A., 1976. Early simian virus 40-specific RNA contains information for tumor antigen formation and chromatin replication. <u>Proc. Natl. Acad. Sci. USA</u>, 73, 366-370.

Graham, F.L., and vander Eb, A.J., 1973. A new technique for the assay of infectivity of human adenovirus 5 DNA. <u>Virology</u>, 52, 456-469.

Guyette, W.A., Matusik, R.J., and Rosen, J.M., 1979. Prolactin-mediated transcriptional and post-transcriptional control of casein gene expression. <u>Cell</u>, 17, 1013-1023.

Hager, L.J., and Palmiter, R.D., 1981. Transcriptional regulation of mouse liver metallothionein-I gene by glucocorticoids. <u>Nature</u>, 291, 340-342.

Hamer, D.H., and Walling, M., 1982. Regulation <u>in vivo</u> of a cloned mammalian gene: Cadmium induces the transcription of a mouse metallothionein gene in SV 40 vectors. <u>J. Mol. Appl. Genet.</u> 1, 273-288.

Hanahan, D., Lane, D., Lipsich, L., Wigler, W., and Botchan, M., 1980. Characteristics of an SV40 plasmid recombinant and its movement into and out of the genome of a murine cell. <u>Cell</u>, 21, 127-139.

Huang, A.L., Ostrawski, M.C., Berard, D., and Hager, G.L., 1981. Glucocorticoid regulation of the Ha-MuSV p21 gene conferred by sequences from mouse mammary tumor virus. <u>Cell</u>, 27, 245-255.

Humphries, R.R., Ley, T., Turner, P., Moulton, A.D., and Nienhuis, A.W., 1982. Differences in human α-, β- and -globin gene expression in monkey kidney cells. <u>Cell</u>, 30, 173-183

Hynes, N.E., Kennedy, N., Rahmsdorf, U., and Groner, B., 1981. Hormone-responsive expression of an endogeneous proviral gene of mouse mammary tumor virus after molecular cloning and gene transfer into cultured cells. Proc. Natl. Acad. Sci. USA, 78, 2038-2042.

Iynedjian, P.B., and Hanson, R.W., 1977. Increase in level of functional messenger RNA coding for phosphoenolpyruvate carboxykinase (GTP) during induction by cyclic adenosine 3':5' monophosphate. J. Biol. Chem., 252, 655-662.

Kelley, P.A., Dijane, J., Houdebine, L.M.. and Teyssot, B., 1981. Evidence for the existence of a prolactin second messenger. J. Cell Biochem. Supplement, 6, 175.

Kioussis, D., Reshef, L., Cohen, H., Tilghman, S.M., Iynedjian, P.B., Ballard, F.J., and Hanson, R.W., 1978. Alterations in translatable messenger RNA coding for phosphoenolpyruvate carboxykinase (GTP) in rat liver cytosol during deinduction. J. Biol. Chem., 253, 4327-4332.

Kressman, A., Clarksen, S.G., Telford, J., and Birnstiel, M.S., 1977. Transcription of Xenopus tDNA$_1$ and sea urchin histone DNA injected into the Xenopus oocyte nucleus. Cold Spring Harbor Symp. Quant. Biol., 42, 1077-1081.

Kurtz, D.T., 1981. Hormonal inducibility of rat α_{2u} globulin genes in transfected mouse cells. Nature, 291, 629-631.

Kurtz, D.T., and Feigelson, P., 1977. Multihormonal induction of hepatic α_{2u}-globulin mRNA as measured by hybridization to complimentary DNA. Proc. Natl. Acad. Sci. USA, 74, 4791-4795.

Kurtz, D.T., Chan, K.M., and Feigelson, P., 1978. Translation control of hepatic α_{2u}-globulin synthesis by growth hormone. Cell, 15, 743-750.

Lamers, W., Hanson, R.W., and Meisner, H., 1982. cAMP stimulates transcription of the gene for cytosolic phosphoenolpyruvate carboxykinase in rat liver nuclei. Proc. Natl. Acad. Sci. USA, 79, 5137-5141.

Lee, F., Mullilgan, R., Berg, P., and Ringold, G., 1981. Glucocorticoids regulate expression of dihydrofolate reductase cDNA in mouse mammary tumor virus chimeric plasmids. Nature, 294, 228-232.

Lipman, M.E., and Thompson, E.B., 1974. Steroid receptors and the mechanism of the specificity of glucocorticoid responsiveness of somatic cell hybrids between hepatoma tissue culture cells and mouse fibroblasts. J. Biol. Chem. 249, 2483-2488.

Manley, J.L., Fire, A., Cano, A., Sharp, P.A., and Gefter, M.L., 1980. DNA-dependent transcription of adenovirus genes in a soluble whole-cell extract. Proc. Natl. Acad. Sci. USA, 77, 3855-3859.

Mantei, N., Boll, W., and Weissman, C., 1979. Rabbit '-globin mRNA production in mouse L cells transformed with cloned rabbit β-globin chromosomal DNA. Nature 281, 40-46.

Maniatis, T., Fritsch, E.F., and Sambrook, J., 1982. Molecular Cloning a Laboratory Manual. Cold Spring Harbor Laboratory.

Martial, J.A., Baxter, J.D., Goodman, H.M., and Seeburg, P.H., 1977. Regulation of growth hormone messenger RNA by thyroid and glucocorticoid hormones. Proc. Natl. Acad. Sci. USA. 74, 1816-1820.

Mathis, D.J., and Chambon, P., 1981. The SV 40 early region TATA box is required for accurate in vitro initiation of transcription. Nature, 290, 310-315.

Mayo, K.E., and Palmiter, R.D., 1981. Glucocorticoid regulation of metallothionein-I mRNA synthesis in cultured mouse cells. J. Biol. Chem. 256, 2621-2624.

Mayo, K.E., and Palmiter, R.D., 1982. Glucocorticoid regulation of the mouse metallothionein-I gene is selectively lost following amplification of the gene. J. Biol. Chem. 257, 3061-3067.

Mayo, K.E., Warren, R., and Palmiter, R.D., 1982. The mouse metallothionein-I gene is transcriptionally regulated by cadmium following transfection into human or mouse cells. Cell, 29, 99-108.

McKnight, G.S., and Palmiter, R.D., 1979. Transcriptional regulation of the ovalbumin and conalbumin genes by steroid hormones in chick oviduct. J. Biol. Chem., 254, 9050-9058.

McKnight, S.L., and Kingsbury, R., 1982. Transcriptional control signals of a eukaryotic protein-coding gene. Science, 217, 316-324.

Meisner, H., Lamers, W.H., and Hanson, R.W., 1983. Cyclic AMP and the synthesis of P-enolpyruvate carboxykinase (GTP) mRNA. Trends in Biochemical Sciences. (in press).

Moreau, P., Hen, R., Wasyluk, B., Everett, R., Gaub, M.P., and Chambon, P., 1981. The SV40 72 base pair repeat has a striking effect on gene expression both in SV40 and other chimeric recombinants. Nuc. Acids Res., 9, 6047-6068.

Mulligan, R.C., and Berg, P., 1980. Expression of a bacterial gene in mammalian cells. Science, 209, 1423-1427.

Mulligan, R.C., Howard, B.H., and Berg, P., 1979. Synthesis of rabbit β-globin in cultured monkey kidney cells following infection of a SV40 β-globin recombinant genome. Nature, 277, 108-114.

Mulvihill, E.R., LePennec, J.P., and Chambon, P., 1982. Chicken oviduct progesterone receptor: Location of specific regions of high affinity binding in cloned DNA fragments of hormone-responsive genes. Cell, 24, 621-632.

Myers, R.M., Rio, D.C., Robbins, A.K., and Tijan, R., 1981. SV40 gene expression is modulated by the cooperative binding of T-antigen to DNA. Cell, 25, 373-384.

O'Malley, B.W., Tsia, M.J., Tsia, S.Y., and Tarle, H.C., 1977. Regulation of gene expression in chick oviduct. Cold Spring Harbor Symp. Quant. Biol. 42, 605-616.

Palmiter, R.D., Chen, H.Y., and Brinster, R.L., 1982. Differential regulation of metallothionein-thymidine kinase fusion genes in transgenic mice and their offspring. Cell, 29, 701-710.

Pithi, P.M., Cinfo, D.M., Kellum, M., Raj, N.B.K., Reyes, G.R., and Hayward, G.S., 1981. Introduction of human β-interferon synthesis with poly (rI-rC) in mouse cells transfected with cloned cDNA plasmids. Proc. Natl. Acad. Sci. USA, 79, 4337-4341.

Robins, D.M., Paek, I., Seeburg, P.H., and Axel, R., 1982. Regulated expression of human growth hormone genes in mouse cells. Cell, 29, 623-631.

Sarrer, N., Gruss, P., Ming, F.I., Khoury, G., and Howley, P., 1981. Bovine papilloma virus DNA - A novel eukaryotic cloning vector. Mol. Cell. Biol 1, 486-496.

Scangos, G., and Ruddle, F.H., 1981. Mechanisms and applications of DNA-mediated gene transfer in mammalian cells - A review. Gene, 14, 1-10.

Shen, Y.M., Hirschhorn, R.R., Mercer, W.E., Surmacz, E., Tsutsui, Y., Soprano, K.J., and Baserga, R., 1982. Gene transfer: DNA microinjection compared with DNA transfection with a very high efficiency. Mol. Cell Biol. 2, 1145-1154.

Shenk, T., 1981. Transcriptional control regions: Nucleotide sequence requirements for initiation by RNA polymerase II and III. Curr. Topics Microbiol. Immunol., 93, 25-26.

Shortle, D., DiMaio, D., and Nathans, D., 1981. Directed mutagenesis. Ann. Rev. Genet., 15, 265-294.

Southern, P.J., and Berg, P., 1982. Transformation of mammalian cells to antibiotic resistance with a bacterial gene under control of the SV40 early region promoter. J. Mol. Appl. Genet. 1, 327-341.

Taylor, R.W., and Smith, R.G., 1982. Effects of highly purified estrogen receptors on gene transcription in isolated nuclei. Biochemistry, 21, 1781-1787.

Teyssot, B., Houdebine, L.M., and Dijane, J., 1981. Prolactin induces release of a factor from membranes capable of stimulating β-casein gene transcription in isolated mammary cell nuclei. Proc. Natl. Acad. Sci. USA, 78, 6729-6733.

Tilghman, S.M., Hanson, R.W., Reshef, L., Hopgood, M.F., and Ballard, F.J., 1974. Rapid loss of translatable messenger RNA of phosphoenolpyruvate carboxykinase during glucose repression in liver. Proc. Natl. acad. Sci. USA, 71, 1304-1308.

Tilghman, S.M., Gunn, J.M., Fisher, L.M., Hanson, R.W., Reshef, L., and Ballard, F.J., 1975. Deinduction of phosphoenol-pyruvate carboxykinase (guanosine triphosphate) synthesis in Reuber H-35 cells. J. Biol. Chem., 250, 3322-3329.

Tilghman, S.M., Hanson, R.W., and Ballard, J.F., 1976. Hormonal regulation of phosphoenolpyruvate carboxykinase (GTP) in mammalian tissues, in Gluconeogenesis, Its Regulation in Mammalian Species (Eds. Hanson and Mehlman), pp. 47-91. Wiley, New York.

Varmus, H.E., Ringold, G., and Yamamoto, R.R., 1979. Regulation of mouse mammary tumor virus gene expression by glucocorticoid hormones, in Monographs on Endocrinology. Glucocorticoid Hormone Action, (Eds. Baxter and Rousseau), pp. 253-278. Springer, New York.

Wagner, E.F., Stewart, T.A., and Mintz, B., 1981. The human
β-globin gene and a functional viral thymidine kinase gene in
developing mice. Proc. Natl. Acad. Sci. USA, 78, 5016-5020.

Weiher, H., and Schller, H., 1982. Segment-specific mutagenesis.
Extensive mutagenesis of a lac promoter/operator element.
Proc. Natl. Acad. Sci. USA., 79 1407-1412.

Weil, P.A., Luse, D.S., Segall, J., and Roeder, R.G., 1979.
Selective and accurate initiation of transcription at the Ad2
major late promoter in a soluble system dependent on purified
RNA polymerase II and DNA. Cell, 18, 461-484.

Wickens, M.P., Woo, S., O'Malley, B.W., and Gurdon, J.B., 1980.
Expression of a chicken chromosomal ovalbumin gene injected
into frog oocyte nuclei. Nature, 285, 628-634.

Wicks, W.D., and McKibbin, J.B., 1972. Evidence of the
translational regulation of specific enzyme synthesis by N^6,
O^2-dibutyryl cyclic AMP in hepatoma cell culture. Biochem.
Biophys. Res. Comm., 48, 205-211.

Wigler, M., Sweet, R., Sim, G.R., Wold, B., Pellicer, A., Lacy,
E., Maniatis, T., Silverstein, S., and Axel, R., 1979.
Transformation of mammalian cells with genes from prokaryotes
and eukaryotes. Cell, 16, 775-785.

Yamamoto, K.T., and Alberts, B., 1976. Steroid receptors for
modulation of eukaryotic transcription. Ann. Rev. Biochem.,
45, 722-746.

Yeung, D., and Oliver, I.T., 1968a. Induction of phosphopyruvate
carboxylase in neonatal rat liver by adenosine 3',5'-cyclic
monophosphate. Biochemistry, 7, 3231-3239.

Yeung, D., and Oliver, I.T., 1968b. Factors affecting the
premature induction of phosphopyruvate carboxylase in
neonatal rat liver. Biochem. J. 108, 325-331.

Yoo-Warren, H., Cimbala, M.A., Felz, K., Monahan, J.E., Leis,
J.P., and Hanson, R.W., 1981. Identification of a DNA clone to
phosphoenolpyruvate carboxykinase (GTP) from rat cytosol. J.
Biol. Chem., 256, 10244-10227.

Zarucki-Schulz, T., Tsai, S.Y., Itakura, K., Soberon, X., Wallace,
R.B., Tsai, M.J., Woo, S.L.C., and O'Malley, B.W., 1982. Point
mutagenesis of the ovalbumin gene promoter sequence and its
effect on in vitro transcription. J. Biol. Chem., 257,
11070-11077.

Zinn, K., Mellon, P., Ptathne, M., and Maniatis, T., 1982.
Regulated expression of an extrachromosomal human β-interferon
gene in mouse cells. Proc. Natl. Acad. Sci. USA, 79,
48-97-4901.

Genes: *Structure and Expression*
Edited by A. M. Kroon
© 1983 John Wiley & Sons Ltd.

STRATEGIES FOR OPTIMIZING FOREIGN GENE
EXPRESSION IN ESCHERICHIA COLI

Herman A. de Boer and H. Michael Shepard

Genentech, Inc.
460 Point San Bruno Boulevard
South San Francisco, California 94080
Genentech Contribution No. 135

CONTENTS

INTRODUCTION

The expression of eukaryotic genes in E. coli is a recent development that has revolutionized molecular biology. It may also have the biggest impact on the pharmaceutical industry since the discovery of the antibiotics. The techniques involved in cloning and expressing eukaryotic genes in bacteria have been given the umbrella-name of genetic engineering.

Important events in the development of this technology include: 1) the discovery of plasmids which are present in many copies within certain bacterial cells, 2) the development of an efficient transformation protocol (to enable the introduction of plasmids into cells with high frequency), 3) the discovery of restriction endonucleases which allow the cutting of DNA at specific sequences, and 4) the discovery of DNA ligases, which allow splicing together of restriction endonuclease-cut heterologous DNA molecules.

In order to express the heterologous DNA sequences into protein in E. coli it is necessary to understand the signals that determine the efficiency of transcription, and those that affect the efficiency of the mRNA translation process. The expression of eukaryotic genes in Escherichia coli was first accomplished when yeast genes cloned into the plasmid Col El were found to complement auxotrophic mutants of E. coli (1,2). Expression of these foreign genes was originally a fortuitous event and was not a result of the precise manipulation of E. coli transcription and translation signals. Since these early experiments, however, our ability to manipulate DNA sequences in vitro has allowed a finer definition of E. coli gene control signals and their use to obtain high level expression of cloned genes. The first E. coli expression systems employed the E. coli lactose (lac) operon promoter and the eukaryotic sequences were expressed as fusion proteins with β-galactosidase. Later "direct" expression experiments, ie., a eukaryotic gene being expressed without the presence of a β-galactosidase leader, were done by Goeddel et al. (5) with the human growth hormone gene. The direct expression technique has now come into general use and its application to the direct expression of human immune interferon (HuIFN-γ) is outlined in the following section.

Many eukaryotic genes have now been expressed in E. coli (see ref. 3 for a current bibliography). These include such divergent coding sequences as the transforming gene of RNA tumor viruses (4), human and animal growth hormones (5,6), neuroactive peptides (7), insulin (8), interferons (9,10,11,12), immunoglobulins (13) and other human serum proteins (14). Some genes are expressed at high levels (up to 10^6 molecules per cell), others are poorly expressed (a few thousand molecules per cell).

In this review we will 1) briefly outline a strategy for the cloning of a gene, 2) briefly outline how a DNA fragment containing the cloned gene can be tailored prior to the fusion with E. coli transcriptional and translation signals, 3) discuss factors that affect the transcriptional efficiency of promoters, and 4) describe some promoters that can be used for foreign gene expression, and

5) describe some studies to increase the translation efficiency of foreign messengers.

Gene Isolation and Tailoring for Expression

The basic steps required to obtain a gene library from messenger RNA of a eukaryotic cell is schematically summarized in Figure 1. Although the example shows the strategy employed in the cloning of the human immune interferon (HuIFN-γ) gene (12), this approach is currently used for the cloning of most genes from

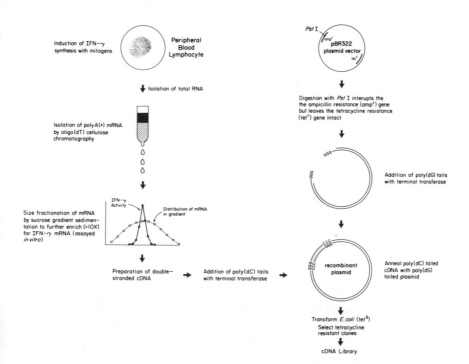

Fig. 1. The Construction of a cDNA Library. The steps involved in the induction of the synthesis of the desired mRNA (IFN-γ) in a specialized cell type, the preparation of double-stranded complementary DNA (cDNA) and the insertion into a plasmid vector are shown. Once the recombinant plasmid is introduced into E. coli in a process called transformation, each plasmid replicates several times giving a final copy number per cell of about 30 plasmids. The transformed cells are spread onto agar dishes containing tetracycline to select for transformants. The colonies are then replica plated onto ampicillin containing plates. Ampicillin sensitive clones are subsequently examined for recombinant plasmids containing the desired cDNA insert.

messenger RNA. In this approach the source of the gene to be cloned is messenger RNA which is in vitro reverse transcribed into a DNA copy which is subsequently cloned into the plasmid pBR322. Thus a library of complementary DNA (cDNA) clones is obtained from which the colony containing the desired gene has to be identified through a screening procedure (see Refs. 10,12,15). After identification of the proper DNA insert the gene has to be tailored to make it suitable for expression in E. coli because the cDNA clone is, if it is a full length clone, a copy of not only the actual coding region of the messenger but also of the untranslated region preceding the translational start codon and the region following the translational stop codon. The region that precedes the translational start codon contains sequences that specify the eukaryotic ribosome binding site. If the messenger codes for a protein that is secreted it also encodes a short stretch of amino acids that constitute the so-called signal peptide which is removed during the process of secretion, thus yielding the mature polypeptide. In this review we will discuss strategies to increase the production of mature protein. To obtain expression of this mature protein in E. coli, all the sequences encoding the signal peptide and the eukaryotic ribosome binding site have to be removed and replaced by the transcriptional and translational regulatory sequences of E. coli, namely promoters and ribosome binding sites, respectively. An example of a gene in the state as it is obtained from a cDNA library is shown in Figure 2. In the case of the HuIFN-γ coding sequence, a BstNI restriction site is present 10 basepairs downstream of the beginning of the sequence encoding the mature polypeptide. This site makes tailoring of the gene simple. If no such conveniently located restriction site is present, a so-called primer-repair reaction has to be done. This elegant method has been described elsewhere (15). In the construction shown in Figure 3 the gene was cut with BstNI and a short synthetic DNA fragment ending with BstNI sticky ends was ligated to the gene. This synthetic DNA replaces not only the first three amino acid codons of the mature protein, but also contains a newly introduced start codon. (Note that no start codon of the eukaryotic mRNA used to be here; it is located in front of the signal peptide codons.) At the left side of this synthetic fragment EcoRI sticky ends are present. The E. coli promoters and ribosome binding sites to be discussed in this paper are all contained on a small EcoRI fragment. When joined to a tailored gene as shown in figure 3 part of the E. coli ribosome binding site (the so-called Shine-Dalgarno sequence) located on the end of the promoter containing fragment, ends up in close proximity to the start codon. Now a hybrid ribosome binding site is formed in which the Shine-Dalgarno (SD) region is prokaryotic and in which the coding sequences are eukaryotic. The last section of this paper discusses some experiments designed to optimize two such chimeric ribosome binding sites.

In the following section we will discuss E. coli promoters. We will describe briefly the nature of promoters and the current state of the research being done on promoters, describe a few promoters that are or can be used for foreign gene expression, and discuss for each promoter some properties that are relevant for that purpose. Finally we will describe some hybrid promoters that have been constructed recently in this laboratory.

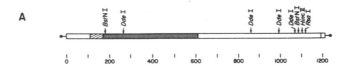

Fig. 2. Restriction endonuclease map and DNA sequence of the human γ interferon gene (HuIFN-γ) cDNA insert. (A) The cDNA insert is bounded by PstI sites (dots at both ends) and poly(dG:dC) tails (single lines; see Fig. 1). The shaded region indicates the coding sequences of the mature protein, the cross-hatched region represents the 20 amino acid signal peptide coding sequence, and the open regions show the 5'- and 3'-untranslated sequences. (B) Nucleotide sequence and deduced amino acid sequence of the HuIFN-γ protein. The sequence of the signal peptide is represented by the residues labeled S1 to S20 (in small type) and the mature sequence of 146 amino acids is presented in large type. Numbers above each line refer to amino acid position and numbers below each line refer to nucleotide position. This figure is reprinted from Gray et al., 1982, Nature 295, 505.

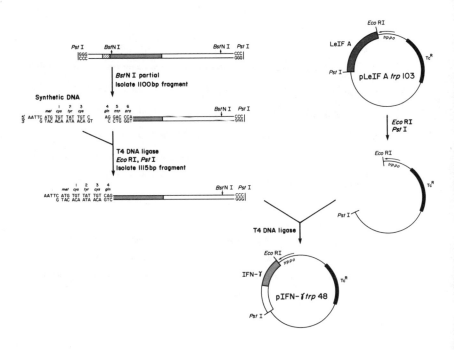

Fig. 3. Construction of a plasmid designed for the direct expression of the HuIFN-γ gene. In the example shown here the gene is partially cut with restriction enzyme (BstNI) which cuts 8 basepairs downstream of the first amino acid (cysteine) of the mature polypeptide. Partial digestion with BstNI of the cDNA sequence was necessary because there is another BstNI site within the 3'-nontranslated region of the gene. After the partial digestion a gene fragment is isolated which extends from the 5'-BstNI site to the PstI site at the 3'-end of the cDNA insert. This fragment is then ligated to a double-stranded synthetic DNA piece containing 1) restriction endonuclease site sticky ends for insertion downstream from a promoter fragment (EcoRI, in this case), 2) an ATG for translational initiation in E. coli, 3) codons for the amino acids upstream from the site of BstNI cleavage (cys, tyr, cys) and overlapping ends that will allow ligation to the BstNI-end of the isolated gene fragment. In order to monitor the ligation of the synthetic piece to the gene fragment the synthetic oligodeoxyribonucleotide is 5'-labeled with ^{32}P-γ-rATP and T4 polynucleotide kinase. After ligation of primer with gene fragment the reaction mixture is digested with EcoRI and PstI to eliminate multimers and the resulting monomeric gene fragment, now containing an ATG immediately preceding the N-terminal cysteine, is isolated following electrophoresis through a polyacrylamide gel. The reconstructed gene is then ligated downstream from a promoter fragment in an expression plasmid and the recombinant is transformed into E. coli. In this case, a promoter fragment derived from the E. coli tryptophan operon (with the trp attenuator deleted) has been used (10,12,15). The structure of the plasmid is confirmed by restriction endonuclease analysis and also by DNA sequencing. These manipulations result in the construction of an expression "cassette", so that the engineered coding sequence can now be easily moved into alternative expression systems. This figure is reprinted with permission from Gray et al., 1982, Nature 295, 506.

PROMOTERS

RNA polymerase starts transcription at promoters.

All the active genes on the chromosome of E. coli are transcribed by RNA-polymerase. Only one type of RNA-polymerase is known to exist in E. coli cells. RNA-polymerase catalyzes the polymerization of nucleotides into RNA molecules whose sequence is dictated by the nucleotide sequence in the DNA template. The core of the enzyme consists of four subunits: two α subunits (MW 40,000 D), one β-subunit (MW 150,000 D) and one β' subunit (MW 160,000 D). This core enzyme is by itself able to catalyze the polymerization reaction. However, it starts this transcription process at random on the DNA template; that is, it lacks specificity. For the initiation of RNA chains at specific sites on the DNA template, i.e. at promoters, the core enzyme requires an additional protein, the sigma factor (MW 86,000 D). The core-enzyme plus the σ factor is called the holoenzyme. The σ factor is present in less than equimolar amounts with respect to the other subunits (reviewed in Ref. 16).

The σ factor releases from the core-enzyme after an RNA chain has been initiated and elongation of the RNA chain occurs without the sigma factor. The region including and preceding the start site for transcription is called the promoter. The sequence of events that occur at a promoter prior to the polymerization of an RNA chain can be summarized in a simplified way as follows:

1. The RNA-polymerase (holoenzyme) binds rapidly at non-specific sites which are randomly distributed throughout the DNA molecule (16,17,20).
2. The RNA polymerase moves along the DNA in search of a promoter sequence (18,19).
3. The RNA polymerase recognizes a promoter sequence and forms a closed binary complex with the DNA duplex (16-20).
4. The RNA polymerase unwinds the DNA in the region around the start site of transcription and forms the so-called open-complex with the DNA duplex (16,21).
5. The RNA polymerase binds the appropriate nucleotide triphosphate and forms a ternary complex (16).
6. A second nucleoside triphosphate is bound and polymerized to the first one and pyrophosphate is released. The release of pyrophosphate and its subsequent hydrolysis provides for energy required for the polymerization process (16).
7. After the chain initiation event, the σ-factor releases (16).

An example of a promoter sequence showing the DNA sequences that are unwound in the open complex is shown in Fig. 4.

A promoter consists of two domains that are in close contact with the RNA polymerase.

How does the RNA polymerase "know" where to start transcription and which are the bases in the DNA to which the enzyme binds specifically? The following three categories of experiments have provided a lot of information concerning this recognition process.

Contact sites with RNA-polymerase

```
══════════        ═══════════════════════
-35 region                 -10 region            lac-repressor
                                                 binding site
                             AATGTGTGGAAT      ───────────────
CCCCAGGCTTTACACTTTATGCTTCCGGCTCGTAT            TGTGAGCGGATAACAATTTCACACAGGAAACAGCTATG
GGGGTCCGAAATGTGAAATACGAAGGCCGAGCATA            ACACTCGCCTATTGTTAAAGTGTGTCCTTTGTCGATAC
                             TTACACACCTTA
```

```
                                                       S.D.
                                                     ────────
                   pppAAUUGUGAGCGGAUAACAAUUUCACACAGGAAACAGCTATG
                                                   ──────
                                                   ribosome
                                                   binding site
```

Fig. 4. Sequences that interact with the RNA polymerase
and lac-repressor. The DNA sequence of the lac-promoter
and the lac-operator are shown. The consensus -35
sequence and the Pribnow box sequence is overlined. The
regions that are protected by the RNA-polymerase against
various treatments are indicated with double lines. The
region that is unwound in the open complex of RNA
polymerase and promoter is indicated by a double space
between both DNA strands. Note that the Pribnow-box area
is in part unwound in the open complex (21). The
lac-operator region to which the lac-repressor binds is
overlined. The 5'-sequence of the RNA strand is shown
beneath the DNA sequence. Note that the first transcribed
base is within the operator and that the RNA-polymerase
and the lac-repressor have overlapping binding sites. The
ribosome binding site on the RNA strand is underlined.
The start codon (ATG) and the Shine-Dalgarno sequence in
the ribosome binding site is overlined.

a.   DNA sequence analysis of promoters. Since only one enzyme
     recognizes all promoters, various investigators have
     determined the sequence of RNA polymerase binding sites
     hoping to find common features shared by all promoters.
     Such common features were indeed discovered. One 6
     basepair region was found (22,23) around position -10 with
     respect to the transcription start site (this start site
     is referred to as +1, the position preceding this start
     site as -1, etc.). The region is referred to as the -10
     region or the Pribnow-box (22). Later (20) a second
     region was found around position -35 (Fig. 4). Based on
     the relative frequencies of occurrence of the four bases
     in these regions a so-called consensus sequence was
     formulated for both domains. The consensus sequence for
     the -35 region is 5'-TTGACA and that for the -10 region is
     5'-TATAAT (20,21). In the various promoters one to three
     deviations from the consensus Pribnow-box and the
     consensus -35 region are quite common. However, no
     naturally occurring promoter exists in which the last T

residue of the Pribnow-box is replaced by a different base
(21). Apparently this T residue plays the most critical
role in promoter function. Mutants in this position
virtually abolish promoter function (24,25). The first
two bases of the Pribnow-box (T and A) also appear to be
highly conserved (21,25). Moreover, the distance between
both domains varies only from 16 bp to 18 bp while 17 bp
is most commonly found in the sequenced promoters (the
distance is measured as the number of basepairs between
both domains, Ref. 21). Below we will discuss further the
effect of this distance on promoter activity and we will
discuss some evidence indicating that a distance of 17 bp
is probably optimal for promoter function. In summary,
the consensus promoter can be presented schematically as
follows:

	-35 region		Pribnow-box	+1
5'	T T G A C A	– 17 bp –	T A T A A T	5-8 bp A or G
3'	A A C T G T	– 17 bp –	A T A T T A	5-8 bp T or C

The DNA sequences of some promoters and promoter mutants
relevant to this paper are shown in figure 5.

b. Analysis of promoter mutants: Most mutations map in or
near the -35 and -10 region or affect the distance between
both regions.
Many promoter mutants were obtained using conventional
genetic techniques. It appears that about 80 percent of
all the promoter mutations are located either in the
Pribnow-box or in the -35 region and a great proportion of
all these mutations map in the most conserved bases within
the Pribnow-box (25), that is in the first two bases and
in the fully conserved 6th base. These mutations affect
the promoter activity adversely; they are so-called "down"
mutations. These base substitutions are indicated in
figure 5 with a downward arrow.

Only a very few mutations are known that increase the
promoter activity (see Fig. 5). Such up-mutations all
increase the extent of homology with the consensus
sequence. For example, in the lac-UV5 promoter the
Pribnow box sequence has changed from TATGTT in the wild
type promoter to TATAAT in the mutant promoter (25). In
the "up"-mutation found in the promoter of the gene for
the lac-repressor (the laci gene) and which causes an
overproduction of the lac-repressor (see below), the -35
sequence has changed from GCGCAA to GTGCAA which is a
change closer to the consensus -35 sequence (26). In the
λPRM promoter "up" mutant (not shown in Fig. 5) the -35
sequence has changed from TAGATA to TAGACA, again
resembling more the consensus -35 sequence (25).

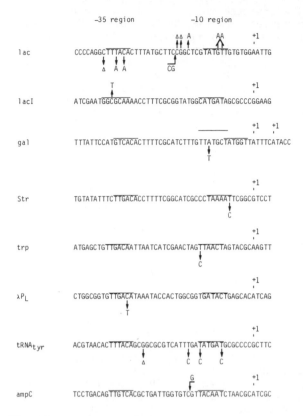

Fig. 5. Promoter sequences and promoter mutations. The DNA sequence (top strand 5'-3') of a few promoters and promoter mutations relevant to this paper are shown. The sequences are aligned at their transcriptional startpoints which are indicated as +1. The –10 and the –35 regions are overlined. Note that the gal-promoter has two overlapping promoters (28). Promoter down mutations are indicated with a downward pointing arrow and promoter-up mutations with an upward pointing arrow. Base changes are indicated with an arrow and the base that has been changed. Deletions are indicated with a Δ. Insertions are indicated with an arrow pointing between two bases. The sequence of these promoters, except the sequence of the amp C promoter, is from a previous review (25). The insertion mutation in the lac-promoter is described in Ref. 29. The deletions in the lac p^S promoter are from Ref. 30. The sequence of the ampC promoter and its deletion is described in Ref. 31. The sequence of the tyr tRNA promoter and its mutations is from Ref. 32. A more complete list of promoter sequences can be found in Ref. 21.

Recently Youderian et al. (27) reported a great number of
mutations in the promoter that directs the transcription
of the antirepressor of the bacteriophage P22. This
promoter (Pant) normally is negatively controlled by two
repressors. In their experimental system the repressor
was, by mutation, made nonfunctional. Under these
conditions the Pant promoter directs the synthesis of
extremely high levels of antirepressor. This
overproduction of antirepressor leads to the inability to
produce progeny phage upon lytic infection. They selected
revertants which have undergone mutational alterations in
the phage Pant promoter. Because of the selection
procedure used, all of the revertants contain promoter
down mutations. DNA sequence analysis of about 70
individual mutants showed that all appeared to have a base
change in either the -35 region or in the -10 region of
the Pant promoter (see Fig. 6). Whereas the natural Pant
promoter differs in only one base from the consensus
sequence (the 5th base is a T instead of an A), in all
mutants this difference appears to be increased.
Strikingly, a revertant of a -35 down mutant was obtained
in which the 5th base in the Pribnow box was altered such
that a consensus Pribnow box was obtained.

From the analysis of these and other promoter mutations
the simple rule emerges that any mutational change which
causes a breakdown of the homology with the consensus
sequence decreases promoter activity. The opposite also
seems to be true. In the few up-mutants found thus far
the mutational change causes an increase in the homology
with the consensus sequence. The greatly increased
activity of the hybrid promoters which will be discussed
below also follow this rule. The effects of these
mutations on promoter activity and their location clearly
confirm the importance of the -10 and the -35 area for
promoter function. However, a promoter with consensus
sequence does not necessarily have maximal efficiency.
Other unknown factors or sequences may also be important
determinants of promoter efficiency.

c. DNA protection experiments; RNA-polymerase binds directly
to the -35 and the -10 region.
In early DNA-RNA polymerase interaction studies (23)
RNA-polymerase was allowed to bind to a DNA fragment known
to contain a transcriptional start site and the complex
was treated with a deoxyribonuclease that cleaves DNA
between any base. The protected DNA was subsequently
analyzed. In such experiments it was found that about 20
bp around the Pribnow-box was protected by the RNA
polymerase. However, after isolation of this fragment the
RNA polymerase was unable to rebind to this short
sequence. Later these kind of experiments were repeated
but instead an exonuclease in combination with a single
strand specific nuclease was used. Under these conditions

a fragment was found to be protected by the RNA-polymerase against exonucleolytic attack, that not only spans the -10 region but also the -35 region (33). These experiments suggest that RNA-polymerase binds to both domains tightly and protects them from cleavage by exonucleases. The first experiment suggests that the endonuclease is apparently able to cleave between both domains suggesting lack of interaction in this region with the RNA polymerase. The picture that evolves from these findings is that the RNA-polymerase has two distinct binding sites that bind to the -35 and the -10 region while the region in between is not engaged in a tight interaction (see Fig. 4). This picture is confirmed in experiments in which DNA fragments containing a promoter were allowed to bind to RNA-polymerase and treated with various chemical reagents that react with unprotected bases leading to breakage of the DNA at the reacted bases. Again, the -10 and -35 regions were protected against such treatments only when the RNA-polymerase was bound to the DNA prior to chemical attack (Ref. 21,34).

Promoters have widely varying efficiencies.

In the following sections we will briefly discuss a few promoters and some of their properties that are relevant for the expression of foreign genes in E. coli. For maximizing the expression of a foreign gene a strong promoter is required and in certain cases (when the product is cytostatic or cytotoxic) a controllable promoter is necessary.

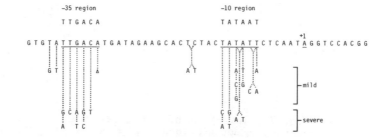

Fig. 6. The DNA sequence of the Pant promoter and various mutants. The Pant promoter directs the transcription of the antirepressor protein of the temperate Salmonella phage P22. Promoter down mutants were selected. The base substitutions are indicated. The two groups of mutations that differ in the extent of loss of promoter function are indicated as "mild" and "severe". These data were taken from Youderian et al. (27).

The efficiency of a promoter is determined by the frequency at which the RNA polymerase starts the transcription cycle. The in vivo efficiency can be measured directly by measurement of the amount of corresponding RNA produced in the cell. However, few such data are available. It is tempting to use, for foreign gene expression, promoters whose natural gene product in E. coli is abundant. Obviously, the amount of gene product synthesized in the cell is determined not only by the strength of the promoter but also by the efficiency of the initiation of translation of the mRNA and its half-life. Therefore, it is important to determine separately the relative contribution of each of these parameters of gene expression.

For this purpose it is necessary to have a system in which the relative efficiencies of promoters can be measured and compared easily and reliably without the requirement to correct the data for differences in translational properties of the mRNAs. Rosenberg and his colleagues (35,36) have developed a plasmid (pKO-1) that contains the entire E. coli galactokinase gene (galK) including its natural ribosome binding site preceded by the natural leader mRNA but lacking its own promoter (see Fig. 7). The galactokinase gene is transcribed only when a promoter is inserted in one of the engineered unique restriction sites located in front of the galK leader mRNA (about 200 bp upstream of the galK start codon; see Figure 7). Between these restriction sites and the leader of the galK gene are three stop codons in all three reading frames (35,36). Thus, any protein initiation that might occur on the promoter containing DNA fragment inserted in one of these restriction sites is aborted and does not contribute to galK expression. The level of galK expression is therefore solely

Fig. 7. Insertion of the EcoRI fragments containing the various promoters into the promoter probe plasmid pKM-1. This plasmid was obtained (37) by insertion of a Rho-dependent terminator (λt_{RI}) in the SmaI site of pKO-1 (35) which reduces the number of transcripts entering the galK gene by at least 70 percent (37,38). The promoter containing pKM derivatives were transformed in either E. coli C600 galK⁻ trpR⁺ or HDB2 galK⁻trpR⁻. Cells were plated on MacConkey galactose indicator plates (35).

determined by the efficiency of the inserted promoter fragment and is independent of the length of the untranslated mRNA and translation signals therein (for details see refs. 35,36).

Constitutive expression of a gene directed by a strong promoter on a high copy number plasmid may reduce the plasmid copy number per cell, or inhibit the growth rate of the cell. To circumvent such complications, Rosenberg et al. (37) constructed pKM-1 (see Figure 7) which they obtained by insertion of a Rho-dependent terminator (λt_{RI}) in the SmaI site of pKO-1. This terminator reduces the number of transcripts entering the galK gene by at least 70 percent (37,38).

In Table I the relative efficiencies as determined in the galactokinase system are listed of some promoters that can be used to express foreign genes. Their sequences are shown in Fig. 8. The lac-UV5 promoter and the gal-promoter are taken as reference. Table I shows that the efficiencies of the promoters tested varies greatly. The order of increasing promoter strength is lac-UV5, gal, trp, tacII, and tacI. λP_L is probably as strong as tacI (37) and rrnE (P1 plus P2) is under certain conditions possibly the strongest of all. The relative efficiency of the lipoprotein promoter has not been published thus far. In the following we will briefly discuss these promoters emphasizing promoter efficiency and promoter regulation.

PROMOTERS USEFUL FOR FOREIGN GENE EXPRESSION

The lac-UV5 promoter: Relatively weak but easy to regulate.
        The lac-UV5 promoter is the weakest promoter listed in Table I. The lac-UV5 promoter has been used quite often for expression purposes, probably because it is the most studied promoter thus far. In wild type E. coli cells the lac-promoter directs the expression of the lac-operon whose first product is β-galactosidase. The wild type lac-promoter has two modes of regulation; it is negatively controlled by the lac-repressor (39,40) and it needs the cyclic-AMP receptor protein for full activity (39,40). The mutant lac-UV5 promoter discussed here does not need this stimulating protein and we will not further discuss this positive mode of regulation. The negative control is exerted by the lac-repressor that binds to the lac-operator. This operator is a 22 bp DNA sequence which is located 6 bp downstream of the lac-Pribnow box (Figure 8). The transcriptional start site of the lac-UV5 promoter is at the beginning of the lac-operator (41). Thus, the lac-repressor prevents the RNA polymerase from binding to the promoter through physical hindrance (Figure 4, 8). Induction occurs after addition of an inducer. The natural inducer is a lactose derivative (allolactose). In the laboratory the synthetically made lactose analog isopropyl-β-D-thiogalactoside (IPTG) which is metabolically stable is routinely used as inducer of the lac-promoter (40). Thus this promoter can be repressed and induced by the lac-repressor and IPTG respectively. In normal cells a few dozen lac-repressor molecules are sufficient to repress a single lac-promoter (42,44).

Table I. Relative Strength of Promoters.

Promoter	Galactokinase units in pKM-1	Galactokinase units in pKO-1	Galactokinase units/mg protein in single copy lysogen	Relative Ratio
lac–UV5	67	--	--	1.0
trp	228	--	--	3.4
tacI	796	--	--	11.8
tacII	472	--	--	7.0
λPL	--	--	--	10–15
lpp	--	--	--	--
lac	--	520	--	1.0
gal	--	680	--	1.3
gal (λgal104)	--	--	1.5	1.0
rrnE (λrrn–gal 1) at μ = 0.5	--	--	7.5	5.0
at μ = 1.0	--	--	15.0	10.0
at μ = 2.4	--	--	(35)	(23.0)

The relative promoter efficiencies are reflected in the relative galactokinase levels. For this purpose the promoters listed here were inserted in pKM-1 (first section), pKO-1 (second section) or fused to the galactokinase gene in a λ lysogen (third section). In section I the galactokinase level of the lac–UV5 promoter is taken as reference and the trp, tacI, tacII (this paper), and λP$_L$ (37) promoters are compared with the lac–UV5 promoter. In section II the lac and the gal promoter are compared in the promoter probe vector pKO (Refs. 35,36; no data using pKM are available). In the third section the rrn promoter is compared with the gal promoter in a λ lysogen. The data in the third section are taken from Ref. 57. The galactokinase level of the rrn–promoter at μ = 2.4 is obtained from linear extrapolation and not a measurement from the graph and is therefore shown in brackets. The galactokinase units are expressed as nanomoles of galactose phosphorylated per minute per ml of cells at OD$_{650}$ = 1.0.

However, when the lac–promoter/operator is present on a high (greater than 30) copy number plasmid (such as pBR322 or any of its derivatives in E. coli), these lac–repressor molecules are titrated out (44) by the abundance of lac–operators. Consequently, derepression occurs even in the absence of inducer (43–46). Mutant strains have been isolated (47) that have a mutation in the –35 region of the laci promoter (26) leading to an overproduction of lac–repressor molecules. E. coli D1210laci^q is an example of such a mutant host. In such laci^q strains the lac–promoter is repressed even when present on a high copy number plasmid, and induction can occur only after addition of an inducer. An experiment illustrating this system will be discussed below (Fig. 11). Colonies of cells that contain β-galactosidase (ie., derepressed lac operons) turn blue on indicator plates containing X-gal (40). In wild type cells the chromosomal lac-operon is repressed hence the colonies appear white on X-gal plates. However, when a plasmid containing a lac-promoter/operator is introduced into wild type cells the chromosomal lac-operon is derepressed due to titration of the lac-repressor. Therefore, such colonies are blue on X-gal plates.

The trp-promoter: Stronger than the lac-UV5 promoter but less easy to control.

The trp-promoter has been used successfully to express foreign genes at very high levels. The trp-promoter is about 4 times stronger than the lac-UV5 promoter (Tables I and II; see below). The trp-promoter is regulated by the trp repressor and by a so-called attenuator (48,49). The attenuator is located downstream of the trp-promoter between the start site of transcription and the translational start site of the first gene of the trp-operon (trpE). Its function is to terminate prematurely a considerable fraction of the transcripts that have started at the promoter (48). The proportion of terminated transcripts varies inversely with the intracellular tryptophan concentration, but attenuation of all transcripts cannot be achieved. In most cases where the trp-promoter has been used to express foreign genes a DNA fragment is used from which this attenuator has been deleted (38) because more efficient expression of genes downstream from the trp promoter is observed in constructions that have such deletions. The attenuator will not be discussed here any further (for a recent review see Ref. 48). The trp repressor binding site, the trp-operator, overlaps with the trp-Pribnow-box (Fig. 8 and Refs. 48,49) thus preventing RNA-polymerase binding. However, the trp-repressor by itself cannot bind to the operator. It needs a so-called aporepressor which is the ultimate product of the trp-operon, namely the amino acid tryptophan (48,49). Thus, the binary complex of the trp-repressor and tryptophan binds to the trp-operator. In order to achieve repression of the trp-promoter, tryptophan has to be added to the growth medium. On the other hand, to derepress the trp-promoter, tryptophan must be removed from the growth medium. In practical terms this means that in order to control the trp promoter an amount of tryptophan must be added to the growth medium which is sufficient to repress the promoter but small enough to be depleted from the medium as it is consumed by the cells for protein synthesis. Full derepression of the trp-promoter cannot be achieved in this way since endogenously synthesized tryptophan inhibits the trp-promoter to a considerable extent (see Table II and Ref. 38). The promoter is fully active when introduced in cells lacking an active trp-repressor (eg. HDB2, see Table II). Clearly, controlled expression using the trp-promoter is much harder to achieve than controlled expression using the lac-UV5 promoter. For this reason we (38,45,46)

```
                                         -35 region            -10 region
Consensus:                    5'        TTGACA.......17 bp......TATAAT

                                             HpaII              +1                           EcoRI
PlacUV5       5' CCAGGCTTTACACTTTATGCTTCCGGCTCGTATAATGTGTGGAATTGTGAGCGGATAACAATTTCACACAGGAAACAGAATTCTATG....HGH

                                        TaqI   HpaI            +1              XbaI  EcoRI
Ptrp          5' GAGCTGTTGACAATTAATCATCGAACTAGTTAACTAGTACGCAAGTTCACGTAAAAAGGGTATCTAGAATTCTATG....HGH

                                                              +1                           EcoRI
PtacI         5' GAGCTGTTGACAATTAATCAT  CGGCTCGTATAATGTGTGGAATTGTGAGCGGATAACAATTTCACACAGGAAACAGAATTCTATG....HGH

                                        TaqI                  +1            HindIII  XbaI  EcoRI
PtacII        5' GAGCTGTTGACAATTAATCATCGAACTAGTT TAATGTGTGGAATTGTGAGCGGATAACAATTAAGCTTAGGATCTAGAATTCTATG....HGH

                                                              +1
λPL    5'  TAACCATCTGCGGTGATAAATTATCTCTGGCGGTGTTGACATAAATACCACTGGCGGTGATACTGAGCACATCAG.........130 bp.........GGATTAGCTGCCAATG....N

                                                              +1
Plpp          CCATCAAAAAAATATTCTCAACATAAAAAACTTTGTGTAATACTTGTAACGCTACATGGAGATTAACTCAATCTAGAGGGTATTAATAATG....lpp

                                        HpaII                 +1
rrn8-P1  TGGTTGAATGTTGCGCGGTCAGAAAATTATTTTAAATTTCCTCTTGTCAGGCCGGAATAACTCCCTATAATGCGCCACCACTGACACGGAACAACGGCAAACACGCCGCCGGGTCAGCGG

                                                              +1
rrn8-P2  GGTTCTCCTGAGAACTCCGGCAGAGAAAGCAAAAATAAATGCTTGACTCTGTAGCGGGAAGGCGTATTATGCACACCCCGCGCCGCTGAGAAAAAGCGAAGCGGCACTGCTCTTT

                                                              +1                           EcoRI
Prac 5-16     5' AAATTTCCTCTTGTCAGGCCGGAATCGGCTCGTATAATGTGTGGAATTGTGAGCGGATAACAATTTCACACAGGAAACAGAATTCTATG....HGH
```

Fig. 8. The DNA sequence of some promoters, and the location of their operators, that are useful for foreign gene expression.

The sequence of the upper strand (5' to 3') is shown of promoters that are used most frequently for foreign gene expression. Also shown are some promoters that are promising for that purpose. The tacI and tacII promoters are hybrids of the trp and the lac-UV5 promoter, and the rac5-16 promoter is a hybrid of the rrnB and the lac-UV5 promoter. The construction and properties of these hybrids are described in this paper and in Refs. 38,45,46. The λP_L promoter is the main leftward promoter of coliphage lambda, directing expression of the gene for the antiterminator protein N (33,50). Plpp is the promoter for the E. coli lipoprotein gene (51,52). In case of the λP_L and Plpp the sequence specifying the ribosome binding site of N-protein and lipoprotein is shown. The sequences of the entire tandem promoter, P_1 and P_2, of the rrnB operon are from Refs. 60-62. In the case of the lac-UV5 promoter the natural Shine Dalgarno sequence of the lac-messenger RNA is shown. This Shine Dalgarno region was fused to the human growth hormone gene (HGH) after conversion of the AluI site (5' AGCT) into an EcoRI site (5' GAATTC) using linkers as described previously (38,65). Similarly, the trp promoter sequence is followed by the Shine-Dalgarno sequence of the trp-leader mRNA. The TaqI site in this region (5' TCGA) was converted into and XbaI and EcoRI site (5' TCTAGAATTC) as described previously (38). The ribosome binding site following the tacI and the rac5-16 promoter is identical to that shown for the lac-UV5 promoter. The ribosome binding site following the tacII promoter is derived from a synthetic DNA sequence. Note that in this case a portable Shine-Dalgarno region is present. This region is flanked by unique HindIII, XbaI and EcoRI restriction sites. Therefore, this segment can be excised easily and replaced by another synthetic fragment encoding a different ribosome binding site. The -10 and -35 consensus sequences in the promoters are underlined. Operator regions are indicated with a broken line beneath the sequence. The Shine-Dalgarno sequence and the start codon is overlined. Relevant restriction sites that are important for fusion to a foreign gene and those used to construct hybrid promoters are also shown.

constructed hybrid promoters that have the combined advantages of the trp and the lac-UV5 promoter. These hybrid promoters, tacI and tacII, are controlled in the same way as the lac-UV5 promoter (see below). They are, surprisingly, not only much stronger than the lac-UV5 promoter but also two to three times stronger than the fully derepressed trp-promoter. The hybrid promoters were designed in such a way that neither one of them is regulated by the trp-repressor and both are therefore insensitive to the intracellular tryptophan concentration.

The λ-P$_L$ promoter, strong and under certain conditions heat inducible.

The λP$_L$ promoter is one of the main promoters of coliphage lambda (λ). This promoter is required for lytic infection (reviewed in Refs. 33 and 50). In a lysogenic bacterium the phage is integrated into the host chromosome and the P$_L$ promoter is repressed by a repressor coded for by the phage cI gene. Mutants in this cI gene have been obtained that code for a repressor which is temperature sensitive (cI857). At 30°C the repressor is able to block transcription from the P$_L$ promoter whereas at 40°C the repressor destabilizes and is no longer able to block the P$_L$ promoter. The λ-repressor prevents transcription from λP$_L$ in a somewhat complicated manner. The mechanism has been elucidated to a large extent and it has been described in an excellent review (33). Whereas the trp-repressor binds to the −10 area and the lac-repressor binds to a region following the start site of transcription, the λ repressor binds to regions upstream of the −10 region. It binds to three distinct regions of about 17 nucleotides long. These three regions are separated by spacers of 3 to 7 bases (see fig. 8).

The efficiency of the λP$_L$ promoter (Table I) is about the same as that of the tacI (37).

The lipoprotein promoter.

Recently the use of the lipoprotein (lpp) promoter for foreign gene expression has been described (51,52). Lipoprotein, an outer membrane protein, is numerically the most abundant protein in E. coli; a single copy of the lpp-gene is responsible for the maintenance of about 750,000 lipoprotein molecules per cell which is equivalent to about 1 percent of the total cell protein (51,52). In other words, per doubling time 750,000 lipoprotein molecules are synthesized. At a doubling time of 25 min. the synthesis rate of lipoprotein from a single copy on the chromosome is about 30,000 protein molecules per minute per cell. If such an expression system is inserted on a high copy number plasmid with 30 copies per cell a theoretical number of lipoprotein molecules per cell of about 22×10^6 can be predicted which is equivalent to 30 percent of total cell protein. The molecular weight of lipoprotein is 7,000D (51,52). For proteins like human growth hormone or the interferons with a molecular weight of about 20,000D the predicted maximal expression level is consequently 60 percent of total cellular protein. Constitutive expression of a gene directed by the lipoprotein translation and transcription signals on a high copy number plasmid will possibly inhibit cell growth and is likely to be lethal. A peculiarity of the lpp gene is that its mRNA is unusually stable; its half life is approximately 12 min (51,52), whereas the average half-life of the E. coli messengers is approximately 3 minutes. At this moment it has not yet been established whether the high yield of the lpp gene is due to efficient transcription, efficient translation, the stability of the messenger RNA, or a combination of these parameters (51,52).

The efficiency of the lipoprotein promoter itself has not been measured. It is believed to be very strong (52). The regulation of this promoter has not yet been elucidated. The utilization of the lipoprotein translational and transcriptional signals is promising, but since such an assembly is likely lethal, an operator must be attached to the promoter. The construction of a hybrid between the lipoprotein promoter and the lac-UV5 promoter, analogous to the construction of the tac promoters (see below) may be useful. Alternatively, if the mRNA translation initiation signals (52) of the lipoprotein operon appear to be very strong, then a fusion of these protein initiation signals to an existing inducible promoter may be very fruitful. It is also interesting to analyze the features that make the lpp-messenger stable and to use the gained knowledge to stabilize the mRNAs of foreign genes in order to enhance their expression levels.

The ribosomal RNA promoter: A superpromoter?

The ribosomal RNA (Prrn) promoter directs the transcription of the ribosomal RNA genes. This RNA is not translated but it is packaged into ribosomes. Ribosomal RNA is completely stable under normal physiological conditions. Therefore, the amount and rate of ribosomal RNA synthesis can be calculated accurately. Such calculations show that the RNA chain initiation occurs at a frequency that is limited by RNA chain elongation rate. In other words these rrn promoters have a maximal efficiency of initiation (53).

In normal cells ribosomal RNA synthesis is controlled by the growth rate of the cells and by the availability of amino acids (53-57). The underlying molecular mechanism is not known (53-58). One peculiar feature of the ribosomal RNA promoter is that two active promoters in a tandem arrangement precede each rRNA gene (58-63). It is not known which regions in the rrn-promoter are responsible for its high efficiency. Neither is it known what the relative contribution of each of the two promoters is to the total strength of the tandem promoter (59).

In a recent paper the growth rate dependence of the rRNA promoter activity has been shown in a system where the rrn-promoter was fused to the galactokinase gene and inserted as a lysogen into the host chromosome (57). From these data the relative activity of the tandem rrn-promoter with respect to the galactokinase promoter (which is about as strong as the lac-UV5 promoter) at various growth rates can be deduced (see Table I). At a doubling time of 120 min. and 60. min the activity is about 5 and 10 times respectively higher than that of the gal-promoter. Since the rRNA synthesis rate is shown to be proportional to the growth rate (54,57) the maximal efficiency at the highest growth rate of a doubling time of 30 min. formally can be extrapolated and we estimate it to be about 20 times higher than that of the gal-promoter. However, this number has not yet been verified experimentally. The use of the rRNA promoter to direct protein synthesis is promising if it can be repressed and induced effectively. The following gives an indication of the theoretical maximal level of a protein in the cell if the transcription of a foreign gene is directed by a ribosomal RNA promoter. At a high growth rate of 2 doublings per hour at least 50 percent of the total RNA synthesized is ribosomal RNA (54,55). E. coli contains 7 rRNA operons per genome (54,56) which means that about 7 percent of total RNA synthesis occurs at each rRNA operon. If a tandem rRNA promoter on a single copy plasmid is fused to a foreign gene such that it directs the synthesis of a protein, then about 12 percent of total mRNA, hence of total protein synthesis, will be mRNA coding for the foreign gene product. Such a fusion is, when inserted into a high copy number plasmid, a lethal situation to the cells. Thus, to utilize the potential of these rrn-promoters they must be converted into inducible promoters.

It should be realized, however, that it is quite possible that the activity of the rRNA promoters, or any other promoter, when inserted in a plasmid is lower than the activity of this promoter when present in its natural location in the chromosome. Also the presence of an excess of these promoters in the cell might diminish the activity of each individual promoter due to shortage of a limiting component in the cell (for example, RNA-polymerase). Several basic questions concerning the rrn-promoters have to be answered prior to developing a sensible strategy to employ these promoters for expression of foreign genes.

Below we will describe the construction of a hybrid between one of the ribosomal RNA promoters and the lac-UV5 promoter (the rac-promoter) analogous to the hybrid promoters tacI and tacII.

HYBRID PROMOTERS

The following section describes in some detail the construction of three different inducible hybrid promoters. The actual construction schemes are shown in order to introduce some procedures routinely used for optimizing the expression of foreign genes and to describe some plasmids that are useful for this purpose. These experiments also demonstrate some interesting applications of the use of synthetic DNA fragments in the construction of regulatory sequences. The three hybrid promoters to be described all contain the lac-operator sequence and therefore can be repressed in lac-repressor overproducing strains by the lac-repressor and they can be induced by isopropyl-β-D-thio-galactoside (IPTG).

The first hybrid promoter (called tacI) is made of natural sequences derived from the trp and the lac-UV5 promoter joined at position -20 with respect to the transcription initiation site. The second hybrid promoter (called tacII) is made of a natural trp-promoter fragment ending within the trp-Pribnow-box which is joined to a synthetic DNA fragment. Thus a hybrid Pribnow-box is formed. It is followed by the lac-operator and an area that codes for a Shine-Dalgarno sequence (see next section) which is surrounded by two unique restriction sites.

The third promoter (called rac5-16) contains the natural sequences of the -35 area derived from a ribosomal RNA promoter (rrnB) and a synthetic DNA fragment that joins the -35 area of the PrrnB part to the lac-UV5 Pribnow-box and operator. This synthetic DNA fragment contains three unique restriction sites which makes it possible to generate various deletion mutants in the area between the Pribnow-box and the -35 sequence.

Construction of tacI and tacII

In this section we will describe the construction of the hybrid trp-lac promoters tacI and tacII. Both hybrid promoters contain the -35 area of the trp-promoter and both contain the lac-operator.

The construction of tacI is shown in Figure 9. The trp-promoter was derived from pHGH207-1 (see Fig. 9) whose construction has been described elsewhere (38). This plasmid contains the entire trp-promoter on a 310 bp EcoRI fragment. This fragment also contains the Shine-Dalgarno (64) area of the trp-leader mRNA. The start codon for initiation of translation is provided by the adjacent human growth hormone gene. The DNA sequence of the sense strand at this junction is:

```
                   XbaI     EcoRI
        S.D.                        start
    5' - G G T A T C T A G A A T T C T A T G ... (HGH)
```

The end of the HGH gene is joined to the HindIII site of the tetracycline resistance gene (tet-gene) of pBR322. The HindIII site and the tet-promoter are destroyed in this process (38,65) but the protein initiation signals on the mRNA of the tet-gene are still intact. Hence the expression of the tetracycline-resistance gene is under direction of the trp-promoter. The 310 bp EcoRI fragment of pHGH207-1 was partially digested with TaqI and the indicated 240 bp fragment was isolated. This fragment contains the -35 sequence of the trp-promoter and it ends at the TaqI site 21 bp upstream of the transcription start-site.

The Pribnow-box of the lac-promoter and the lac-operator sequences were derived from pHGH107-11 (Fig. 9). The structure of this plasmid, except the 90 bp EcoRI fragment that contains the lac-UV5 promoter and the Shine-Dalgarno area of the lac-mRNA, is identical to that of pHGH207-1. Plasmid pHGH107-11 has a single lac-promoter fused to the HGH-gene. It is derived as described previously (38) from pHGH107 (65) which has two lac-promoters in a row. The sequence at the junction of the lac-promoter/operator and the HGH gene is:

```
                   EcoRI
        S.D.                        start
    A G G A A A C A G A A T T C T A T G ... (HGH)
```

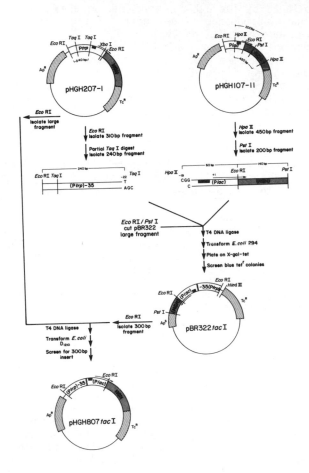

Fig. 9. Construction scheme of pHGH807tacI containing the trp-lac hybrid promoter tacI. The trp and the lac portion of tacI is indicated between brackets as Ptrp and Plac, respectively; their junction is indicated with a vertical wavy line. The location of the −35 consensus sequence is indicated as −35. The Pribnow box is shown as a black box. The direction of transcription is indicated with an arrow. The human growth hormone gene is indicated with HGH.

From pHGH107-11 the indicated 200 bp HpaII PstI fragment was isolated. This HpaII site is at position –19 with respect to the transcriptional start of the lac-UV5 promoter. The 240 bp trp-fragment, the 200 bp lac-fragment and the large EcoRI-PstI fragment of pBR322 were covalently joined with T4 ligase. After transformation of E. coli 294 and plating on X-gal (5-bromo-4-chloro-3-indolyl-β-D-galactoside) tetracycline plates, plasmid DNA from blue, tetracycline resistant colonies was analyzed. Thus pBR322 tacI was obtained. Its structure is shown in Figure 9. From pBR322 tacI the 300 bp EcoRI fragment that contains the newly constructed tacI promoter was purified and inserted into the large EcoRI fragment of pHGH207-1. After transformation of E. coliD$_{1210}$laci^q the plasmid pHGH807tacI was obtained. In this expression plasmid the tacI containing fragment provides for the Shine-Dalgarno sequence allowing protein synthesis to start at the ATG of the adjacent HGH gene. The DNA sequence of the tacI promoter/operator and ribosome binding site is shown in Figure 8.

The EcoRI fragments that contain the trp and the tacI promoter also code for the Shine-Dalgarno area of the ribosome binding site. For a systematic analysis of ribosome binding sites in the future it would be very useful to be able to vary the nucleotide sequence in the ribosome binding site area. Therefore, a portable Shine-Dalgarno element was included in the design of the synthetic part of the second hybrid promoter called tacII. The construction of pHGH907tacII is shown in Figure 10. In the trp-promoter, the trp-repressor binding site includes the Pribnow-box which in this case also contains a HpaI site (see figure 8 and ref. 49). Sequences downstream of this HpaI site were replaced by a synthetic

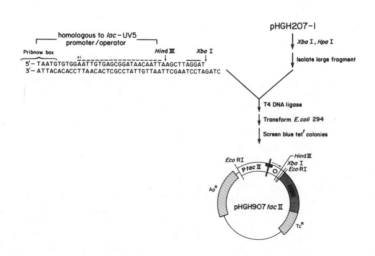

Fig. 10. Construction scheme of pHGH907tacII containing the half synthetic trp-lac hybrid promoter tacII and the portable Shine-Dalgarno area.

DNA fragment, thus destroying the trp-repressor binding site. Plasmid pHGH207-1 was opened with HpaI and XbaI and a synthetic DNA fragment of 46 bp, whose sequence is shown in figure 10, was inserted. The sequence of this synthetic fragment is identical to that found in the homologous area of the lac-UV5 promoter/operator. Thus the first 2 bases of the trp-Pribnow-box (TTAACTA) are joined to the last 5 bp of the lac-UV5 Pribnow-box (TATAATG) resulting in a hybrid Pribnow-box (TTTAATG). Downstream of this hybrid Pribnow-box is a 5 nucleotide sequence which is identical to that found in the lac-UV5 promoter/operator. This 5 base long sequence is the left arm of the region of symmetry that flanks the core of the natural lac-operator (39). It is followed by nucleotides that specify the core of the lac-operator. The lac-operator is followed by a HindIII site and a Shine-Dalgarno sequence with a 4 basepair homology with 16s ribosomal RNA (64). The synthetic fragment ends with an XbaI site by which it is fused to the HGH gene. Thus, the Shine-Dalgarno area is flanked by two different restriction sites which are unique to the plasmid and will allow it to be used in alternative expression systems. It should be noted that the tacII promoter/operator lacks the right arm of the region of symmetry that flanks the natural lac-operator.

Measurement of the Relative Efficiency of the Natural and the Hybrid Promoters

All the natural and hybrid promoters described above are contained on EcoRI fragments which also encode a Shine-Dalgarno region but which lack a start codon. Because the DNA sequences downstream of the position -20 of the PtacI promoter are identical to those of the parental lac-UV5 promoter, it is likely that transcription starts in the same region as that in the lac-UV5 promoter (41). The length and nucleotide sequence of the 5' untranslated region of each mRNA is likely to be the same. The 5'-untranslated region contains the Shine-Dalgarno area, which precedes the start codon of the HGH gene in the same way in each case. Therefore, the entire ribosome binding site is similar on the mRNA of the tac and the lac-UV5 promoter. Thus, the initial rate of HGH accumulation after induction of these promoters in cells of E. coliD1210laci^q must reflect the relative efficiencies of transcription. In the experiment of Figure 11, cells of E. coliD1210laci^q containing pHGH807tacI or pHGH107-11 were grown in M9 medium supplemented with 0.2 percent casamino-acids and induced with 1.0 mM IPTG. Samples were taken before and after induction and the HGH level was determined using a standard radioimmune assay (65). The HGH concentration per optical density unit of cells was calculated and plotted versus time. Figure 11 shows that the initial rate of HGH accumulation after induction of the tacI promoter is about ten-fold higher than that of the induced lac-UV5 promoter. This means that the tacI promoter is about ten times as efficient as the lac-UV5 promoter. Before induction the HGH level in repressed cells with the tacI promoter is significantly higher than that in repressed cells with the lac-UV5 promoter. This probably reflects the difference in promoter strength.

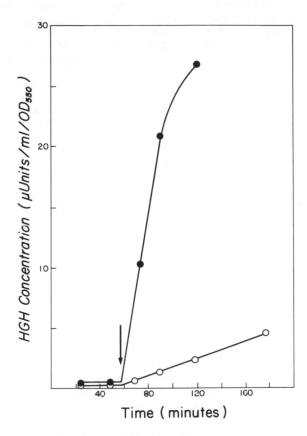

Fig. 11. HGH production directed by the lac-UV5 and the tacI promoter. Overnight cultures were used to inoculate 50 ml LB-ampicillin (20 μg/ml) to a cell-density of 0.03 (OD$_{550}$). After one hour the first sample was taken (t=40 min). Induction was done by addition of 1.0 mM IPTG at t=80 minutes. HGH levels were determined in a radioimmunoassay as outlined previously (88). Symbols: O-O: E. coli D1210/pHGH107-11 (single lac-promoter);●--● : E. coli D1210/pHGH 807 tacI (tacI-promoter).

The tacI and the lac-UV5 promoter could be compared by measuring the expression of the HGH gene since their ribosome binding sites are identical. However, the relative efficiency of these promoters cannot be compared in this way with the trp and the tacII promoter since they are contained on EcoRI fragments that code for different ribosome binding sites. In order to measure

reliably relative efficiencies of any promoter we have used the
plasmid system which has been designed by Rosenberg and coworkers
(35,36) to circumvent such complications.

From the plasmids pHGH107-11, pHGH207-1, pHGH807tacI (Figure
9), pHGH907tacII (Figure 10) the small EcoRI fragments containing
the lac, trp, tacI and the tacII promoter respectively, were
isolated and inserted into pKM-1 (see Figure 7). Each plasmid was
introduced in E. coli C600 galK⁻ and E. coli HDB2
galK⁻trpR⁻. Cells were grown in minimal medium containing 0.5
percent tryptophan-free casamino acids and 0.2 percent fructose.
The galactokinase levels were determined as described (35).

The lac-promoter and the hybrid promoters on pKM-1 are
derepressed since the lac-repressor in the host cells (E. coli
C600) is titrated by the abundance of lac-operator sequences on the
high copy number plasmids (38,44). In addition, any residual
lac-repressor activity was counteracted by the addition of 1 mM
IPTG to the growth medium. The trp-promoter on these plasmids is
derepressed due to depletion of any residual tryptophan in the
medium. The galactokinase levels in cells with pKM-1 harboring the
various promoters are shown in Table II (left panel). The
galactokinase level in C600/pKM-lac is about 67 units/ml/OD. The
galactokinase level in C600/pKM-trp is approximately 2.2 times
higher than that in C600/pKM-lacUV5, implying that the trp-promoter
under these conditions is about two times more active than the
lac-UV5 promoter.

The host used in this assay (E. coli C600) has a functional
trp-repressor. Previously we found that the HGH levels in hosts
that have a functional trp-repressor (trpR⁺) are lower than those
in trpR⁻ hosts containing the plasmid pHGH207-1 (unpublished
observation). It is likely that endogenously synthesized
tryptophan causes some repression of the trp-promoter. Hence, the
relative trp-promoter strength as measured in C600/pKMtrp is
underestimated. To determine the relative trp-promoter strength
accurately a derivative of E. coli C600 galK⁻ was constructed
that lacks a functional trp-repressor (HDB2 galK⁻trpR⁻). The
pKM-1 derived plasmids containing the various promoters were
introduced into HDB2. In order to confirm once more that HDB2
indeed lacks a functional trp-repressor we measured the
galactokinase levels in C600/pKM-trp and HDB2/pKM-trp in the
absence and in the presence of 10 µg/ml tryptophan in the growth
medium. In C600/pKM-trp the galactokinase level had dropped to 2
percent of the level measured in the absence of exogenously added
tryptophan. In contrast, in HDB2 the trp-promoter activity was
barely affected by 10 µg/ml tryptophan (93 percent residual
level). These observations clearly confirm the absence of
functional trp-repressor in HDB2.

The data in Table II (left panel) show that the actual
trp-promoter strength as measured in HDB2/pKMtrp is at least three
times higher than that of the lac-UV5 promoter. Table II (right
panel) also shows that the galactokinase level in HDB2/pKM-trp is
about 1.6 times higher than that in C600/pKMtrp. This increased
trp-promoter activity must be due to the absence of functional
trp-repressor in HDB2. No significant difference in promoter

Table II   Comparison of the relative strength of the natural and the hybrid promoters in pKM-1.

Host	Plasmid/ Promoter	Galactokinase Units	Ratio Relative to pKM-lac	Relative Promoter Activity		Galactokinase activity in HD82 / Galactokinase activity in C600	
C600 trpR+	pKM-lac	67	1.0	$\frac{tacI}{tacII}$	1.7 (0.2)	lac	1.1 (0.1)
C600 trpR+	pKM-trp	144	2.2 (0.2)	$\frac{tacI}{trp}$	3.5 (0.4)	trp	1.6 (0.1)
C600 trpR+	pKM-tacI	796	11.8 (1.5)	$\frac{tacII}{trp}$	2.1 (0.3)	tacI	0.9 (0.2)
C600 trpR+	pKM-tacII	472	7.0 (0.9)			tacII	1.2 (0.1)
HD82 trpR-	pKM-trp	228	3.4 (0.5)				

Cells were grown in M9 minimal medium supplemented with 0.5 percent casamino acids and 0.2 percent fructose to an OD_{650} of 0.6 and assayed for galactokinase activity as described (22). The galactokinase units (left panel) are expressed as nanomoles of galactose phosphorylated per minute per ml of cells at OD_{650} = 1.0. For each experiment two appropriate dilutions of the cultures were made and the galactokinase activity was determined in triplicate. The average value was taken of the dilutions of which the total counts incorporated in galactose phosphate were less than 25 percent of the input counts. The results shown are the average values of 8 or more independent experiments. The standard error of the mean of all experiments is indicated between brackets. In the middle panel the relative strength of the hybrid promoters and the trp promoter are compared. Between brackets is shown the standard error of the mean of 8 such ratios of 8 independent experiments. The right panel shows the ratio of the galactokinase activity in HD82 and C600 harboring the various promoters in pKM-1. The mean value of 5 independent measurements are shown. The numbers between brackets represent the standard error of the mean.

activity is observed in both hosts with the lac-UV5 promoter (right panel).

The relative efficiencies of the hybrid promoters tacI and tacII were measured in a similar fashion in pKM-1 in both hosts. The tacI promoter appeared to be at least 10 times more efficient than the lac-UV5 promoter and at least three times as strong as the fully derepressed trp-promoter. The tacII promoter is about seven times stronger than the lac-UV5 promoter, i.e. the tacI promoter is 1.5 times stronger than the tacII promoter (middle panel). The activity of the tacI and tacII promoter is not affected by the presence or absence of the trp-repressor (right panel) which is consistent with the absence of an intact trp-repressor binding site in both hybrid promoters.

Hybrid promoters are stronger than the parental promoters.

The trp-promoter in the trpR- host is approximately three times stronger than the lac-UV5 promoter. The tacI and the tacII promoters are about eleven and seven times, respectively, stronger than the lac-UV5 promoter. Thus the hybrid promoters are more efficient than either one of the parental promoters. The differences between the hybrid promoters and each one of the parental promoters are too extensive to attribute the cause of this increased activity to one particular aspect of the hybrid promoter sequence.

There are several features in the DNA sequence of a promoter that affect its efficiency: The nucleotide sequence of the Pribnow-box (20,21,35); the nucleotide sequence of the -35 area (20,21,35); the distance between these two domains (29,30); and possibly the AT-richness of areas upstream of the -35 sequence (51,52,58,63).

The consensus sequence for the Pribnow-box is 5'-TATAAT and that for the -35 sequence is 5'-TTGACA. The distance between the Pribnow-box and the -35 area is in most cases 17 bp (21); a distance of 18 bp or 16 bp occurs far less frequently and the effect on promoter efficiency of various mutants in which this distance has been changed suggests that the consensus distance is 17 bp and that this distance is possibly optimal for maximal promoter activity. The highly efficient ribosomal RNA promoters that all have a distance of 16 bp seem to be exceptional (21,59-63).

It seems likely that a promoter with a consensus -35 and a consensus -10 region and a distance of 17 bp between these domains is efficient. The hybrid promoters described here confirm this apparent rule as summarized in Table III. The lac-UV5 promoter has a consensus Pribnow-box (TATAAT) but no consensus -35 sequence (TTTACA). The trp-promoter does not have a consensus Pribnow-box sequence (TTAACT) but it does have a consensus -35 sequence (TTGACA). Consequently, the tacI-promoter not only has a consensus -35 sequence (TTGACA) but also has a consensus Pribnow-box sequence (TATAAT). In tacI the distance between both domains is 16 bp which, by comparison with the mutant β-lactamase promoter (31) and the mutant lac p^S promoter (30), may be suboptimal.

TABLE III. Comparison of the DNA sequences in the -10 and in
the -35 area of the various promoters with the
consensus sequence.

	-35 region	distance in basepairs	-10 region
Consensus	TTGACA	17	TATAAT
lac-UV5	TTTACA	18	TATAAT
trp	TTGACA	17	TTAACT
tacI	TTGACA	16	TATAAT
tacII	TTGACA	17	TTTAAT
rac5-16	TTGTCA	16	TATAAT
rrnB-P1	TTGTCA	16	TATAAT

The tacII promoter has a consensus -35 sequence 5'-TTGACA but no consensus Pribnow-box sequence (TTTAAT). Although an A residue is highly preferred at the second position in the Pribnow-box a T residue at this position is the second most preferred base (21,25). The distance between both domains in tacII is 17 bp. It is possible that, in this promoter, the optimal distance compensates for a suboptimal Pribnow-box sequence, resulting in a promoter with strength comparable to tacI. It should be noted that these deviations from the consensus sequence as they occur in the hybrid promoters may be much less important for promoter activity and that the increased efficiency of the hybrid promoter compared to the parental lac–UV5 promoter is merely due to an optimization of the distance between both domains from 18 bp in the parental lac–UV5 promoter to 17 bp and 16 bp in the tacII and tacI promoter respectively. We will describe below a hybrid of the ribosomal RNA promoter and the lac–UV5 promoter which contains unique restriction sites in the –20 area. This will allow us to vary the distance between the –35 area and the –10 area to any desired length in order to study the effect of spacer length on promoter efficiency.

The rac Promoter: A Hybrid of the Ribosomal–RNA Promoter and the lac–UV5 Promoter

We argued earlier that the ribosomal RNA promoters are among the most efficient promoters of E. coli (53–55) and that each rrn–operon sequenced thus far is transcribed by a tandem promoter (58–63). Each of these two promoters are located about 200 and 300 basepairs upstream of the sequences that code for mature 16s rRNA (Fig. 8). It has been shown that both promoters are active in vitro (66) and in vivo (59) but it is not known what the relative contribution is of the first (P_1) and the second (P_2) promoter to the overall efficiency (59). Neither is it known which features in the ribosomal RNA promoters give them high efficiency. Since the substitution of the sequences upstream of position –20 of the lac–UV5 promoter by homologous sequences of the stronger trp–promoter greatly enhance the efficiency of the promoter we designed a similar hybrid of the rrnB and the lac–UV5 promoter.

The construction of this hybrid promoter (the rac promoter) is shown in Figure 12. Because we thought that the hybrid promoter might be lethal when transformed into cells on a high copy number plasmid, the strategy was to construct first an inactive promoter and to activate it later. This plan involves the subcloning of the –35 area and the AT–rich domains upstream of the –35 area of the promoter of the rrnB operon from plasmid pKK3535 (60,62). This fragment ends at a HpaII site at position –28 with respect to the start site of transcription of the first ribosomal RNA promoter P_1 (83,85), and it begins at the AluI site at position –154 which was converted into an EcoRI site using an EcoRI linker. The lac–part is the same as described for the construction of the tacI promoter (see figure 9). Thus pBR322–Prrnlac was obtained (Fig. 12). The sequence of Prrnlac I is shown in Figure 13. The distance between the Pribnow–box and –35 area is 10 bp which makes the promoter inactive (data not shown). At the HpaII site at the junction between the rrn part and the lac part of PrrnlacI a synthetic DNA fragment was inserted according to a scheme which will be described elsewhere. This insert increases the length between the Pribnow–box and the –35 sequence from 10 bp to 31 bp (see Figures 12 and 13). This promoter (Prrn–lacII) is also inactive. Thus, the assembly of a potentially lethal promoter was avoided and after each ligation and transformation blue colonies on X–gal plates could be selected and screened. The synthetic insert contains three unique restriction sites, namely SacI, XhoI, and

XbaI. In order to activate the rrn-lacII promoter the plasmid (pHGH-Prrnlac) was digested with SacI and XbaI and the protruding ends were removed with the single strand specific nuclease S_1. The plasmid was reclosed with T_4 DNA ligase. At this stage the hybrid promoter was expected to be active. Therefore, cells that are lac-repressor overproducers ($D_{1210}lacI^q$) were transformed with this DNA. At this stage the activated promoter is repressed by the lac-repressor. In this activated promoter (Prac5-16) the distance between the -35 sequence and the Pribnow-box is 16 bp which is the same (Table III and Fig. 13) as that in the parental ribosomal RNA promoter. The rac-promoter can be induced with IPTG in a similar fashion as tacI (not shown). The activity of this rac5-16 promoter is about the same as that of the tacI promoter (data not shown). The growth rate dependency of this rac-promoter has not yet been established.

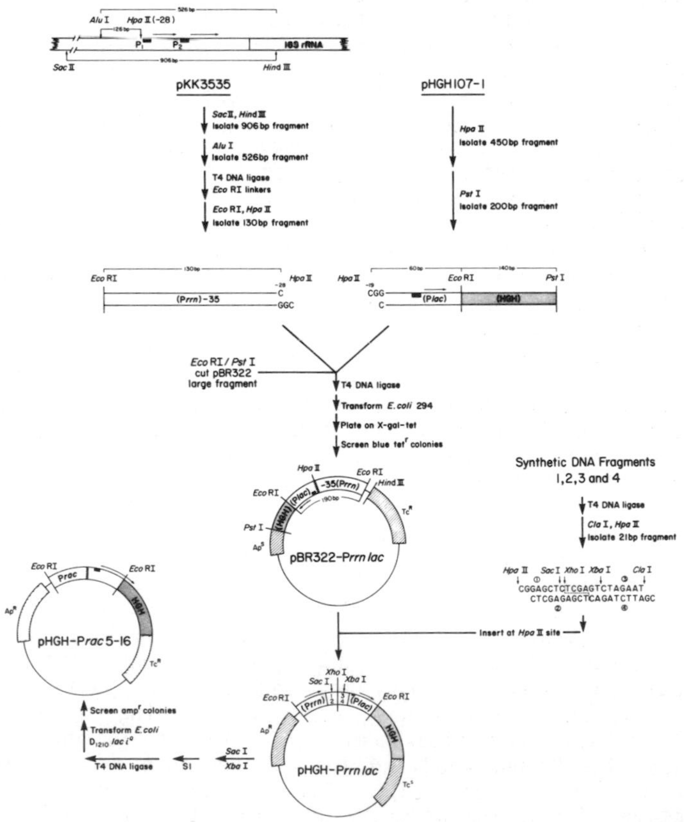

Fig. 12. Construction scheme of the rac5-16 promoter and its precursors PrrnlacI and PrrnlacII.

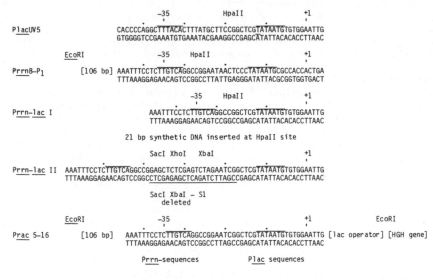

Fig. 13. The DNA sequence of the lac-UV5, the rrnB-P₁, the shortened (Prrn-lacI), the lengthened (Prrn-lacII) and the activated (rac5-16) hybrid promoter. The -35 sequence and the Pribnow-box sequence is overlined. The synthetic insert of Prrn-lacII is underlined. Its sequence is also shown in Figure 12. The transcription start site of Plac (41) and Prrn (59,66) or the nucleotide corresponding to that start site in the case of the Prrn-lacI, the Prrn-lacII and the Prac5-16 is indicated.

TRANSLATION OF CHIMERIC MESSENGER RNAs IN E. COLI

Summary of the important parameters for translation.

The initiation of translation in E. coli is a multicomponent process involving base-pairing between three different mRNA species (16s rRNA, mRNA and fmet-tRNA$^{met}_f$), the three initiation factors IF1, 2, and 3, and the ribosomal protein S1 (67). The efficiency with which the initiation complex forms, in general, regulates the rate of formation of product in E. coli (64,67,68). Of these components the most variable is the mRNA. This variability can be represented as the result of differences in the intracellular concentration of a particular messenger species, and in its sequence. There are several variables contained within the parameter of sequence variation, which will occupy the greatest part of this discussion. During the formation of the translational initiation complex a short sequence of 3-12 nucleotides (the Shine-Dalgarno (S/D) region; ref. 64,69) forms a duplex with the 3'-end of 16s rRNA. For some messengers it seems that the stability of the hybrid (and therefore the length of the complementary region) can be related to the frequency of initiation (67,69). Several other sequence related parameters are of importance. First, if the nucleotide sequence surrounding the Shine-Dalgarno (S/D) and/or the initiating AUG can form a hybrid with some sequence either 5' or 3' within the mRNA, then this duplex will inhibit or prevent the initiation process (68,70,71). In addition, the distance between the S/D sequence and the AUG initiator can be optimized to increase the frequency of translational initiation (71-73). One final parameter which has recently received considerable attention is the effect of the nucleotide which occurs on either side of the AUG initiation codon. In this respect, Ganoza et al. (74) have shown that a pyrimidine residue immediately preceding the AUG start may enhance initiation complex formation, while Taniguchi and Weissmann (75) have reported that a transition in the Qβ coat protein ribosome binding site that changes the sequence from AUGG to AUGA increases the yield of initiation complex in vitro.

The effects of spacing and sequence in the ribosome binding site.
We have explored most recently the effect of varying the distance between the Shine-Dalgarno sequence of the E. coli trp ribosome binding site and the AUG initiator of mature human leukocyte interferon (IFN-αA) and fibroblast interferon (IFN-β1; ref. 71). The original expression plasmids directing the synthesis of these polypeptides were constructed similarly to the IFN-γ expression plasmid discussed earlier. A diagram of each and a description of the protocol used to alter the spacing between the trp S/D (GGT) and the initiating ATG is shown (Figure 14). Using this approach we were able to alter the lengths of the ribosome binding sites of these genes stepwise (Table IV) from the natural trp length of 7 nucleotides (pLeIF A7) to a minimum of 2 nucleotides (pFIFtrp 29) and a maximum of 15 (pLeIF A15). What we found during these experiments was that in the case of plasmids directing the expression of IFN the natural trp spacer length of 7 nucleotides was less optimal for the expression of either IFN-α or IFN-β1 than a spacing of 9 nucleotides (Figure 15). Because the optimal spacing is identical for both plasmids, despite differences in the ribosome binding site sequence between the S/D sequence and the ATG start, the "spacer" region, as well as downstream from the site of translational initiation, it seems that spacing and not sequence is probably the predominant factor in determining the optimal efficiency of translation of these transcripts. Expression of IFN-α decreases linearly as the length of the spacer region is increased. Two parameters are being changed simultaneously in this experiment. The first is spacing and the second is nucleotide sequence. From the linear kinetics of the decrease in IFN-αA production as the spacing becomes longer, despite changes in spacer sequence, it seems that within the limits imposed by this experimental design the sequence composition of this spacer does

not affect translation. This is not the case for IFN-β₁ expression. In this example abrupt changes in the production of IFN-β₁ are observed with changes in the spacer length of one or two nucleotides (Table IV, Fig. 15; compare spacer lengths of 8, 9 and 10 nucleotides). Also, those plasmids encoding spacers of only two or four nucleotides synthesize very little or undetectable amounts of IFN-β₁.

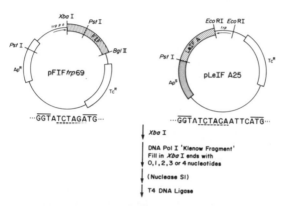

Series of expression plasmids with altered
spacing between GGT and ATG

Fig. 14.   Construction of IFN-β₁ and IFN-αA expression plasmids with altered ribosome binding sites.   Only the portion of the DNA sequence containing the Shine–Dalgarno sequence, spacer region, and initiating ATG is shown.   To the left of the SD sequence (GGT) of either plasmid is the trp promoter, and to the right of the ATG is the interferon coding sequence.   ApR and ApS indicate intact or partially deleted β-lactamase genes (respectively).   TcR designates tetracycline resistance.   The arrow at the top of each diagram indicates the direction of transcription of the E. coli tryptophan promoter–operator (trp p o) fragment used in these experiments.

TABLE IV. Spacer length variations in the trp-ribosome binding site.

Plasmid	Sequence	Spacing[a]	Change from Natural trp Leader	Relative Interferon Production
1. FIFtrp29	GGTATATG	2	-5	0.004
2. FIFtrp49	GGTATCGATG	4	-3	0.007
3. FIFtrp69	GGTATCTAGATG	6	-1	0.37
4. FIFtrp89	GGTATCTATAGATG	8	+1	0.49
5. FIFtrp99	GGTATCTACTAGATG	9	+2	1.00
6. FIFtrp109	GGTATCTAGCTAGATG	10	+3	0.16
7. LeIF A7	GGTATAATTCATG	7	0	0.44
8. LeIF A9	GGTATCGAATTCATG	9	+2	1.00
9. LeIF A25	GGTATCTAGAATTCATG	11	+4	0.70
10. LeIF A13	GGTATCTATAGAATTCATG	13	+6	0.37
11. LeIF A15	GGTATCTAGCTAGAATTCATG	15	+8	0.16

Table IV. Effect of varying the length of the trp leader ribosome binding site on expression of cloned IFN-β_1 and IFN-αA. The plasmids were constructed as described in Figure 3 from the expression plasmids pFIFtrp69 and pLeIF A25 (lines 3 and 9). Cell lysates were prepared using the Triton X-100 procedure (ref. 15). Interferon content was determined when the cultures reached a density of 3.5×10^8 cells/ml using the CPE inhibition assay. Escherichia coli 294/pFIFtrp99 contains about 2×10^4 molecules of mature IFN-β_1 per cell, and cells with the pLeIF A9 plasmid contain about 1×10^5 molecules per cell of mature IFN-αA. Results are given as the fraction of maximal expression.

[a]The distance between the end of the Shine-Dalgarno sequence and the AUG start of the interferon coding region. The natural spacing is 7 nucleotides in the trp leader peptide ribosome binding site.

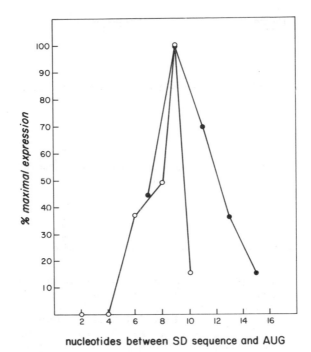

Fig. 15. Effect of spacing on efficiency of IFN synthesis. Assays were performed as described in the legend to Table IV. Data for IFN-β_1 (O) and IFN-αA (●) are expressed as the percent of maximal synthesis observed for the various spacings between the Shine-Dalgarno and AUG sequences.

Messenger RNA secondary structure.

Data obtained using different plasmids directing the synthesis of IFN-β_1 suggest that factors other than spacing (perhaps secondary structure) may be of importance. Figure 16 shows the results of a computer analysis of possible secondary structures formed by the ribosome binding site region of three of the IFN-β_1 expression plasmids (pFIFtrp 69 and pFIFtrp 89, which have secondary structures identical to pFIFtrp 99, are not shown). Even though pFIFtrp 99 directs the synthesis of considerably more IFN-β_1 than does pFIFtrp 69, the secondary structure in the ribosome binding site region of their respective mRNAs is identical. Therefore some factor other than secondary structure (probably spacing) must determine the differences between

242

Plasmid	Structure	ΔH

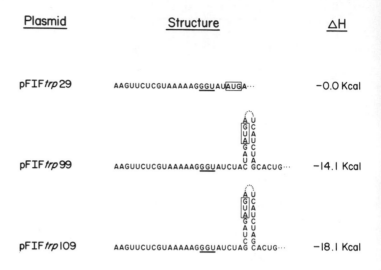

pFIF*trp*29   AAGUUCUCGUAAAAAGGGUAUAUGA···   −0.0 Kcal

pFIF*trp*99   AAGUUCUCGUAAAAAGGGUAUCUAC GCACUG···   −14.1 Kcal

pFIF*trp*109   AAGUUCUCGUAAAAAGGGUAUCUAG CACUG···   −18.1 Kcal

Fig. 16. Predicted secondary structures of trp-IFN fusion transcripts. The sequence shown includes the 5'-untranslated region from the E. coli trp operon, the trp leader Shine-Dalgarno sequence (GGU), and the spacer region (between the GGU and AUG sequences). The sequence continues into the remainder of the coding region for mature IFN-β1 (ref. 15). The loop at the top of each stem is 216 bases long. No stable stem structures can be drawn for pFIFtrp29 or 49. The stability (ΔH) of the stems was calculated according to Tinoco et al. (79). The stem structures of pFIFtrp69, 89, and 99 all have identical stabilities.

expression levels of pFIFtrp 69, 89 and 99. A similar analysis of the pFIFtrp 29 and pFIFtrp 49 interferon transcripts (the least efficient of the IFN-β1 expression plasmids) either upstream or downstream from the S/D or AUG sequences shows no regions that would include them in a stable stem structure.

In an attempt to determine whether the drastically lower yield of IFN-β1 from the pFIFtrp 29 plasmid resulted from its lack of secondary structure in the spacer region and therefore potentially increased susceptibility to intracellular nucleases compared to the pFIFtrp 99 and pFIFtrp 109 transcripts, all three plasmids were transformed into E. coli K12, strain E. coli PR13 (pnp 13, rna 19; refs. 76,77). This strain is deficient in polynucleotide phosphorylase (pnp13) and an E. coli ribonuclease (rna 1) and has been shown to allow longer mRNA half-lives for some prokaryotic-eukaryotic fusion messenger RNAs in E. coli (77). The results (Table V) indicated that the amount of IFN-β_1 produced by PR13/pFIFtrp 29 was about eight-fold greater than 294/pFIFtrp 29 (the strain normally employed), but still less than 20 percent of that synthesized by pFIFtrp 99. IFN-β_1 synthesis directed by pFIFtrp 99 (which seemed to have the optimal structure in the wild type strain, 294) was increased by less than two-fold. This result suggests that the pFIFtrp 29 transcript is probably several-fold more labile toward intracellular nucleases than the pFIFtrp 99 transcript, perhaps due to its lack of secondary structure in the ribosome binding site region. Because the amount of IFN-β_1 synthesis in strain PR13 directed by pFIFtrp29 remains 5-fold lower than that directed by pFIFtrp99 it could be that the very short spacing (2 bases) between the pFIFtrp29 S/D sequence and its AUG initiator is the major factor determining low levels of interferon production from this plasmid. A similar observation has been made by Singer et al. (73) who have obtained deletions resulting in a T4 rIIB mRNA spacer region less than 5 nucleotides in length. The conclusion that lack of secondary structure within the initiation region of the pFIFtrp 29 transcript leads to greater lability in E. coli 294 compared to the other plasmids (pFIFtrp 69, 89, 99 and 109; Figure 16) is apparently contradicted by the similar increase in interferon synthesis in E. coli PR13 directed by pFIFtrp 109 (Table V), which has the most stable predicted messenger RNA secondary structure. It is possible that in this latter case the stem structure of the pFIFtrp 109 transcript may inhibit the movement of initiated ribosomes past the AUG into the elongation step of translation. This explanation has been convincingly offered by Iserentant and Fiers (70) to explain radical changes in production of the lambda cro protein in some lac-cro fusions (72).

TABLE V.   Interferon levels in E. coli 294 and E. coli PR13.

| Plasmid | Molecules per Cell of IFN-β_1 | | PR13:294 |
	294	PR13	
1. pFIFtrp99	1.8×10^4 (1.0)	2.9×10^4 (1.0)	1.6
2. pFIFtrp29	2.1×10^2 (0.01)	1.7×10^3 (0.06)	8.0
3. pFIFtrp109	1.7×10^3 (0.09)	1.8×10^4 (0.64)	10.8

Table V. Stabilization of pFIFtrp29 and pFIFtrp109 transcripts in PR13. Cells (E. coli 294 or PR13) were grown and assayed for IFN-β_1 content as described in the legend to Table IV. Values given in parentheses represent the fraction of IFN-β_1 directed by either pFIFtrp29 or pFIFtrp109 compared to pFIFtrp99.

We have proposed that the pFIFtrp 109 mRNA may be less protected by ribosomes and therefore more susceptible to intracellular nucleases than the more efficiently translated fibroblast interferon mRNA of pFIFtrp 99. At any rate it seems that transformation of a plasmid encoding a chimeric mRNA of interest into the PR13 strain and then comparing expression between the PR13 strain and a wild type E. coli might be a very useful test of whether or not the structure of the messenger is optimal for expression.

In the experiments discussed above the effect of sequence on translation was not clearly observed (except in the case of the pFIFtrp 109 transcript where a stable stem structure probably inhibits translation). An extensive search of many E. coli gene sequences has been carried out by Gold and co-workers (69,73). They have found that throughout the ribosome binding region (which includes about 10-20 nucleotides on either side of the initiating AUG) there are preferred nucleotides for many positions. It is likely that their analysis will be very useful to our understanding of how ribosome binding sites have evolved and in the design of efficient ribosome binding site sequences upstream from the initiating AUG. However, because the ribosome binding site extends to either side of the initiating AUG, and the sequence of each cloned eukaryotic gene is unique, it is difficult to plan optimal sequences downstream from the initiator. In most cases this has not presented much of a problem. However, the expression of some viral surface proteins does seem to be limited by the coding sequence downstream from the initiating AUG. This is especially apparent in the case of the expression of the antigenic protein (VP$_3$) of the Foot and Mouth Disease Virus (80). In this case appreciable quantities of VP$_3$ can be made only when the sequence encoding the N-terminal portion of the protein is fused in vitro to the E. coli sequence encoding the trpΔLE 1413 protein (which is coded from the 5'-end of the trp leader gene and the last third of the trpE protein gene; ref. 80). This construction therefore has provided natural E. coli ribosome binding site sequence downstream from the initiating AUG. Other evidence that sequences within the ribosome binding site (other than the Shine-Dalgarno and ATG initiator) are important is now becoming available. For example, it has been shown that the nucleotide composition immediately following the Shine Dalgarno region affects the translation efficiency (81). A dramatic increase in the synthesis of HuIFN-β$_1$ in E. coli has recently been observed when the nucleotide just preceding the AUG is altered (78) in a way not possible in earlier experiments (71). An interesting possibility which begins to suggest itself from this work, and the data of Gold et al. (69,73), is that messenger sequences downstream from the initiating AUG may in some cases dictate an acceptable sequence within the ribosome binding site upstream from the initiator.

FUTURE PROSPECTS

The current array of cloning techniques makes it possible to clone almost any gene of interest for human health care, animal health care, or of industrial and agricultural interest. Once a gene has been cloned it is a great challenge to obtain a high level of expression either in a homologous or in a heterologous host. Some of the first heterologous genes were from the very beginning expressed at high levels (eg. human growth hormone; ref. 5) when introduced into E. coli, others were very poorly expressed. The rules that govern the expression levels are still poorly understood. The need for high level expression, be it for academic or commercial purposes, has stimulated greatly the amount of basic research on promoters and ribosome binding sites not only in E. coli but also in yeast and mammalian cells. At the same time the methods to synthesize DNA fragments have made great strides. Now, DNA fragments of up to 50 base pairs long can be synthesized in a few days. Synthetic promoters (82,83) and ribosome binding sites (81,84) have been made already. Variants can be designed and made readily, opening up a whole new array of possibilities to generate mutations at predetermined sites. In principle, any base in a regulatory sequence can be altered either through direct synthesis of the region or through modification of the region, using synthetic primers in site directed mutagenesis (reviewed in ref. 85). Thus, in the near future we may expect considerable progress in the understanding of the role of particular nucleotide sequences in regions that determine the level of gene expression in any system.

REFERENCES

1. Walz, A., Ratzkin, B. and Carbon, J. (1978). Proc. Natl. Acad. Sci. (U.S.A.) 75:6172–6176.
2. Struhl, K. and Davis, R.W. (1977). Proc. Natl. Acad. Sci. (U.S.A.) 74:5255–5259.
3. Wetzel, R. and Goeddel, D.V. In press. In The Peptides. Eds. Gross, E. and Meienhofer, J. Publ. Academic Press, Inc., N.Y., N.Y.
4. McGrath, J.P. and Levinson, A.D. (1982). Nature 295:423–425.
5. Goeddel, D.V., Heyneker, H.L., Hozumi, T., Arentzen, R., Itakura, K., Yansura, D.G., Ross, M.J., Miozzari, G., Crea, R. and Seeburg, P.H. (1979). Nature 281:544–548.
6. Seeburg, P. (1982). Personal communication.
7. Shine, J., Fettes, I., Lan, N.C.Y., Roberts, J.L. and Baxter, J.D. (1980). Nature 285:456–461.
8. Goeddel, D.V., Kleid, D.G., Bolivar, F., Heyneker, H.L., Yansura, D.G., Crea, R., Hirose, T., Kraszewski, A., Itakura, K. and Riggs, A.D. (1979). Proc. Natl. Acad. Sci. (U.S.A.) 76:106–110.
9. Taniguchi, T., Guarente, L., Roberts, T.M., Kimelman, D., Douhan, J. and Ptashne, M. (1980). Proc. Natl. Acad. Sci. (U.S.A.) 77:5230–5233.

246

10. Goeddel, D.V., Yelverton, E., Ullrich, A., Heyneker, H.L., Miozzari, G., Holmes, W., Seeburg, P.H., Dull, T., May, L., Stebbing, N., Crea, R., Maeda, S., McCandliss, R., Sloma, A., Tabor, J.M., Gross, M., Familletti, P.C. and Pestka, S. (1980). Nature 287:411-416.
11. Streuli, M., Nagata, S. and Weissmann, C. (1980). Science 209:1343-1347.
12. Gray, P.W., Leung, D.W., Pennica, D., Yelverton, E., Najarian, R., Simonsen, C.C., Derynck, R., Sherwood, P.J., Wallace, D.M., Berger, S.L., Levinson, A.D. and Goeddel, D.V. (1982). Nature 295:503-508.
13. Amster, O., Salomon, D., Zemel, O., Zamir, A., Zeelon, E.P., Kantor, F. and Schechter, I. (1980). Nucleic Acids Res. 8:2055-2065.
14. Lawn, R.M., Adelman, J., Bock, S.C., Franke, A.E., Houck, C.M., Najarian, R.C., Seeburg, P.H. and Wion, K.L. (1981). Nucleic Acids Res. 9:6103-6128.
15. Goeddel, D.V., Shepard, H.M., Yelverton, E., Leung, D. and Crea, R. (1980). Nucleic Acids Res. 8:4057-4074.
16. Chamberlin, M.J. (1979) In: RNA polymerase. Eds. R. Losick and M.J. Chamberlin, pp. 17-67, Cold Spring Harbor Laboratory.
17. Chamberlin, M.J. (1979) In: RNA polymerase. Eds. R. Losick and M.J. Chamberlin, pp. 159-191, Cold Spring Harbor Laboratory.
18. Park, C.S., Hillel, Z. and Wu, C.W. (1982) J. Biol. Chem. 257, 6944-6949.
19. Park, C.S., Wu, F.Y.H. and Wu, C.W. (1982) J. Biol. Chem. 257, 6950-6956.
20. Gilbert, W. (1979) In: RNA polymerase. Eds. R. Losick and M.J. Chamberlin, pp. 193-205, Cold Spring Harbor Laboratory.
21. Siebenlist, U., Simpson, R.B., and Gilbert, W. (1980) Cell 20, 269-281.
22. Pribnow, D. (1975b) J. Mol. Biol. 99, 419-443.
23. Schaller, H., Gray, C., and Hermann, K. (1975) Proc. Natl. Acad. Sci. USA 72, 737-741.
24. Post, L.E., Arfsten, A.E. and Nomura, M. (1978a) Cell 15, 231-236.
25. Rosenberg, M. and Court, D. (1979) Ann. Rev. Genet. 13, 319-353.
26. Calos, M.P. (1978) Nature 274, 762-765.
27. Youderian, P., Bouvier, S. and Susskind, M.M. (1982) Cell, in press.
28. de Crombrugghe, B. and Pastan, I. (1978) In: The Operon. Eds. J.H. Miller and W.S. Reznikoff, pp. 303-324.
29. Mandecki, W. and Reznikoff, W.S. (1982) Nucleic Acids Res. 10, 903-912.
30. Stefano, J.E. and Gralla, J.D. (1982) Proc. Natl. Acad. Sci. USA 79, 1069-1072.
31. Jaurin, B., Grundstrom, T., and Normark, S. (1981) Nature 290, 221-225.
32. Berman, M.L. and Landy, A. (1979) Proc. Natl. Acad. Sci. USA 76, 4303-4307.
33. Ptasne, M. (1978) In: The Operon. Eds. J.H. Miller and W.S. Reznikoff, pp. 325-343, Cold Spring Harbor Laboratory.

34. Johnsrud, L. (1978) Proc. Natl. Acad. Sci. USA 75, 5314–5318.
35. McKenney, K., Shimatake, H., Court, D., Schmeissner, U., Brady, C., and Rosenberg, M. (1981) Gene Amplification and Analysis. Vol. II: Structural Analysis of Nucleic Acids by Enzymatic Methods (Elsevier-North Holland) Jack G. Chirikjian and Takis S. Papas, Eds., 384–408.
36. Rosenberg, M. (1981) Promoters, Structure and Function. Praeger Scientific Publishing Co., M.J. Chamberlin and R. Rodriguez, Eds., in press.
37. Rosenberg, M. (1982) Personal communication.
38. de Boer, H.A., Comstock, L.J., Yansura, D., and Heyneker, H. (1981) Promoters Structure and Function, Praeger Scientific Publishing Co., M.J. Chamberlain and R. Rodriguez, Eds., in press.
39. Reznikoff, W.S. and Abelson, J.N. (1978) In: The Operon. Eds. J.H. Miller and W.S. Reznikoff, pp. 221–243.
40. Miller, J.H. (1972) In: Experiments in Molecular Genetics, Cold Spring Harbor Laboratory, pp. 407–411.
41. Carpousis, A.J., Stefano, J.E. and Gralla, J.D. (1982) J. Mol. Biol. 157, 619–633.
42. Miller, J.H. (1972) In: Experiments in Molecular Genetics, Cold Spring Harbor Laboratory, pp. 356–359.
43. Miller, J.H. (1972) In: Experiments in Molecular Genetics, Cold Spring Harbor Laboratory, pp. 385–393.
44. Heyneker, H.L., Shine, J., Goodman, H.M., Boyer, H.W., Rosenberg, J., Dickerson, R.E., Narang, S.A., Itakura, K., Lin, S.Y., and Riggs, A.D. (1976) Nature 263, 748–752.
45. de Boer, H.A., Heyneker, H., Comstock, L.J., Wieland, A., Vasser, M. and Horn, T. (1982) Miami Winter Symposium, in press.
46. De Boer, H.A., Comstock, L.J. and Vasser, M. (1982) Proc. Natl. Acad. Sci. U.S.A., in press.
47. Sadler, J.R., Tecklenburg, M. and Betz, J.L. (1980) Gene 8, 279–300.
48. Yanofski, C. (1981) Nature 289, 751–758.
49. Platt, T. (1978) In: The Operon. Cold Spring Harbor Laboratory, p. 263–302.
50. Rosenberg, M., Court, D., Shimatake, H., Brady, C. and Wulff, D. (1978) In: The Operon. Eds. J.H. Miller and W.S. Reznikoff, Cold Spring Harbor Laboratory, pp. 345–369.
51. DeRienzo, J.M., Nakamura, K. and Inouye, M. (1978) Ann. Rev. Biochem. 47, 481–532.
52. Nakamura, K. and Inouye, M. (1979) Cell 18, 1109–1117.
53. Hamming, J., Gruber, M., and Ab, G. (1979) Nucleic Acids Res. 7, 1019–1033.
54. Nomura, M. (1976) Cell 9, 633–644.
55. Kjeldgaard, N.O. and Gausing, K. (1979) In: Ribosomes. Cold Spring Harbor Laboratory, pp. 369–392.
56. Ellwood, M., and Nomura, M. (1980) J. Bacteriol. 143, 1077–1080.
57. Miura, A., Krueger, J.H., Itoh, S., de Boer, H.A., and Nomura, M. (1981) Cell 25, 773–782.
58. De Boer, H.A., Gilbert, S.F., and Nomura, M. (1979) Cell 17, 201–209.

248

59. De Boer, H.A., and Nomura, M. (1979) J. of Biol. Chem. 254, 5609–5612.
60. Brosius, J., Dull, T.J., Sleeter, D.D., and Noller, H.F. (1981) J. Mol. Biol. 148, 107–127.
61. Young, R.A., and Steitz, J.A. (1979) Cell 17, 225–234.
62. Brosius, J., Palmer, M.L., Kennedy, P.J., and Noller, H.F. (1978) Proc. Natl. Acad. Sci. USA 75, 4801–4805.
63. Kiss, I., Boros, I., Udvardy, A., Venetianer, P., and Delius, H. (1980) Biochim. Biophys. Acta 609, 435–447.
64. Shine, J., and Dalgarno, L. (1974) Proc. Natl. Acad. Sci. USA 71, 1342–1346.
65. Goeddel, D.V., Heyneker, H.L., Hozumi, T., Arentzen, R., Itakura, K., Yansura, D.G., Ross, M.J., Miozzari, G., Crea, R., and Seeburg, P.H. (1979) Nature 281, 544–548.
66. Gilbert, S.F., de Boer, H.A., and Nomura, M. (1979) Cell 17, 211–224.
67. Steitz, J.A. (1979). In: Biological Regulation and Development, I. (Gene Expression): pp. 349–399. Ed. R.F. Goldberger, Plenum Press, New York.
68. Lodish, H.F. (1976). Ann. Rev. Biochem. 45:39–72.
69. Gold, L., Pribnow, D., Schneider, T., Shinedling, S., Singer, B.S. and Stormo, G. (1981). Ann. Rev. Microbiol. 35:365–403.
70. Iserentant, D. and Fiers, W. (1980). Gene 9:1–12.
71. Shepard, H.M., Yelverton, E. and Goeddel, D.V. (1982). DNA 1:125–131.
72. Roberts, T.M., Kacich, R. and Ptashne, M. (1978). Proc. Natl. Acad. Sci. (U.S.A.) 76:760–764.
73. Singer, B.S., Gold, L., Shinedling, S.T., Hunter, L.R., Pribnow, D., Nelson, M.A. (1981). J. Mol. Biol. 149:405–432.
74. Ganoza, M.C., Fraser, A.R. and Neilson, T. (1978). Biochemistry 17:2769–2775.
75. Taniguchi, T. and Weissman, C. (1979). J. Mol. Biol. 128:481–500.
76. Reiner, A.M. (1969). J. Bacteriol. 97:1437–1443.
77. Hautala, J.A., Bussett, C.L., Giles, N.H. and Kushner, S.R. (1979). Proc. Natl. Acad. Sci. (U.S.A.) 76:5774–5778.
78. Matteuci, M. Unpublished observations.
79. Tinoco, I., Borer, P.N., Dengler, B., Levine, M.D., Uhlenbeck, O.C., Crothers, D.M., and Gralla, J. (1973). Nature New Biol. 246:40–41.
80. Kleid, D.G., Yansura, D., Miozzari, G., Heyneker, H., European Patent Application publication number 0036776, 30 September 1981.
81. De Boer, H.A., Comstock, L.J., and Hui, A. To be published.
82. Dobrynin, V.N., Korobdo, V.G., Severtsova, I.V., Bystrov, N.S., Chuvpilo, S.A. and Kolosov, M.N. (1980) Nucl. Acids Res. Symposium Series No. 7, 365–376.
83. Dunn, R.J., Belagaje, R., Brown, E.L. and Khorana, G.H. (1981) J. Biol. Chem. 256, 6109–6118.
84. Jay, G., Khoury, G., Seth, A.K. and Jay, E. (1981) Proc. Natl. Acad. Sci. 78, 5543–5548.
85. Shortle, D., DiMaio, D. and Nathans, D. (1981) Ann. Rev. Genet. 15, 265–294.

Genes: *Structure and Expression*
Edited by A. M. Kroon
© 1983 John Wiley & Sons Ltd.

INTERPLAY BETWEEN DIFFERENT GENETIC SYSTEMS IN
EUKARYOTIC CELLS: NUCLEOCYTOPLASMIC-MITOCHONDRIAL
INTERRELATIONS

Hans de Vries and Peter van 't Sant

Laboratory of Physiological Chemistry, State
University at Groningen, Bloemsingel 10,
9712 KZ GRONINGEN, The Netherlands

INTRODUCTION

All eukaryotic cells contain mitochondria. These cell organelles are
the sites of tricarboxylic acid cycle reactions, fatty acid oxidation,
electron transport and oxidative phosphorylation. Mitochondria consist of two membranes which enclose the matrix space (Fig. 1). The
two membranes, which differ in chemical composition, permeability
properties and enzyme content, are largely separated by the intermembrane space. The invaginations of the inner membrane are called
cristae. Mitochondria are the major source of energy for the cells.
Energy (ATP) is generated in a process called oxidative phosphorylation: metabolites are oxidized by oxygen to carbondioxyde and water

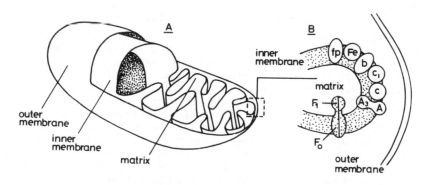

Fig. 1. Schematic representation of a mitochondrion
a. Three dimensional model, showing outer membrane, inner
   membrane and matrix space.
b. Enlargement of part of crista, showing the respiratory
   chain (schematically) and the ATPase complex. Explanation of the symbols: fp: flavoprotein; Fe: iron-sulfur
   protein; b,c_1,c,aa_3, cytochromes b,c_1,c, and aa_3, resp.;
   F_1F_0: parts of the ATPase complex.

and ATP is formed from ADP and phosphate. The enzymes involved in this process are located in the mitochondrial inner membrane: oxidation proceeds via the electron transport chain, composed of flavoproteins, iron-sulfur proteins, ubiquinone and several cytochromes. The component reacting with molecular oxygen is usually called cytochrome c oxidase. The ATP synthesizing system, which is driven by the electron flux through the electron transport chain, is called the mitochondrial ATPase complex. It is composed of the F_o part, located in the membrane and the F_1 part protruding into the matrix (for a recent monograph, see Tzagoloff, 1982).

Mitochondria, like chloroplasts (cf. the chapter by Bogorad), contain a genetic-biosynthetic system of their own. This means that eukaryotic cells do not only have genes in their nuclei, but also on their mitochondrial DNAs. It appears that, different as the sizes of DNA species are, all mtDNAs[*] contain roughly the same information: for a set of proteins and for mitochondrial transfer RNAs and ribosomal RNAs. The DNAs do not harbour a sufficient number of genes to secure independence of the organelles from the nuclei: the majority of mitochondrial proteins is synthesized outside the mitochondria, and encoded in genes located in the nucleus. Even the biosynthetic machinery of the mitochondria is of hybrid origin.

Obvious questions are: how is the balance between the two systems kept? What information, if any, is exchanged between the nucleocytoplasmic and the mitochondrial genetic systems? Questions like this have been posed ever since the presence of a mitochondrial biosynthetic system was discovered around 1960, and the answer or answers are still very incompletely known. These questions, of course, can be regarded as deduced from the more general problem of regulation: what keeps the concentration of cell components constant?

In principle, regulation can occur at several levels: at the levels of transcription, of transcript-processing, of translation and of post-translational modification. For complex structures like ribosomes and multi-subunit protein complexes, assembly is another possible level of regulation. Also the amount of DNA, both the nuclear and the mitochondrial, has to be controlled by regulation of replication. Moreover, the presence of two separate biosynthetic systems, one inside, the other one outside the mitochondria, imparts the extra dimension of mutual regulation to all of these processes. Finally, since the mitochondrion is physically separated from the rest of the cell, import of extramitochondrially made molecules through the mitochodrial membranes is in principle subject to regulation. The import mechanism of cytosolically synthesized proteins has been elucidated to a great extent. Essentially, this mechanism implies that most mitochondrial proteins are synthesized outside the organelle as precursors which are usually larger than the mature proteins found within the mitochondria. The precursors are imported by a process, dependent on energy-rich bonds inside the mitochondria. During import the N-terminal extensions are proteolytically removed from the precursors. For further details on this subject we refer to Neupert's

---

[*] Abbreviations: mt = mitochondrial; kD, MD = kiloDalton, MegaDalton; (k)bp = (kilo)basepairs

chapter in this volume. The purpose of the present essay is to describe the current knowledge, and lack of knowledge, regarding these
interrelations. We will try to answer some of the following questions:
How are the concentrations of components of mitochondrial complexes
kept constant? Is there direct coupling between the two biosynthetic
systems? What are the interactions between the two genetic systems?
What are the responses of the cell to disturbance of mitochondrial
function? Before answering these questions, we will pay attention to
the components and the products of the mitochondrial biosynthetic
system

THE MITOCHONDRIAL BIOSYNTHETIC SYSTEM: ITS COMPONENTS AND ITS PRODUCTS

The first reports showing the ability of mitochondria to synthesize
proteins came in 1955-1958 from Simpson and his coworkers (Simpson
et al. 1957). A few years later several other authors confirmed their
findings and extended these from skeletal muscle to heart, liver and
later on, to yeast, *Neurospora* and *Tetrahymena* (for a review, see
Kroon *et al*. 1972). These early investigators sometimes had difficulties in convincing their colleagues that this protein-synthetic activity really was mitochondrial, and did not originate from contaminating microsomes or, even worse, bacteria. The latter suspicion was
inspired by the discovery that chloramphenicol, a specific inhibitor
of prokaryotic protein synthesis, inhibits protein synthesis by isolated mitochondria. Nevertheless, by 1965 it was generally accepted
that mitochondria do synthesize proteins, in a process sensitive to
chloramphenicol and several other inhibitors of bacterial protein
synthesis, and insensitive to cycloheximide, an inhibitor of microsomal protein synthesis. By that time also mitochondrial DNA was discovered (Nass and Nass, 1963). This finding indicated that mitochondria not only translate messages into protein, but that the very
genes for these proteins are present in the organelles. A short
summary of the major differences, as we know now, between the mitochondrial and cytoplasmic biosynthetic systems is given in Table 1.

Mitochondrial DNA. The mtDNAs vary considerably in size among different groups of organisms. As shown in Table 2, mitochondria from all
animals contain small circular molecules of about 5 μm contour length
(16 kbp). Such a homogeneity in size is not found in lower eukaryotes
and plants: among fungi lengths vary from 6 μm (19 kbp) in *Torulopsis
glabrata* to 25 μm (75 kbp) in *Saccharomyces*; most fungal mtDNAs are
circular, but *Hansenula mrakii* mtDNA is linear. The protozoa *Tetrahymena* and *Paramecium* have 15 μm linears, and in plant mitochondria
very large mtDNAs, strongly differing in size, were found. For a
recent symposium the reader is referred to Kroon and Saccone, 1980.

Several investigators have looked at the concentration of mtDNA in
the cell. The ratio of mtDNA to nuclear DNA is generally fairly
constant, as is the number of mitochondria. Conflicting results have
been reported on the influence of the cell cycle on the mtDNA concentration. For mammalian cells it was reported that mtDNA synthesis
occurs discontinuously in S and G_2 phase (Pica-Mattoccia and Attardi
1971), but for yeast it appears that mtDNA synthesis takes places
uniformly throughout the cell cycle (Cottrell 1981). It has been

TABLE 1. Some differences between the mitochondrial and nucleo-cytoplasmic protein-biosynthetic systems

	Mitochondrial	Nucleo-cytoplasmic
– DNA:	generally circular	linear
Histones bound	no	yes
– DNA replication sensitive to:		
Actinomycin D	yes	yes
ethidium bromide	yes	no
– DNA transcription sensitive to:		
α-amanitin	no	yes
rifampicins	yes	no
ethidium bromide	yes	no
– Protein synthesis:		
initiating methionine formylated	yes	no
cycloheximide	no	yes
chloramphenicol	yes	no
tetracycline	yes	no
erythromycin	varying with organism	no

shown, furthermore, that mtDNA synthesis is not coupled to nuclear DNA synthesis: when cytosolic protein synthesis is blocked by cycloheximide, nuclear DNA synthesis stops almost completely, while the synthesis of mtDNA continues (Grossman *et al.* 1969).

The data reviewed in this section do not give many answers about the mechanism of control of mtDNA replication. Actually, only moderate progress has been made in about 10 years (*cf.* Borst 1972).

The genes on mitochondrial DNA. As shown in Table 2, mitochondrial DNAs range in size from 15,000 to 2,400,000 basepairs. Nevertheless, the major gene products seem to be the same for all eukaryotic organisms investigated so far (for a recent review, see Tzagoloff *et al.* 1979). First, the protein-synthetic machinery is partly composed of mitochondrially-coded components: 2 mt ribosomal RNAs, a set of mt transfer RNAs and (at least for yeast and *Neurospora*) one single mt ribosomal protein. Second, 5 genes coding for polypeptide subunits of enzymes involved in oxidative phosphorylation are located on mtDNAs: for the 3 largest subunits of cytochrome *c* oxidase, for cytochrome *b* which is a component of the bc_1 complex and for one, and in some organisms two, subunits of the mitochondrial ATPase complex. The gene for the smallest ATPase subunit, the DCCD-binding proteolipid

TABLE 2. Size of a few mitochondrial DNAs

Species	Length in basepairs
Animals	14,000 - 18,000
Fungi	
yeasts:	
Saccharomyces cerevisiae	75,000
Kluyveromyces lactis	33,000
Torulopsis glabrata	19,000
Neurospora crassa	63,000
Aspergillus nidulans	35,000
Protozoa	
Acanthamoeba	36,000
Tetrahymena	45,000
Higher plants	
tobacco	60,000
Musk melon	2,400,000

(8 kD) is located on mtDNA in *Saccharomyces* and plants and on nuclear DNA in most other organisms. A most intriguing finding has been the recent discovery by Agsteribbe and coworkers (Van den Boogaart *et al.* 1982) of a sequence on *Neurospora* mtDNA, which also potentially codes for a DCCD-binding proteolipid. The recent avalanche of sequencing data has culminated in the complete elucidation of the sequence of several mammalian mtDNAs, human being the first (Anderson *et al.* 1981). A great surprise was that besides the genes for the gene products mentioned above, several reading frames potentially coding for proteins were found. These Unassigned Reading Frames (URF's) are highly conserved in human, rat, mouse, bovine mtDNAs and must, therefore, have important functions. In lower eukaryotes URF's have also been found, some of which are homologous to the URF's on mammalian mtDNA.

A striking feature of the genes listed above is that all of their gene products, except for the tRNAs, are components of complexes. These complexes are composed of parts fabricated inside and of parts manufactured outside the mitochondrion. The mitochondrial ribosomes are made of ribosomal RNAs (made inside) and of ribosomal proteins (all imported, except for one in fungi). Cytochrome *c* oxidase, the bc_1 complex and the ATPase complex again are of mixed composition: part made inside, part imported. Hence the options for the cell to regulate mitochondrial biogenesis appear almost infinite in number.

TABLE 3. The major mitochondrial gene products

	Animals	Fungi	Protozoa	Plants
ribosomal RNAs	2	2	2	3
transfer RNAs	22	25	10	not determined
cytochrome c oxidase subunits	3	3	3	3
cytochrome b				
ATPase subunits	1	1 or 2	?	2

Mitochondrial ribosomes, tRNAs and other components of the mitochondrial biosynthetic system. The mt-ribosomes are composed, like all
other ribosomes, of RNAs and proteins. This simple statement conceals
an astonishing variety in types of mitochondrial ribosomes. Until
some 10 years ago ribosomes used to be classified in two types:
a) the 70S, prokaryotic type, consisting of rRNAs of 1.2, 0.6 and
0.04 MD (24S, 16S and 5S, resp.) and of about 55 ribosomal proteins,
having a total molecular weight of about 0.9 MD. b) the 80S, eukaryotic type, containing rRNAs of 1.4, 0.7 and 0.04 MD and about 80
ribosomal proteins, adding up to some 2 MD. Both types of ribosomes
dissociate into unequal subunits: 80S yields 60S and 40S, 70S gives
50S and 30S. For more complete and expert information, we refer the
reader to the chapters by Noller and by Garrett. The sensitivity of
mitochondrial protein synthesis to chloramphenicol led to the expectation that mt ribosomes would be of the 70S type. Nevertheless, the
first sedimentation coefficient found for mitochondrial ribosomes,
from bovine liver, was 55S (O'Brien and Kalf, 1967). After the initial disbelief of most workers in the field, their findings were confirmed for HeLa cells (Attardi and Ojala, 1971; Brega and Vesco,
1971) and for rat liver (De Vries and Van der Koogh-Schuuring, 1973).
It is now established that all animal mitochondrial ribosomes
contain very small rRNAs, 16S and 12S (0.6 and 0.3 MD), no 5S RNA and
90 ribosomes proteins (O'Brien et al., 1980). It seems likely that a
small part of the sequence of 5S rRNA has been incorporated into the
16S rRNA (Nierlich, 1982). The total molecular weight of these ribosomes is about 0.7 MDal. Hence, the protein: RNA ratio is inversed as
compared to bacterial ribosomes. The subunits are 40S and 30S. To
add further confusion, fungal mt ribosomes are somewhat larger than
bacterial ribosomes (73-75S) and contain 25S and 17S rRNAs (no 5S
RNA again), Tetrahymena mt ribosomes are 80S, and plant mt ribosomes
are 78S particles. Plant mitoribosomes, as an exception, contain 5S
RNA (cf. Leaver and Gray, 1982). Evidently, there is no uniformity at
all in mitochondrial ribosomes, which indicates that ribosomal functions can be performed by widely differing structures. For all mitochondria, the rRNAs are coded by mtDNA whereas in general all ribosomal proteins are imported. Import of these proteins and assembly of
ribosomes potentially offers possibilities for regulation.

Mitochondrial tRNAs are fewer in number than one would expect from
the genetic code: according to Crick's wobble hypothesis a minimum of
32 species is required to translate all possible triplet codons.
Mitochondria can synthesize proteins with a lower number because the
genetic code as well as the codon-anticodon interactions are special
(*cf.* the chapter by Kroon). In most species about 25 tRNA species have
been found. This number generally equals the number of tRNA genes
present on the mtDNA. Only in *Tetrahymena pyriformis* a significant
discrepancy between the number of genes and of species has been found
(Suyama, 1982): on the mtDNA 10 tRNA genes seem to be present, whereas 26 tRNA species could be detected. This hardly leaves open possibilities other than that of tRNA import. Import of a single, but probably inactive, tRNA species has been reported for *Saccharomyces*
(Martin *et al.*, 1979). *Tetrahymena*, however, is the only known species
where import of tRNAs active in mitochondrial protein synthesis into
the mitochondrion seems to take place. Although the significance of
this finding for mitochondrial biogenesis in general is not clear
yet, it once more points to the peculiarity of *Tetrahymena* which has
a linear instead of a circular mtDNA (only a few other organisms have
linear mtDNAs) and contains two copies of the gene for the large rRNA.
Furthermore, it shows that generalizations drawn from studies on
mitochondria of one particular species are sometimes unjustified.

Regarding the factors and enzymes involved in translation, transcription and replication it can be said that, like most other mitochondrial proteins, these are nuclear-coded, cytoplasmically synthesized and imported into the mitochondrion. These include DNA polymerase, RNA polymerase, initiation and elongation factors.

THE FATE OF MITOCHONDRIAL GENE PRODUCTS

We have already pointed out above that most mitochondrial gene products are part of complexes and, therefore, have to be assembled
correctly. Moreover, many primary gene products are not ready for
immediate use, but have to be processed first. For the rRNAs processing includes splicing and trimming, for several mRNAs also
splicing is necessary; for some proteins proteolytic cleavage of an
N-terminal part is required.

The processing of mitochondrial gene products. The fate of mitochondrial primary transcripts is entirely different in animals and in
lower eukaryotes. It has been found about a decade ago by Attardi's
group (Aloni and Attardi, 1971) that in HeLa cells mtDNA is transcribed symmetrically. Only one initiation site of transcription for
each strand is present. The transcripts of the strand containing the
majority of the genes is probably processed while the chain is growing, since no giant RNA from this strand has been found (Montoya
et al., 1981). This is different from the transcription of the other
strand; large transcripts have been found covering about 70% of this
strand. The RNAs from both strands are processed by precise endonu-
cleolytic cleavage at very specific sites. In general, tRNA sequences
are used as punctuation marks for this cleavage (Ojala and Attardi,
1981). Messenger RNAs are polyadenylated during or immediately after
cleavage. In a number of cases the UAA stopcodon is generated only by
this polyadenylation. Usually animal mt mRNAs have no leader sequence

but start at their 5' end with the AUG (or AUA) for the initiating methionine. Moreover, also the rRNAs are ready for use after the first cleavage. Finally, no intervening sequences are present in animal mtDNA, so processing of transcripts is comparatively simple in animal mitochondria. It is not known with certainty, but it seems likely that the cleavage enzyme(s) are of extramitochondrial origin. No mutants are available, however, to prove or disprove this notion. As compared to the transcript processing in animals, the lower euka-ryotes offer a very complicated picture. Intervening sequences are often found in their mitochondrial genes. Furthermore, the genes are never adjacent or surrounded by punctuation-mark tRNAs. Sometimes long stretches of non-coding DNA are interspersed between the genes. Many data, mainly from Lambowitz' lab for *Neurospora* and from the groups of Rabinowitz and of Borst and Grivell for yeast, have been gathered on the transcription of Ascomycete mtDNA and the processing of the transcripts (Tabak *et al.*, 1982; Bertrand *et al.*, 1982). It is difficult to sketch a general picture. Nevertheless, it can be said: a) that most genes are transcribed separately, except perhaps for some closely packed tRNA genes; b) that no polyadenylation of mRNAs occurs; c) that usually transcripts have to be trimmed to their mature size. An example of this is the small rRNA, which is synthe-sized as a molecule slightly longer at the 5' end (Osinga and Tabak, 1982); d) initiation of transcription of both rRNA genes probably proceeds via a 17-nucleotide promotor sequence (Osinga and Tabak, 1982); e) the primary transcripts of those genes that contain inter-vening sequences have to be spliced to yield mature transcripts. This holds for the large rRNAs of most fungi (one intervening sequen-ce of 1 to 2.3 kbasepairs) and for the genes for cytochrome *b* and for subunit 1 of cytochrome *c* oxidase of *Saccharomyces*. The latter protein genes are extremely mosaic and their transcript processing follows a complicated path of splicing, in which both cytosolically and mitochondrially made splicing enzymes play their roles. We would like to refer to the chapter by Grivell for detailed information on these mosaic genes. For the splicing of the large rRNA precursor only nuclear-coded enzymes are used. In *Neurospora* the primary trans-cript is a 5.2 kbase molecule, which accumulates in certain mutants (Bertrand *et al.*, 1982). It is interesting that the mitochondria of these mutants still have large ribosomal subunits. These subunits, however, are aberrant and do not seem to be functional in protein synthesis, although they contain a full complement of large subunit proteins.

Not only transcripts, but also some mitochondrially coded polypep-tides have to be processed. At least for subunit 2 of cytochrome *c* oxidase in yeast (Sevarino and Poyton, 1980) and *Neurospora crassa* (Werner *et al.*, 1980) and for subunit 1 of the same enzyme complex in *Neurospora crassa* it has been demonstrated that precursor proteins with an N-terminal extension exist. In yeast Sevarino and Poyton were able to partially block the processing of the precursor of subunit 2 of cytochrome *c* oxidase *in vitro*. For *Neurospora* it was shown that both subunit 1 and subunit 2 of cytochrome *c* oxidase have lost their original N-termini (Werner *et al.*, 1980; Van 't Sant *et al.*, 1981). By the use of anti-sera a 45-kD precursor of subunit 1 was identified, first in a mutant and later in wild type. Finally amino acid sequences deduced from the genes for subunit 1 and 2 of *Neurospora* cytochrome *c* oxidase (unpu-

blished observations of our group) were compared with
the known N-terminal sequences of the proteins. This revealed that
the precursors are 26 and 12 amino acids longer resp. than the
mature products. The processing of these precursors and perhaps also
the integration of the polypeptides into the mitochondrial inner membrane are steps in the biogenesis of the mitochondria which allow regulation. It was demonstrated by Van 't Sant et al., (1981) that after inhibition of cytoplasmic protein synthesis of Neurospora cells with cycloheximide
for more than 1 hour, mitochondrially synthesized precursor proteins are
no longer processed, resulting in an accumulation of these polypeptides. These data suggest that the processing machinery is of nucleocytoplasmic origin and that the enzyme(s) involved have a high turn-
over rate. So probably a continuous cytoplasmic protein synthesis is
necessary for the proper assembly of these mitochondrial translation
products.

The assembly of respiratory chain complexes and of mt ribosomes. The
final stage for the mitochondrial gene products is assembly into
multicomponent complexes. The respiratory chain complexes finally
find their location in the mitochondrial inner membrane. Little is
known about the mechanism or control of the assembly of these complexes. Specific inhibition of either cytoplasmic or mitochondrial
protein synthesis should ultimately result in shortage of part of
the components of these complexes and, consequently, in impaired
assembly. Furthermore, impaired processing of precursors of subunits
of the complexes should lead to defective assembly. An example, already given above, is the accumulation in Neurospora of the precursor
of subunit 1 of cytochrome c oxidase in unassembled form after inhibition of the formation of the enzyme(s) by cycloheximide.

Another factor playing a role in the assembly of cytochrome oxidase,
is heme. A yeast mutant incapable of synthesizing heme cannot synthesize cytochromes in the absence of heme or heme precursors. Nevertheless, three subunits of cytochrome c oxidase, the mitochondrially made
subunits 2 and 3 and the cytoplasmically made subunit 6, are
present in an unassembled form. (Saltzgaber-Müller and Schatz, 1978).
Under non-restrictive conditions, the enzyme is active and fully
assembled. It is still not known what the precise effect of heme on
assembly is. The fact that subunit 1 appears to carry heme (Winter
et al., 1980) may be of importance in this respect, since it is the
only mitochondrial subunit which is not present in the absence of
heme.

Both in yeast and in Neurospora many mutants deficient in cytochrome
c oxidase assembly have been described. Studies with these mutants,
mainly performed in the laboratories of Bertrand and Schatz, indicate
that in the assembly of cytochrome c oxidase both mitochondrial and
nuclear genes play key roles. Moreover, a regulation by the electron
flux through the respiratory chain appears to exist. The genes
governing correct assembly are not only the structural genes, but
also (and more interestingly) non-structural genes. Many assembly-
deficient mitochondrial mutations have nuclear suppressors or can be
suppressed by inhibition of respiration. This means that the information for the structural genes itself is present on the mtDNA, but
that some steps on the route to correct assembly are impaired in the

mutants, and can be repaired by the suppressor gene product (*e.g.*, mutant ribosomal proteins or splicing enzymes or by the "message" originating from respiratory inhibition (see below). An example of an assembly deficiency is the mitochondrial deletion mutant of *Neurospora*, E35. This mutant contains all subunits of cytochrome *c* oxidase, but in unassembled form (De Vries *et al.*, 1981). Possibly, the part of the mtDNA which has been deleted in this mutant, contains a sequence playing a role in the assembly. The assembly of the mitochondrial ATPase has been studied in a yeast nuclear mutant by Todd *et al.* (1981). In this mutant, pet936, the cytosolically synthesized subunits, or their precursors, accumulate outside the inner membrane. In this case, the defect seems to be in transport rather than in assembly proper.

The mitochondrial ribosome is another complicated multicomponent complex, composed of rRNAs, synthesized inside, and ribosomal-proteins, generally synthesized outside. It might be expected that the synthesis and import of the ribosomal proteins is tuned to the rate of rRNA synthesis and processing. In *E. coli* a very exciting regulatory system has been discovered (Nomura *et al.*, 1982). There certain key ribosomal proteins inhibit the translation of their own mRNA. This simple and effective negative feedback system will probably not be operative for the mitochondrial ribosomal proteins, because their presence in free form inside the mitochondria will have no consequences for their translation which occurs outside. For fungi there is, however, the possibility that the one ribosomal protein synthesized inside the mitochondrion, *var*-1 for yeast (Terpstra and Butow, 1979) and S5 for *Neurospora* (La Polla and Lambowitz, 1981), performs such a role. It is known from the work of Lambowitz' group (La Polla and Lambowitz, 1977) that inhibition of mitochondrial protein synthesis by chloramphenicol induces the formation of aberrant small mt ribosomal subunits, deficient in S5 ("chloramphenicol particles"). Mutations in the gene for S5 or in genes regulating its expression conceivably could have effects on synthesis or assembly of all mitochondrial proteins. The poky mutant of *Neurospora* appears to be such a mutant. Indeed the manyfold defects in poky all seem to be ultimately caused by the (partial) deficiency of S5: the poky mutants have a very low concentration of small ribosomal units, and these subunits have the appearance of "chloramphenicol-particles". Hence, for *Neurospora* a role in the assembly of ribosomal subunits appears to exist for the S5 r-protein. A similar role has been found by Terpstra and Butow (1979) for the yeast *var*-1 protein.

Examining what is known about assembly of the complexes it is apparent that regarding most of the molecular mechanisms underlying assembly knowledge is lacking. Nevertheless, even without knowing these mechanisms in detail, the contours of the processes which govern assembly are gradually emerging. In the next sections we will try to describe these processes in more detail.

THE INTERPLAY BETWEEN THE TWO BIOSYNTHETIC-GENETIC SYSTEMS

We have indicated in the preceding paragraphs that targets for regulation of mitochondrial biosynthesis are transcription, translation, processing, import and assembly. It should be emphasized that regulation can serve two purposes. First, the balance between components from inside and outside the mitochondria must be kept. Second, there must be mechanisms by which the cell reacts on impairment of mitochondrial function, ultimately resulting in lack of energy for the cell. The major approaches used to dissect these two types of regulation have been: the use of inhibitors of mitochondrial protein synthesis, of oxidative phosphorylation and of cytoplasmic protein synthesis; and the use of mutants with disturbed synthesis or regulation.

Keeping the balance. Since the respiratory chain complexes and the mitochondrial ribosomes are composed of both mitochondrially and cytoplasmically synthesized parts, one would expect that a very precise regulation will be necessary to produce corresponding amounts of polypeptides of the two translational origins. A direct way in which an interdependence might be arranged could be a tight coupling between the two protein synthesizing systems. Inhibition of one system should then immediately block the counterpart. Experiments to investigate this kind of mutual regulation are rather simple to perform since both polypeptide synthesizing systems can specifically be inhibited *in vivo*; the cytoplasmic translation by cycloheximide and the mitochondrial protein synthesis by chloramphenicol. In this way it was demonstrated by Weiss and colleagues that in *Neurospora* cytoplasmic subunits of the "hybrid" enzyme complexes accumulate in the presence of chloramphenicol (Weiss *et al.*, 1975). Such an accumulation has also been found for enzyme complexes not containing mitochondrial translation products. Hence the cytoplasmic translation continues when mitochondrial protein synthesis is inhibited. Similar results were obtained for many other organisms. On the other hand, for *Kluyveromyces lactis* is has been reported (Zennaro *et al.*, 1977) that chloramphenicol treatment causes an immediate arrest of cytosolic protein synthesis. This organism may therefore have, by way of exception, a stringent coupling between the two systems. The complementary experiment, inhibition of cytoplasmic protein synthesis by cycloheximide, has also been performed. For *Neurospora*, Neupert and Rücker (1976) demonstrated that under these conditions mitochondrial protein synthesis *in vivo* continues for at least two hours (*i.e.*, one generation period). In fact, it had been demonstrated even earlier that mitochondrial protein synthesis does not stop immediately in the absence of cytoplasmic translation: isolated mitochondria are capable to incorporate radioactive amino acids into proteins for at least 30 min (Kroon, 1963). Actually, it was this finding that established, about a quarter of a century ago, that mitochondria have their own protein-synthetic system (Simpson *et al.*, 1957; Kroon, 1963). Depending on the intactness of the mitochondria, on the abilities of the experimenter, and on the particular organism, organ or growth stage, the pioneer "mitochondriacs" (term coined by Borst, 1969) reported that mitochondrial protein synthesis was either completely independent or completely dependent – or anything in between – on: added ATP; respiratory substrates; Mg^{2+}; cytoplasmic protein factors. The role of cytosolic proteins has been reinvestigated recently by Finzi *et al.* (1981): addition of a low-molecular weight protein fraction isolated from a cytosolic postpolysomal supernatant stimulates yeast mitochondrial protein synthesis several-fold. Whether these polypeptides are really regulatory proteins or simply enzymes

involved in mitochondrial protein synthesis has not yet been elucidated. Until then, the claim that we deal here with a way of regulating mitochondrial protein synthesis has to be regarded with skep-
ticism.

Summarizing, it is evident that the cytoplasmic and mitochondrial
translation systems are not very tightly coupled to say the least,
except perhaps for *Kluyveromyces*, "not coupled" may even be the real
situation. Tight coupling of a different kind, however, has been found
by Ray and Butow (1979 a,b): inhibition of yeast cytoplasmic protein
synthesis leads to rapid inhibition of mt ribosomal RNA synthesis.
The authors suggest that the cessation of cytosolic protein synthesis,
more specifically the "stalling" of ribosomes as polysomes, is the
source of the blocking signal to mt transcription. In other words,
specific regulatory signals are required here.

Responses of the cell to impaired mitochondrial function. Most eukaryotic organisms are strictly aerobic. Hence, their survival depends
very much on the undisturbed functioning of their mitochondria. Only
a few eukaryotes have mechanisms to escape from inhibition of respiration or of the mitochondrial biosynthetic system. One particularly
inventive organism, in many respects, is the yeast *Saccharomyces
cerevisiae*, which can live without oxygen and therefore without
functional mitochondria. Its peculiarities in this and other respects
will be treated briefly in the paragraph on "Facultatively aerobic
yeasts" and in the chapter by Grivell.

For higher plants and moulds like *Neurospora crassa* the escape route
is the induction of an alternate oxidase. Inhibition of respiration
by, *e.g.*, antimycin A or cyanide induces an alternate electron
transport pathway which is sensitive to salicylhydroxamic acid (SHAM)
and insensitive to cyanide (for a review, see Lambowitz and Zannoni,
1978). This oxidase is also induced by inhibition of the mitochondrial biosynthetic system or by mutations causing mitochondrial
lesions. The alternate oxidase is still poorly characterized. Little
is known about the identity of the oxidase or the mechanism of oxygen
reduction. The only known aspects are that it is less efficient in
ATP synthesis and that cytochromes do not participate in this pathway. The induction of the alternate oxidase requires transcription
of nuclear genes and cytosolic protein synthesis (Edwards and Unger,
1978).

Animals do not appear to have many options to cope with dysfunction
of mitochondria. This may be the reason that mitochondria in rat
liver and intestine have a large surplus of mitochondrial respiration
complexes (De Vries and Kroon, 1970; De Jong *et al.*, 1978): despite
the continuous presence in the blood of chloramphenicol at maximally
inhibiting concentrations regeneration of liver tissue after partial
hepatectomy proceeds normally during three days at least: the weight
and protein content double in that period, whereas no new-formation
of cytochrome *c* oxidase can occur. Only prolonged treatment with antibiotics inhibiting mitochondrial protein synthesis result in inhibition of cell growth. Hence, these antibiotics (tetracycline for instance) are effective as antiproliferative agents in experimental
tumor systems (Van den Bogert *et al.*, 1981).

In most organisms non-lethal (or not immediately lethal) inhibition
of mitochondrial functions results in overproduction of several cytoplasmically synthesized mt proteins. To explain this phenomenon, a
dynamic model for the regulation of synthesis of mitochondrial proteins in *Neurospora crassa* has been proposed by Barath and Küntzel
(1972). These authors suggested that a mitochondrially coded and synthesized protein is transported to the nucleus and represses coordinately the synthesis of the enzymes involved in mitochondrial replication, transcription and translation. The model implies that the
mitochondrial repressor is diluted out during nuclear DNA replication.
As a consequence, transcription of the formerly repressed genes becomes possible. Proteins are then synthesized and transported into
the mitochondrion. Hence, mitochondrial transcription and translation
start and newly synthesized repressor molecules are exported to the
nucleus to block transcription of the genes again. This again causes
a switch-off of the mitochondrial genetic system and the cycle is
closed. The hypothesis was based on the observation that the synthesis of several nuclear-coded mitochondrial proteins, such as elongation factors, increases when mitochondrial protein synthesis is
blocked. This phenomenon was also observed in other organisms and
for other enzymes like cytochrome *c* and the alternate oxidase. All
these observations suggested that a mitochondrial gene product might
be involved in keeping the correct balance between the two trans-
cription-translation systems. Fig. 2 gives a schematic description
of the model.

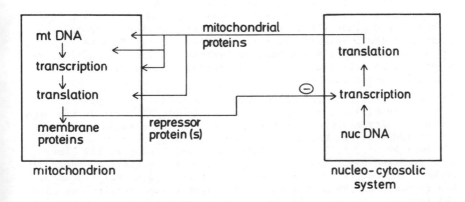

Fig. 2. The repressor model of Barath and Küntzel (1972).

However, in the ten years since the model was published, only one isolated report has been made on the export of a mitochondrial translation product (Macklin *et al.*, 1977). Other investigators were not able to trace even one exported mitochondrial translation product (Van 't Sant *et al.*, 1980). Furthermore, an argument against the hypothesis is that no overall induction of nuclear-coded mitochondrial proteins was found, *e.g.* the aminoacyl-tRNA synthetases showed highly different degrees of induction (Van 't Sant *et al.*, 1980). A different model for regulation seems more adequate to explain the observations which supported the repressor idea. This alternative model, first brought forward by Lambowitz and Zannoni (1978), implies that a signal for synthesis of nuclear-coded mitochondrial proteins is not generated by a shortage of a specific mitochondrial translation product but by an impairment of oxidative phosphorylation. The controlling agent might be a metabolite, the concentration of which is depending upon the activity of oxidative phosphorylation. Lines of evidence, given by Lambowitz and Zannoni, and mainly obtained in *Neurospora*, are:

1. All kinds of mutants, both cytoplasmic and nuclear, with an incomplete respiratory chain have an induced alternate oxidase and an elevated cytochrome c concentration. Even mutants with a normal activity of the mitochondrial protein synthesis, but unable to assemble one of the enzyme complexes of the respiratory chain, contain higher amounts of cytochrome c and the alternate oxidase.
2. Copper deficiency also causes alternate oxidation and increased cytochrome c contents, even in wild type. As a consequence of this deficiency active cytochrome c oxidase cannot be formed whil the mitochondrial protein synthesis is unimpaired (Schwab, 1973).
3. Metabolic inhibition of oxidative phosphorylation by antimycin, KCN, oligomycin or CCCP, also induces the alternate oxidase and a high cytochrome c concentration.
4. Certain culture conditions, not directly affecting the mitochondrial functions, cause the increased activities as well.

Because transcription of nuclear DNA is required for the synthesis of the cyanide-insensitive respiration, derepression is necessary. In most of the possibilities for induction described above, a shortage of mitochondrially synthesized repressor is not a plausible explanation. An effect caused by a disturbance of oxidative phosphorylation is more suitable.

Barath and Küntzel supposed in their hypothesis that regulation occurs predominantly at transcriptional level. Although it has been shown that for synthesis of the alternate oxidase in *Neurospora crassa* (Edwards and Unger, 1978) and of cytochrome c in yeast (Zitomer *et al.*, 1979) transcriptional control does exist, other data revealed that substantial regulation of synthesis of nuclearly coded mitochondrial proteins occur at other levels. For instance, the increase of the phenylalanyl-tRNA synthetase activity caused by chloramphenicol can not be explained by derepression of transcription because inhibition of the cytoplasmic ribosomes by cycloheximide hardly influenced the induction (Beauchamp and Gross, 1976). This suggests that post-translational effects are involved in the regulation. One may think, for instance, of the transport of proteins across the mitochondrial membranes accompanied by processing of precursors (*c.f.* the chapter by Neupert). Summarizing, the repressor hypothesis in its simple form cannot be operative. Of course it cannot be excluded completely that synthesis of one or only a few proteins is regulated in this way.

Nuclear and mitochondrial genes affecting mitochondrial function. The vast majority of mt proteins is encoded in the nucleus. An almost infinite range of mutants is therefore possible with mutations in nuclear structural genes. More pertinent to the subject of this chapter are the mutants with mutations in the steps leading to the assembly of hybrid complexes. Most of the mutants useful in this respect were isolated in *Saccharomyces* and in *Neurospora*. Due to the talent of *Saccharomyces* to survive easily defects in the respiratory chain, attention has focused on mutants of this organism lacking functional cytochrome *b*, cytochrome *c* oxidase or ATPase. Different authors have isolated nuclear mutations modifying the expression of mitochondrial genes. Due to the recent work of mainly Slonimski's group on the cytochrome *b* gene and its expression and of Grivell and Borst's group on the cytochrome *c* oxidase subunit 1 gene and its expression, it has become evident that, for this yeast, sophisticated networks exist within the mitochondrial genes and between the genes. In these interactions, signals of the introns, many of which code for splicing enzymes (maturases), play a major role. The chapter by Grivell gives a full display of these interrelations, which are probably unique to yeast. An interesting phenomenon, from a regulatory viewpoint, is the suppression of mit⁻ mutations. These are mutations characterized by the absence of one mitochondrial protein gene product not caused by deficiency of mitochondrial protein synthesis (Slonimski and Tzagoloff, 1976). Both nuclear and mitochondrial suppressors of mit⁻ mutations were found (Coruzzi and Tzagoloff, 1980; Groudinsky *et al.*, 1981). The mechanism of suppression is still far from clarified, but in general they seem to cure deficient splicing of mutated introns. The introns often carry the genetic information for their own splicing enzymes and sometimes also for splicing of other introns. It is possible that the suppressor mutations cause (non-specific) alterations in the translation machinery or that they generate aberrant components of the splicing complexes. In the latter case, the presence of such an aberrant component would functionally (conformationally?) alleviate the defect caused by the primary mutation. Most interestingly, strains carrying the nuclear NAM2-1 or the mitochondrial mim2-1 suppressor are able to restore the synthesis of cytochrome *c* oxidase subunit 1 even when the cytochrome *b* gene is completely absent. The cytochrome *b* intron-4 product is indispensable for correct splicing of the cytochrome *c* oxidase subunit 1 pre-mRNA. It was hypothesized, therefore, (Groudinsky *et al.*, 1981) that these mutations switch on the synthesis of silent genes for an active maturase, either mitochondrial or nuclear. The nuclear yeast mutant, pet494-1, completely lacks the mitochondrially synthesized cytochrome *c* oxidase subunit (Ono *et al.*, 1975). In this case, a nuclear-coded splicing enzyme may be absent. Recent results from Tzagoloff's lab (Dieckman *et al.*, 1982) indicate that the expression of cytochrome *b* in respiration-deficient mutants of yeast can be restored by transformation with a recombinant plasmid. The nuclear gene present in this plasmid can code for a 70 kD basic protein. Probably, also in this case the nuclear gene product is an RNA-processing enzyme. A final example of the influence of processing steps on mitochondrial expression in yeast is given in a recent paper by Martin and Underbrink-Lyon (1981). They present strong evidence that a mitochondrial locus is indispensable for the processing of mitochondrial transcripts to mature tRNAs. Complementation experiments suggest that this locus acts in trans and may code for a processing enzyme.

Also in *Neurospora crassa* nucleocytoplasmic-mitochondrial interactions have been studied through the analysis of mutants. In this ascomycete it is much more difficult than in yeast to isolate mutants in mitochondrial biogenesis. Until now, for instance, nobody has been able to isolate mutants with defects in the structural protein genes. This indicates that at least during some stage in the life cycle, the alternate oxidase is not sufficient to sustain life, but that a functional respiratory chain is necessary then (*cf.* "Mitochondria and development"). The mutants of *Neurospora* of interest are both mitochondrial and nuclear. Mitochondrial mutants, all of which have abnormal respiratory systems, have been classified by Bertrand *et al.* (1976). Within the scope of this chapter, the following are important:
1) "poky" mutants, defective in mt ribosome assembly;
2) "stopper" mutants, which have deletions in their mtDNAs and have disturbed assembly of cytochrome *c* oxidase and cytochrome *b*;
3) mutants deficient in cytochrome *c* oxidase only. One of these, *mi*-3, accumulates the precursor protein to subunit 1 of cytochrome *c* oxidase, which suggests that the defect is primarily in protein processing (Werner *et al.*, 1980).

These mutants appear deficient in regulation, and some of the defects can be overcome. The defects of the poky mutants can be suppressed by a nuclear suppressor (Bertrand, 1980). Since the primary defect here is supposed to be in the S5 ribosomal protein, the suppressor is likely to be an abnormal mitochondrial ribosomal protein or one of the factors involved in mitochondrial protein synthesis. Most interesting is the effect of inhibition of cytochrome *b* function in some mutants deficient in cytochrome *c* oxidase production. Either by addition of the cytochrome *b* inhibitor antimycin, or by the genetic introduction of cytochrome *b* deficiency, these mutants are forced to form this enzyme complex. It is very likely that this phenomenon reveals a direct influence of the electron flux through the respiratory chain on cytochrome *c* oxidase formation. It may be, as hypothesized by Bertrand (1980), that an effector molecule is either generated or ceased to be generated when electron transport is blocked. The precursor to subunit 1 of cytochrome *c* oxidase, accumulated in *mi*-3 is also processed under the influence of antimycin inhibition or by the introduction of cytochrome *b* deficiency through crossing with a suitable mutant.

The stopper mutants seem to be assembly mutants: all components of cytochrome *c* oxidase are present in the stopper E35, whereas only minimal (but probably essential) amounts of the assembled complex are present. It is possible that an 11 kD mt translation product, which is absent in E35, plays a role in assembly. A nuclear mutant lacking the mitochondrially made subunit 1 of cytochrome *c* oxidase has also been described (Nargang *et al.*, 1978). It is not clear as yet what nuclear gene product is missing here, but it is unlikely that it is a splicing enzyme like in yeast pet 494-1: the *Neurospora* gene for subunit 1 appears to be uninterrupted by intervening sequences.

Finally, an important group of nuclear mutants has been studied jointly by the groups of Lambowitz and Bertrand: these temperature-sensitive mutants are defective in the splicing of the mitochondrial large rRNA which has a 2.3 kbp intron. Three different nuclear genes are required for this splicing reaction, one of which is phenotypically suppressed by antimycin (Bertrand *et al.*, 1982). Hence, at least one component required for mitochondrial RNA splicing is regulated by electron transport.

In summary, much of the regulation of mitochondrial gene expression seems related to processing phenomena. We anticipate that recombinant DNA technology such as used by Dieckman *et al.* (1982) with yeast, will be of extreme value for the elucidation of the interactions between nucleus and mitochondrion.

Mitochondrial genes affecting non-mitochondrial functions. Although it may seem unexpected, mitochondrial loci can influence properties of the cell or organism which appear not mitochondrial at all. The first example were found in the plant kingdom: the economically very important trait of maize, male-sterility, was found to be cytoplasmically inherited. Other plants also show this phenomenon. Recent investigations have shown that mitochondria from such strains contain different types of plasmid-like DNA molecules besides the normal mtDNA (for a review, see Leaver and Gray, 1982). The most likely explanation seems that these plasmid-like molecules act like transposons, because they have several structural features of transposons (for a review see Cohen & Shapiro, 1980). Experiments by the group of Levings (Levings *et al.*, 1980) showed that in revertants to male-fertility the plasmid-like molecules have integrated into mtDNA. The mechanism of action of these plasmids on sexual function is a complete mystery still. There are, however, some hints that synthesis of aberrant mitochondrial translation products is the cause of the trait.

Neurospora crassa exhibits another type of sexual defect caused by a mitochondrial mutation. Stopper mutants which are characterized by an irregular growth, a cyanide-insensitive respiration and deficiency of cytochromes b and aa_3, are also female-sterile: they make no or only few protoperithecia. The conidia, however, are perfectly capable of fertilizing other strains' protoperithecia. The molecular abnormality of the stopper mutants is the deletion of part (not always the same part) of the mtDNA, usually followed by formation of smaller circular DNAs, consisting of a segment of the wild-type mtDNA.

A comparable phenomenon appears to be the so-called senescence in *Podospora* (*e.g.* Kück *et al.*, 1981) and in some mutants of *Aspergillus amstelodami* (Lazarus *et al.*, 1980): after a certain period of growth, the mycelium becomes abnormal and dies. It seems that also here the mitochondrial DNA is the cause: plasmid-like molecules, possibly of a transposon-like nature, are found in all senescent mitochondria. These molecules again are derived from parts of the mtDNA.

Summarizing, although the molecular explanation of these extra-mito-chondrial effects is lacking still, in all cases known so far additional populations of mtDNA play a role. In some cases these additional mtDNAs have the features of transposons.

Mitochondria and development. It may be deduced from the preceding paragraph that in plants and fungi the sexual reproductive cycle can be negatively controlled by mutations in mitochondrial DNA. Comparatively little is known, however, about the possible role of mitochondrial DNA during the life or cell cycle. In general, major increases in mitochondrial activities (respiration, transcription and translation) occur when conidia, ascospores or plant seeds germinate. It is in no way clear here what goes first. Several reports by Brambl and his group on the fungus *Botryodiplodia theobromae* (Brambl, 1980) have made clear that germination of conidia is under the control of mitochondrial function: germination is strictly dependent upon cyanide-sensitive respiration. Nevertheless, the mitochondria in the dormant conidia of this organism do not possess a complete respiratory chain: no heme a or cytochrome c oxidase are present. This enzyme appears very rapidly during germination in a cycloheximide sensitive, chloramphenicol-insensitive process. Brambl showed that in mitochondria of dormant spores only the three mitochondrial subunits of cytochrome c oxidase are present. Precursors of the cytoplasmically synthesized subunits are present in the cytosol. Evidently, as soon as germination starts the cytosolic subunits are processed and imported to assemble the functional enzyme. The cycloheximide-sensitivity of this assembly shows that a component of the import/processing/assembly pathway has to be synthesized before germination. For *Neurospora* it was also found (Stade and Brambl, 1981) that cyanide-sensitive respiration is essential for germination. These authors state that in this fungus, unlike the situation in *Botryodiplodia*, cytochrome c oxidase is already present in assembled form in the mitochondria of conidia, even in strains which appear cytochrome c oxidase-deficient during log phase. However, their evidence that indeed the oxidase is assembled is not very convincing.

Although at this moment it is only speculation, it may well be that the presence on *Neurospora* mtDNA of a gene potentially coding for a proteolipid subunit of the mitochondrial ATPase complex also has a function in development (Van den Boogaart *et al.*, 1982). It may be that during spore germination the mtATPase complex contains mitochondrially synthesized proteolipid subunits, instead of the regular imported ones.

In yeast, a role for mitochondrial DNA in germination was established by Hartig *et al.* (1981). Germination of ascospores is under the influence of a mitochondrial gene, which is not identical to any known structural gene. It might be that the mutants, described by these authors showing blocked ascospore germination have a mutation in a regulatory region on the mtDNA necessary for the reinitiation of mitochondrial genetic activity during germination.

Finally, interesting observations have been made on the biogenesis of cytochrome c oxidase and ATPase during the cell cycle of synchronized yeast cells (Somasundaram and Jayaraman, 1981 a,b). For both enzymes it was found that cytosolically synthesized subunits accumulate in the cytosol during G_1 and early-S phases. The mitochondrially made subunits are synthesized during late-S phase, and only then the cytosolically made subunits move to the inner membrane. This interesting experimental system holds promises for future elucidation of regulatory processes.

SOME SPECIAL CASES

Most of the experiments reported in the preceding paragraphs have been performed with only a few selected organisms: rat, mouse and HeLa cells; yeast, *Neurospora* and some other fungi; a few protozoa. We pointed to the differences between the various species wherever useful, and also noticed that extrapolations from one organism to another are not always allowed.

In this paragraph we want to elaborate some more on the pecularities of *Saccharomyces*, the organism most amenable to the study of mitochondrial biogenesis and, nevertheless, in many cases an exception rather than an example. Furthermore, we will briefly summarize the differences between higher and lower eukaryotes. Finally, some attention will be paid to the mitochondria of plants.

Facultatively aerobic yeasts. The interrelations between the mitochondrial genome and the nuclear genome in facultatively aerobic yeast strains (*Saccharomyces*) are much more complex than in other eukaryotes. The cause of this complexity is the ability of these strains to live without functional mitochondria. The cells are then characterized by the absence of a functional respiratory chain and the absence of mitochondrial protein synthesis. This situation can be induced by environmental conditions: high concentrations of glucose, or the absence of oxygen. Also deletions in the mitochondrial genome or mutations in nuclear genes can cause the same phenotypical changes. Such mutants are easily recognizable on plates since their colonies are smaller ("petite") than normal colonies (Roodyn and Wilkie, 1968). It appears that only strains in which petites can be induced (petite-positive yeasts) can live without functional mitochondria. About the mechanisms responsible for the adaptations to high glucose concentrations or to anaerobic growth, little is known. It seems that mitochondrial protein synthesis is blocked in some way since the concentration of messenger RNAs and the ratio of RNAs are about the same in glucose-repressed and in derepressed mitochondria (Jakovcic *et al.*, 1979).

A further unique aspect of *Saccharomyces* is the existence of both haploid and diploid cells in stable form. Haploid cells are descendants from single ascospores, and can live stably until they can mate with haploid cells of the opposite mating type. The resulting diploid zygotes can divide and generate diploid progeny indefinitely. Only under special conditions diploid cells go into meiotic division to yield an ascus containing 4 haploid spores.

In general, all mitochondrial DNA is inherited strictly uniparentally both for animals, plants and other fungi. In some way, paternal mtDNA is always excluded from or degraded in the zygote. So the cells derived from the zygote, either by meiotic or by mitotic division, only contain the maternal mtDNA (Kroon *et al.*, 1978). *Saccharomyces*, however, contains mixed populations of mtDNA in the diploid cells arising after copulation (Coen *et al.*, 1969). Usually, after a few generations these populations segregate mitotically to yield varying

percentages of the parental phenotypes. The progeny cells are mixed
then, but the cells themselves contain only one type of mtDNA. A
further peculiarity of yeast is the frequent recombination occurring
in the zygotic stage between different mtDNAs. All these specialities
of *Saccharomyces* have made it the favorite organism of mitochondrial
geneticists. Fungi like *Neurospora* are only haploid, except for the
zygote: this diploid cell immediately goes into meiotic division to
yield ascospores again. Nonetheless, mitochondrial recombination could
be shown to occur in heterokaryons (Mannella and Lambowitz, 1978).
Heterokaryons are cells arisen by fusion between two different
mycelia. These heterokaryons have two different types of nuclei, and
in principle also two different types of mitochondria. These authors
showed that a limited degree of recombination also occurs in
Neurospora heterokaryons.

The differences between higher and lower eukaryotes. In this paragraph we will discuss some differences between the mitochondrial
genetic systems of mammals of lower eukaryotes which might have consequences for the way in which mitochondriogenesis is regulated. Some
of the points have already been raised above.
A. The size of the mtDNA molecules. The mammalian mtDNAs are circles
   of about 16,000 bp. Although some lower eukaryotes have a mt
   genome of about the same size, most have a much larger genome
   (Table 2). About the same number of genes are located on the
   genomes of both origins. The difference in length is mainly
   caused by the presence of intron sequences in genes and long inter-
   spacements between genes in lower eukaryotes. Therefore, the processing of mtRNAs in mammals seems a much simpler process than
   that in lower eukaryotes.
B. Replication of mtDNA. The way in which the mtDNA is replicated in
   mammals is quite different from the situation in *e.g.* yeast. The
   mammalian mtDNA has one defined origin of replication for each
   strand (Battey *et al.*, 1980). The replication is initiated by
   RNA-priming at the origin of the strand carrying most of the genes
   (the heavy strand) and proceeds unidirectionally. The starting
   point of the replication of the other (light) strand is located
   at about two-thirds of the genome length downstream from the
   heavy strand origin. The initiation of replication of the light
   strand starts only after passage of the newly synthesized heavy-
   strand beyond this origin. The replication is also unidirectional.

   About the replication of the mtDNA in lower eukaryotes little is
   known. On yeast mtDNA probably multiple origins exist. From the
   retention of specific wild-type sequences in different petites the
   conclusion has been drawn that seven or eight origins of replication are located on the mtDNA of *Saccharomyces cerevisiae* (De
   Zamaroczy *et al.*, 1980).

   Little research has been performed on the replication in other
   lower eukaryotes. From the composition of mtDNA of several mitochondrial mutants of *Aspergillus* it can be concluded that also in
   this organism multiple origins of replication exist (Lazarus
   et al., 1980).

C. The gene organization of mtDNA. One way to regulate gene
   expression is a specific gene organization. A well-known example

of this is the lactose operon in *Escherichia coli*. Three genes
for enzymes, involved in the catabolism of lactose, are located
close together and are regulated by one operator. Binding of a
repressor to a site on the DNA, called operator, inhibits the
transcription of all three genes. Although no evidence has been
found, it could be that the complete mt genome of mammals is like
an operon. Transcription of both strands is initiated at one point
allowing a simple repression of all genes. In lower eukaryotes a
"Jacob and Monod"-like repressor regulation of all mt genes is not
possible since transcription is initiated at different sites (see
3.1). This could be inferred from 5' end group determination of
yeast mtRNAs (Levens *et al.*, 1981) and from binding of RNA polymerase to *Paramecium* mtDNA (Seilhamer and Cummings, 1980)

It is striking that the gene order is strictly conserved among
mammals. It has been postulated that this gene order plays an
important role in the regulation of transcription. The rRNA genes
are close together and tandemly transcribed. In this way equal
amounts of both molecules are synthesized. Moreover these genes
are the first genes, larger than tRNA genes, to be transcribed
after initiation of transcription of the H-strand. Attardi (1980)
suggested that premature termination of transcription beyond the
rRNA cistron could be a mechanism to obtain higher amounts of the
rRNAs. Such a mechanism could probably not account for a more than
10-fold larger molar amount of rRNA as compared to mRNAs. Nevertheless, at least more than a 100 times higher concentration is
found. This indicates that other regulatory processes might also
be occurring. One could think of differences in metabolic stability of the mtRNAs.

In lower eukaryotes there are no indications that the gene order
is involved in regulation of transcription. In the first place
the arrangement of the mt genes in the lower eukaryotes is highly
divergent, even among yeasts. In the second place, the ribosomal
RNA genes are not adjacent and separately transcribed (Tabak
et al., 1982).

D. The mRNAs. Two important differences exist between the mature mt
mRNAs of both origins. The mRNAs of mammals start at their 5' end
with the startcodon AUG or only a few basepairs more (Montoya
et al., 1981). In this way specific regulation at the level of
protein synthesis initiation seems hardly possible. In contrast,
the mt mRNAs of lower eukaryotes contain considerably longer 5'
ends. On the other hand, the mRNAs of mammals have poly-A tails
of 35 to 55 residues at their 3' ends. In yeast and *Neurospora*
polyadenylated mRNAs have not been found (Groot *et al.*, 1974;
H. de Vries and J.C. de Jonge, unpublished results).

E. The protein products. Two differences are conspicuous when the mt
protein products of both origins are compared. First, in yeast and
Neurospora one ribosomal protein has been identified as mt translation product. In higher eukaryotes there are no indications for
a ribosome-associated mt translation product. The other difference
between the mt synthesized proteins from both origins is that in
Neurospora crassa and in yeast some of these proteins are synthesized as elongated precursor proteins. These fungi have, there-

fore, .one more opportunity to regulate the biogenesis of mitochondria, *viz*. via the proteolytic cleavage of these precursors.

Taking the data mentioned in this paragraph, it seems that in mammals
both the mtDNA and the expression of this genome are considerably
simpler than in lower eukaryotes. It may be that a more sophisticated
regulatory system is needed in the lower eukaryotes, in view of the
much larger variation in conditions in the life cycle and environment
of these cells.

Plant mitochondria. The investigation   of plant mitochondria has
always been hampered, and overshadowed, by the presence of the other
organelles, the chloroplasts. Only in the last decade or so, plant
mitochondria have become subject of much research. A comprehensive
review has been presented by Leaver and Gray (1982). In general,
plant mitochondria have larger DNAs than those from animals and fungi.
The size estimates range from 60 kbp for tobacco to 2,400 kbp for musk-
melon. Moreover, restriction enzyme analysis shows that considerable
variation in stoechiometry of fragments exists within one type of
mitochondria. The most probable explanation is that mtDNA populations
in plants are heterogeneous themselves. This heterogeneity is intramitochondrial, rather than intermitochondrial, since the restriction
patterns are strictly reproducible. It may be, as proposed by
Levings *et al*. (1979) that the plant mitochondrial genome is distributed among individual chromosomes. An alternative explanation arose
from size fractionation experiments, followed by cross-hybridization
(Dale, 1981). The homology between different size classes found in these
experiments, points to derivation of plant mtDNA from a single circular molecule followed by duplication, deletion rearrangements,
recombinations, etc.

Plant mtDNAs contain genes for tRNAs and for ribosomal RNAs. Remarkable here is that a 5S rRNA gene was found, which is not present, or
perhaps present in a completely different form, in other mitochondria. In general, there is the impression that the gene content of
plant mtDNA is higher than that of other mtDNAs.

Concluding this very brief summary, plant mitochondria are rather
different in many respects and promise to offer a fertile field of
research in the near future. Especially the intriguing male-sterility
phenomenon seems to be near its explanation soon.

CONCLUSIONS AND PROSPECTS

It will be clear to the reader of this chapter that, notwithstanding
the impressive knowledge gained in the last 5 to 10 years, many
questions regarding the interactions of the nuclear and the mitochondrial genetic systems remain unanswered as yet.

On of the problems being solved at this moment is the mechanism of
import of nuclear coded and cytoplasmically synthesized proteins. As
described at large by Neupert in this volume, many aspects of this
import process have been clarified by now. The assembly mechanism of
the multicomponent complexes of oxidative phosphorylation, on the
other hand, is not elucidated at all. Heme certainly plays a role, as
far as cytochromes are concerned. Also the electron flux through the
respiratory chain is of importance in the assembly, but the mediator

signal has to be found yet. Processing of mitochondrially synthesized
RNAs and proteins by nuclear coded enzymes appears to play a crucial
role, at least in fungi, in keeping the balance between the two systems, much more than direct coupling between cytoplasmic and mitochondrial protein synthesis.

The function of introns in mitochondrial genes of many lower eukaryotes may partly be to offer the possibility of regulation by
splicing. Animal mitochondria apparently do not have such possibilities, perhaps because the animal cell is an extremely constant environment to its mitochondria. Nevertheless, also in animals regulation is necessary to keep the total amount of mitochondrial protein
per cell constant. Future research, we expect, will be directed
towards:
1) the elucidation of the nature of the signal(s) from the electron
   transport chain to the nucleus which result in compensatory production of mitochondrial components;
2) the characterization of the nuclear genes regulating mitochondrial function;
3) the further understanding of phenomena like cytoplasmically determined sterility;
4) the control of mitochondrial biogenesis during development. The
   plants with their three genetic systems in one cell appear the
   most difficult to tackle in this respect.

ACKNOWLEDGEMENTS

Research carried out in the authors' laboratory was supported by a
grant from the Dutch Organization for the Advancement of Pure
Research (ZWO) to A.M. Kroon. We are indebted to Drs. A.M. Lambowitz,
C.J. Leaver, P.P. Slonimski, Y. Suyama and G. Schatz who kindly
provided us with copies of manuscripts still in press. Karin van Wijk
typed the manuscript and assisted with the lay-out, Theo Deddens
drew the figures. Finally, we thank Etienne Agsteribbe and Ab Kroon
for their critical reading of the manuscript.

REFERENCES

Aloni, Y. and Attardi, G., 1971. Symmetrical in vivo transcription
   of mitochondrial DNA in HeLa cells. Proc. Natl. Acad. Sci. U.S.,
   16, 1757-1761.
Anderson, S., Bankier, A.T., Barrell, B.G., De Bruijn, M.H.L.,
   Coulson, A.R., Drouin, J., Eperon, I.C., Nierlich, D.P., Roe, B.A.,
   Sanger, F., Schreier, P.H., Smith, A.J.H., Staden, R. and
   Young, I.G., 1981, Sequence and organization of the mammalian
   mitochondrial genome. Nature, 290, 457-464.
Attardi, G., Cantatore, P., Ching, E., Crews, S., Gelfand, R.,
   Merkel, C., Montoya, J. and Ojala, D. 1980. The remarkable features
   of gene organization and expression of human mitochondrial DNA, in:
   The Organization and Expression of the Mitochondrial Genome
   (Eds. Kroon A.M. and Saccone, C.) Elsevier/North-Holland, Amster-
   dam, 103-120.
Attardi, G. and Ojala, D. 1971. Mitochondrial ribosomes in HeLa
   cells. Nature, 229, 133-135.

Barath, Z. and Küntzel, H., 1972. Cooperation of mitochondrial and nuclear genes specifying the mitochondrial genetic apparatus in *Neurospora crassa*. Proc. Natl. Acad. Sci. USA, 69, 1371–1374.

Battey, J., Nagley, P., Van Etten, R.A., Walberg. M.W. and Clayton, D.A., 1980. Expression of the mouse and human mitochondrial DNA, in: The Organization and Expression of the Mitochondrial Genome (Eds. Kroon, A.M. and Saccone, C.) Elsevier/North-Holland, Amsterdam, 277–286.

Beauchamp,P.M. and Gross, S.R., 1976. Increased mitochondrial leucyl- and phenylalanyl-tRNA synthetase activity as a result of inhibition of mitochondrial protein synthesis. Nature, 261, 338–340.

Bertrand, H., 1980. Biogenesis of cytochrome *c* oxidase in *Neurospora crassa*: interactions between mitochondrial and nuclear regulatory and structural genes, in: The Organization and Expression of the Mitochondrial Genome (Eds. Kroon, A.M. and Saccone, C.) Elsevier/North-Holland, Amsterdam, 325–332.

Bertrand, H., Bridge, P., Collins, R.A., Garriga, G. and Lambowitz, A.M., 1982. RNA splicing in *Neurospora* mitochondria. Characterization of new nuclear mutants with defects in splicing the mitochondrial large rRNA. Cell, in press.

Bertrand, H., Szakacs, N.A., Nargang, F.E., Zagozeski,C.A., Collins, R.A. and Harrigan, J.C., 1976. The function of mitochondrial genes in *Neurospora crassa*. Can. J. Genet. 18, 397–409.

Borst, P., 1969. Biochemistry and function of mitochondria, in: Handbook of Molecular Cytology (Ed. Lima-de-Faria, A.) North-Holland, 914–942.

Borst, P., 1972. Mitochondrial nucleic acids. Annu. Rev. Biochem., 41, 333–376.

Brambl, R., 1980, Mitochondrial biogenesis during fungal spore germination: biosynthesis and assembly of cytochrome *c* oxidase in *Botryodiplodia theobromae*. J. Biol. Chem., 255, 7673–7680.

Brega, A. and Vesco, C., 1971, Ribonucleoprotein particles involved in HeLa mitochondrial protein synthesis. Nature New Biol., 229, 136–139.

Coen, D., Deutsch, J., Netter, P., Petrochilo, E. and Slonimski, P.P., 1969. Mitochondrial genetics: I. Methodology and phenomenology, in: Control of Organelle Development (Ed. Miller, P.L.) Cambridge University Press, 449–496.

Cohen, S.N. and Shapiro, J.A., 1980. Transposable genetic elements. Sci. Amer., 242, 36–45.

Coruzzi, G. and Tzagoloff, A., 1980. Assembly of the mitochondrial membrane system: nuclear suppression of a cytochrome *b* mutation in yeast mitochondrial DNA. Genetics, 95, 891–903.

Cottrell, S.F., 1981. Mitochondrial DNA synthesis in synchronous cultures of the yeast *Saccharomyces cerevisiae*. Exptl. Cell Res., 132, 89–98.

De Jong, L., Holtrop, M. and Kroon, A.M., 1978. Oxidative phosphorylation in mitochondria isolated from small intestinal epithelium of thiamphenicol-treated rats and control rats. Biochim. Biophys. Acta, 501, 405–414.

De Vries, H., De Jonge, J., Van 't Sant, P., Agsteribbe, E. and Arnberg, A., 1981, A "stopper" mutant of *Neurospora crassa* containing two populations of aberrant mitochondrial DNA. Curr. Genet., 3, 205–211.

De Vries, H. and Kroon, A.M., 1970. On the effect of chloramphenicol and oxytetracycline on the biogenesis of mammalian mitochondria. Biochim. Biophys. Acta, 204, 531-541.

De Vries, H. and Van der Hoogh-Schuuring, R., 1973. Physicochemical characteristics of isolated 55-S mitochondrial ribosomes from rat-liver. Biochem. Biophys. Res. Commun., 54, 308-314.

Dieckman, C.L., Pape, L.K. and Tzagoloff, A., 1982. Identification and cloning of a yeast nuclear gene (CBP1) involved in expression of mitochondrial cytochrome b. Proc. Natl. Acad. Sci., 79, 1805-1809.

Edwards, D.L. and Unger, B.W., 1978. Induction of hydroxamate-sensitive respiration in Neurospora mitochondria. Transcription of nuclear DNA is required. FEBS Lett., 85, 40-42.

Finzi, E., Sperling, M. and Beattie, D.S., 1981. Partial purification of cytosolic proteins which control yeast mitochondrial protein synthesis. J. Biol. Chem., 256, 11914-11922.

Groot, G.S.P., Flavell, R.A., Van Ommen, G.J.B. and Grivell, L.A., 1974. Nature, 252, 167-169.

Grossman, L.I., Goldring, E.S. and Marmur, J., 1969. Preferential synthesis of yeast mitochondrial DNA in the absence of protein synthesis. J. Mol. Biol., 46, 369-376.

Groudinsky, O., Dujardin, G. and Slonimski, P.P., 1981. Long range control circuits within mitochondria and between nucleus and mitochondria. 2. Genetic and biochemical analyses of suppressors which selectively alleviate the mitochondrial intron mutations. Mol. & Gen. Gen., 3, 493-504.

Hartig, A., Schroeder, R., Mucke, E. and Breitenbach, M., 1981. Isolation and characterization of yeast mitochondrial mutants defective in spore germination. Current Genet., 4, 29-35.

Jakovcic, S., Hendler, F., Halbreich, A. and Rabinowitz, M., 1979. Transcription of yeast mitochondrial DNA. Biochem. 18, 3200-3220.

Kroon, A.M., 1963. Protein synthesis in heart mitochondria. I. Amino-acid incorporation into the protein of isolated beef-heart mitochondria and fractions derived from them by sonic oscillation. Biochim. Biophys. Acta, 72, 391-402.

Kroon, A.M., Agsteribbe, E. and De Vries, H., 1972. Protein synthesis in mitochondria and chloroplast, in: The Mechanism of Protein Synthesis and its Regulation (Ed. Bosch, L.) North-Holland, Amsterdam, 539-589.

Kroon, A.M., De Vos, W.M. and Bakker, H., 1978. The heterogeneity of rat-liver mitochondrial DNA. Biochim. Biophys. Acta, 519, 269-273.

Kroon, A.M. and Saccone, C., Eds., 1980. The Organization and Expression of Mitochondrial Genome., Elsevier/North-Holland, Amsterdam.

Kück, U., Stahl, U. and Esser, K., 1981. Plasmid-like DNA is part of mtDNA in Podospora anserina. Curr. Genet., 3, 151-156.

Lazarus, C.M., Earl, A.J., Turner, G. and Küntzel, H., 1980. Amplification of a mitochondrial DNA sequence in the cytoplasmically inherited "ragged" mutant of Aspergillus amstelodami. Eur. J. Biochem., 106, 633-643.

Leaver, C.J. and Gray, M.W., 1982. Mitochondrial genome organization and expression in higher plants. Annu. Rev. Plant Physiol., in press.

Levens, D., Ticho, B., Ackerman, E. and Rabinowitz, M., 1981. Transcriptional initiation and 5' termini of yeast mtRNA. J. Biol. Chem., 256, 5226-5232.

Levings, C.S., Kim, B.D., Pring, D.R., Conde, M.F., Mans, R.J., Laughnan, J.R. and Gabay-Laughnan, S.J., 1980. Cytoplasmic reversion of cms-S in maize: association with a transpositional event. Science, 209, 1021-1023.

Levings, S.C.S., Shah, D.M., Hu, W.W.L., Pring, D.R. and Timothy, D.H., 1979. Molecular heterogeneity among mitochondrial DNAs from different maize cytoplasms, in: Extrachromosomal DNA, ICN-UCLA Symposium on Molecular and Cellular Biology (Eds. Cummings, D., Borst, P., Dawid, I.B., Weissmann, S. and Fox, C.F.) 63-73.

Lambowitz, A.M. and Zannoni, D., 1978. Cyanide-insensitive respiration in Neurospora, genetic and biophysical approaches, in: Plant Mitochondria (Eds., Ducet, G. and Lance, C.) Elsevier/North-Holland, 283-291.

LaPolla, R.J. and Lambowitz, A.M., 1977. Mitochondrial ribosomes assembly in Neurospora crassa. Chloramphenicol inhibits the maturation of small ribosomal subunits. J. Mol. Biol., 116, 189-206.

LaPolla, R.J. and Lambowitz, A.M., 1981. Mitochondrial ribosome assembly in Neurospora crassa. Purification of the mitochondrially synthesized ribosomal protein S-5. J. Biol. Chem., 256, 7064-7067.

Macklin, W.B., Meyer, D.J., Woodward, D.O. and Erickson, S.K., 1977. Chloramphenicol-sensitive labelling of protein in microsomes of Neurospora crassa. Nature, 269, 447-450.

Mannella, C.A. and Lambowitz, A.M., 1978. Interaction of wild type and poky mitochondrial DNA in heterokaryons of Neurospora. Biochim. Biophys. Res. Commun., 80, 673-679.

Martin, N.C. and Underbrink-Lyon, K., 1981. A mitochondrial locus is necessary for the synthesis of mitochondrial tRNA in the yeast Saccharomyces cerevisiae. Proc. Natl. Acad. Sci. US, 78, 4743-4747.

Martin, R.P., Schneller, J.-M., Stahl, A.J.C. and Dirheimer, G., 1979. Import of nuclear deoxyribonucleic acid coded lysine-accepting transfer ribonucleic acid (anticodon C-U-U) into yeast mitochondria. Biochemistry, 18, 4600-4605.

Montoya, J., Ojala, D. and Attardi, G., 1981. Distinctive feature of the 5' terminal sequences of the human mitochondrial mRNAs. Nature, 290, 465-570.

Nargang, F.E., Bertrand, H. and Werner, S., 1978. A nuclear mutant of Neurospora crassa lacking subunit 1 of cytochrome c oxidase. J. Biol. Chem., 253, 6364-6369.

Nass, M.M.K. and Nass, S., 1963. Intramitochondrial fibers with DNA characteristics. J. Cell Biol., 19, 593-629.

Neupert, W. and Rücker, A.V., 1976. Coördination of mitochondrial and cytoplasmic protein synthesis in Neurospora crassa, in: Genetics and Biogenesis of Chloroplasts and Mitochondria (Eds. Th. Bucher et al.) 231-238.

Nierlich, D.P., 1982. Fragmentary 5S rRNA gene in the human mitochondrial genome. Mol. and Cell Biol., 2, 207-209.

Nomura, M., Dean, D. and Yates, D.L., 1982. Feedback regulation of ribosomal protein synthesis in Escherichia coli. Trends in Biochem. Sci., 7, 92-95.

O'Brien, T.W., Denslow, N.D., Harville, T.O., Hessler, R.A. and
  Mathews, D.E., 1980. Functional and structural roles of proteins in
  mammalian mitochondrial ribosomes, in: The Organization and
  Expression of Mitochondrial Genome (Eds., Kroon, A.M. and
  Saccone, C.) Elsevier/North-Holland, Amsterdam, 301-305.
O'Brien, T.W. and Kalf, G.F., 1967. Ribosomes from rat liver mitochondria. J. Biol. Chem., 242, 2172-2185.
Ojala, D., Montoya, S. and Attardi, G., 1981. tRNA punctuation model
  of RNA processing in human mitochondria. Nature, 290, 470-475.
Ono, B., Fink, G. and Schatz, G., 1975. Mitochondrial assembly in
  respiration-deficient mutants of Saccharomyces cerevisiae. Effects
  of nuclear amber suppressors on the accumulation of a mitochondrially made subunit of cytochrome c oxidase. J. Biol. Chem.,
  250, 775-782.
Osinga, K.A. and Tabak, H.F. 1982. Initiation of transcription of genes for
  mitochondrial ribosomes RNA in yeast: comparison of the nucleotide
  sequence around the 5'-ends of both genes reveals a homologous
  stretch of 17 nucleotides. Nucl. Acids Res., in press.
Pica-Mattoccia, L. and Attardi, G., 1971. Expression of the mitochondrial genome in HeLa cells. V. Transcription of mitochondrial DNA
  in relationship to the cell cycle. J. Mol. Biol., 57,
  615-621.
Ray, D.B. and Butow, R.A., 1979. Regulation of mitochondrial ribosomal RNA synthesis in yeast. I. In search of relaxation of
  stringency Mol. Gen. Gen., 173, 227-238.
Ray, D.B. and Butow, R.A., Regulation of mitochondrial ribosomal
  RNA synthesis in yeast. II. Effects of temperature sensitive
  mutants defective in cytoplasmic protein synthesis. Mol. Gen. Gen.,
  173, 239-248.
Roodyn, D.B. and Wilkie, D., 1968. The biogenesis of mitochondria,
  Methuen, London.
Saltzgaber-Müller, J. and Schatz, G., 1978. Heme is necessary for the
  accumulation and assembly of cytochrome c oxidase subunits in
  Saccharomyces cerevisiae. J. Biol. Chem., 253, 305-310.
Schwab, A., 1973. Mitochondrial protein synthesis and cyanide-
  resistant respiration in copper depleted cytochrome c oxidase
  deficient Neurospora crassa. FEBS Lett., 35, 63-66.
Seilhamer, S.I. and Cummings, D.J., 1980. Determination of E. coli
  RNA polymerase binding sites in Paramecium mitochondrial DNA.
  Biochem. Intern., 1, 331-338.
Sevarino, K.A. and Poyton, R.O., 1980. Mitochondrial biogenesis:
  identification of a precursor to yeast cytochrome c oxidase sub-
  unit II, an integral polypeptide. Proc. Natl. Acad. Sci. US, 77,
  142-146.
Simpson, M.V., McLean, J.R., Cohen, G.I. and Brandt, I.K., 1957.
  In vitro incorporation of leucine-1-C^{14} into the protein of liver
  mitochondria. Fed. Proc., 16, 249-250.
Slonimski, P.P. and Tzagoloff, A., 1976. Localization in yeast mito-
  chondrial DNA of mutations expressed in a deficiency of cytochrome oxidase and/or coenzyme QH$_2$-cytochrome c reductase.
  Eur. J. Biochem., 61, 27-42.
Somasundaram, T. and Jayaraman, J., 1981. Synthesis and assembly of
  cytochrome c oxidase in synchronous cultures of yeast.
  Biochemistry, 20, 5369-5372.

Somasundaram, T. and Jayaraman, J., 1981. Synthesis and assembly of adenosinetriphosphate in synchronous cultures of yeast. Biochemistry, 20, 5373-5379.

Stade, S. and Brambl, R., 1981. Mitochondrial biogenesis during fungal spore germination: Respiration and cytochrome c oxidase in *Neurospora crassa*. J. Bacteriol., 147, 757.

Suyama, Y., 1982. Native and imported tRNAs in *Tetrahymena* mitochondria: two dimensional polyacrylamide gel electrophoresis, in: Mitochondrial Genes (Slonimski, P.P., Borst, P. and Attardi, G., Eds.) Cold Spring Harbor Press, 449-456.

Tabak, H.F., Grivell, L.A. and Borst, P., 1982. Transcription of mitochondrial DNA, in: CRC Critical Reviews in Biochemistry (Ed., Fasman, G.D.) CRC Press, Boca Raton, in press.

Terpstra, P. and Butow, R.A., 1979. The role of var 1 in the assembly of yeast mitochondrial ribosomes. J. Biol. Chem., 254, 12662-12669.

Todd, R.D., Buck M.A. and Douglas, M.G., 1981. Localization of unassembled subunits of the mitochondrial ATPase in an assembly-defective yeast nuclear mutant. J. Biol. Chem., 256, 9037-9048.

Tzagoloff, A., 1982. Mitochondria, Plenum Press.

Tzagoloff, A., Macino, G. and Sebald, W., 1979. Mitochondrial genes and translation products. Annu. Rev. Biochem. 48, 419-442.

Van den Bogert, C., Dontje, B.H.J., Wybenga, J.J. and Kroon, A.M., 1981. Arrest of *in vivo* proliferation of Zajdela tumor cells by inhibition of mitochondrial protein synthesis. Cancer Res., 41, 1943-1947.

Van den Boogaart, P., Samallo, J. and Agsteribbe, E., 1982. Similar genes for a mitochondrial ATPase subunit in the nuclear and mitochondrial genomes of *Neurospora crassa*. Nature, 298, 187-189.

Van 't Sant, P., Mak, J.F.C. and Kroon, A.M., 1980. Regulation of the synthesis of mitochondrial proteins: is there a repressor? in: The Organization and Expression of the Mitochondrial Genome (Eds., Kroon, A.M. and Saccone, C.) Elsevier/North-Holland, Amsterdam, 387-390.

Van 't Sant, P., Mak, J.F.C. and Kroon, A.M., 1981. Larger precursors of mitochondrial translation products in *Neurospora crassa*. Indications for a precursor of subunit 1 of cytochrome c oxidase. Eur. J. Biochem., 121, 21-26.

Weiss, H., Schwab, A.J. and Werner, S., 1975. Biogenesis of cytochrome oxidase and cytochrome b in *Neurospora crassa*. in: Membrane Biogenesis (Ed., Tzagoloff, A.) Plenum Press, New York, 125-153.

Werner, S., Machleidt, W., Bertrand, H. and Wild, G., 1980. Assembly and structure of cytochrome oxidase in *Neurospora crassa*, in: The Organization and Expression of the Mitochondrial Genome (Eds., Kroon, A.M. and Saccone, C.) Elsevier/North-Holland, Amsterdam, 399-411.

Winter, D.B., Bruyninckx, W.J., Foulke, F.G., Grinich, N.P. and Mason, H.S., 1970. Location of heme on subunits I and II and copper on subunit II of cytochrome oxidase. J. Biol. Chem. 255, 11408-11419.

Zamaroczy, M., Marotta, R., Faugeron-Fonty, G., Goursot, R., Mangin, M., Baldacci, G. and Bernardi, G., 1981. The origins of replication of the yeast mitochondrial genome and the phenomenon of suppressivity. Nature, 292, 75-78.

Zennaro, E., Falcone, C., Frontali, L. and Puglisi, P.P., 1977.
Dependence of cytoplasmic on mitochondrial protein synthesis in
Kluyveromyces lactis CBS2360. Mol. Gen. Gen., 150, 137-140.
Zitomer, R.S., Montgomery, D.L., Nichols, D.L. and Hall, B.D., 1979.
Transcriptional regulation of the yeast cytochrome *c* gene.
Proc. Natl. Acad. Sci. US, 76, 3627-3631.

Genes: *Structure and Expression*
Edited by A. M. Kroon
© 1983 John Wiley & Sons Ltd.

MOSAIC GENES AND RNA PROCESSING IN MITOCHONDRIA

L.A.Grivell, L.Bonen and P.Borst

Section for Molecular Biology, Laboratory of Biochemistry,
University of Amsterdam, Kruislaan 318, 1098 SM Amsterdam,
The Netherlands

INTRODUCTION

Mitochondria, like chloroplasts but unlike every other cytoplasmic
organelle in the eukaryotic cell, contain DNA and a genetic system
capable of expressing the genes within it. This DNA codes for
various components of the mitochondrial translation system and for a
limited number of proteins which go to make up certain respiratory
enzymes in the mitochondrial inner membrane. It, therefore, makes a
small, but highly essential, contribution to the process of mitochondrial biogenesis.
    Since the mid 1960's, when many mitochondrial DNAs (mtDNAs)
were physically characterized for the first time, a powerful combination of biochemical and molecular genetics, coupled recently with
rapid DNA sequencing techniques has led to the identification of
genes in mtDNA, to a reasonably complete picture of their structure
and to the first inklings of how their expression may be regulated.
Some of these genes are split and the purpose of this brief essay
will be to review what is now known about the structure of such
genes, their mode of expression and their evolution. Other aspects
of mitochondrial gene organization have been dealt with extensively
in recent reviews (Dujon, 1981; Borst et al., 1982; Grivell, 1982a;
Tabak et al., 1982) and form the subject of a recent symposium
volume (Slonimski et al., 1982).

FEATURES OF MITOCHONDRIAL GENE ORGANIZATION

Contrary to what this heading may imply, there is no single typical
form of mitochondrial gene expression. For all their relative
simplicity, mitochondrial genomes are diverse in form, complexity
and mode of gene organization. Overall sizes range from 14-18 kb in
higher animals and insects, through 19-108 kb in various fungi, to
upwards of 250 kb in higher plants (Borst et al., 1982). Despite
their 5-fold differences in size, the mitochondrial genomes of man
and the yeast Saccharomyces contain roughly the same basic information and the difference in size can be almost entirely accounted for
by differences in gene organization (Borst and Grivell, 1981a).
Genes in mammalian mtDNAs are tightly packed on the genome and they
do not contain introns (Anderson et al., 1981). The same genes in
yeast mtDNA are separated by vast expanses of AT-rich DNA and some
of them are multiply split so that they extend over inordinately

280

large segments of the DNA (Fig. 1).

    Study of plant mitochondrial genes is still in its infancy, so that it is still anyone's guess how much of the large genome size is due to the presence of genes absent from other mtDNAs. A slightly higher coding potential is suggested by the complexity of the pattern of plant mitochondrial translation products, compared with yeast or man (Leaver and Gray, 1982), but this still leaves much of the genome unaccounted for. Moreover, there are wide variations in mtDNA complexity between even closely related plants and this has

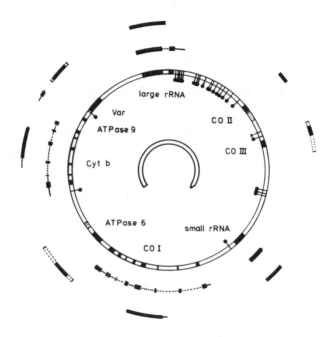

Fig. 1. Gene organization and transcription of yeast mtDNA. Genes for the two mitochondrial rRNAs and major proteins are shown as solid blocks on the main circle and are accentuated by shading. The transcripts of these genes are shown on the outer circles, with the likely structures and map positions of the primary transcripts and mature RNAs given on the middle and outer circles, respectively. Uncertainties in sequence composition and position are indicated by dotted segments. ⌶ represents a tRNA gene and ⅄ is the gene for tRNA^thr located on the DNA strand opposite to that containing all other known genes. Stippled blocks in the main circle are unassigned reading frames outside genes (those in introns are not shown). Co I, II, III: genes for subunits I-III of cytochrome c oxidase; ATPase 6, 9: genes for subunits 6 and 9 of the ATPase complex; Cyt. b: the gene for cytochrome b; Var: the gene for a ribosome-associated protein, defined by the var1 locus. Taken from Tabak et al. (1982).

prompted the suggestion that much of the plant mitochondrial genome may either fulfil a sequence-independent function, or even have no function at all (Ward et al., 1981).

SPLIT GENES IN FUNGAL mtDNAs: UNUSUAL ASPECTS

a) Optional introns. Although at least one gene in a plant mtDNA is known to be split (Fox and Leaver, 1981), attention so far has been focussed almost exclusively on the unusual features of introns in fungal mtDNAs. In yeast, the best studied organism, introns are restricted to only three genes, those for the large ribosomal RNA (rRNA), cytochrome b (cob) and subunit I of cytochrome c oxidase (oxi3) (Table I) and this is also true for Aspergillus (Waring et al., 1981; Netzker et al., 1982). Whether Neurospora will also conform to this pattern is not certain, since sequence analysis of this mtDNA is far from complete. In the only strain examined so far, the gene for the large rRNA was found to be split (Hahn et al., 1979; Yin et al., 1982), but that for subunit I was continuous (De Jonge and De Vries, 1982).

In Saccharomyces (Table I) and probably Aspergillus too (Earl et al., 1981), the exact structure of the split genes is strain-dependent, with certain introns being present in some strains but not in others (Fig. 2). Such introns have been termed 'optional' (Borst, 1980) and their presence or absence is largely responsible for the strain-dependent differences in the size of yeast mtDNA observed in early work (Sanders et al., 1977). Whether all introns in mtDNA are optional is unresolved as yet, since definition of an intron as such is dependent on finding a strain that lacks it. It is possible, however, that some introns - due to the information they contain and the role their translation products may play in RNA processing - are optional only in certain contexts i.e. in a

TABLE I

MITOCHONDRIAL GENES AND THEIR INTRONS IN YEAST

Adapted from Borst (1981) and based on sequence data compiled by Grivell (1982b). Information on var1 is from Hudspeth et al. (1982).

Gene	Number of introns	
	Total	Optional
Large rRNA	1	1
Cytochrome b	5*	5
Cytochrome c oxidase, subunit I	9*	5*
Small rRNA	0	0
tRNAs (25 genes analysed)	0	0
ATPase complex, subunits 6 and 9	0	0
Cyt. c oxidase, subunits II and III	0	0
Ribosome-associated protein (var1)	0	0

* Minimal estimate.

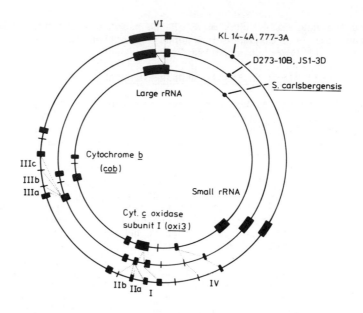

Fig. 2. The optional introns in yeast mtDNA. The three
concentric circles represent the mitochondrial genomes of
a number of commonly used yeast strains, that differ in
the organization of their split genes as the result of the
presence or absence of optional introns. These introns
correspond to the major insertions in yeast mtDNA described originally by Sanders et al. (1977). Nomenclature
is according to these authors, modified in the light of
more recent mapping or DNA sequence data (Grivell, 1982b).
The yeast strains shown are S. carlsbergensis and the S.
cerevisiae strains KL14-4A, 777-3A, D273-10B and JSI-3D.

particular nuclear background, or when other mitochondrial introns
are modified or absent (see below and Turner et al., 1982).

b) Circular RNAs. An unusual phenomenon, associated with the splicing
out of certain introns in yeast mitochondria, is that of RNA circularization (Arnberg et al., 1980). The circular RNAs represent the
greater part, if not the entire excised intron, and they are derived
from three introns in the gene for cytochrome c oxidase subunit I
(I1, I2 and I5) and the first intron of the long form of the cytochrome b gene (Halbreich et al., 1980; Hensgens et al., 1982a). They
appear to be covalently closed, but closure may not be achieved by a
normal phosphodiester bond because the enzyme reverse transcriptase
is unable to proceed across the joint (Bonen, L., unpublished
observation). Whether the circles fulfil an independent function, or
whether they are merely side products of an unusual splicing mechanism, is not yet clear. In Tetrahymena the single intron in the gene
for the large cytoplasmic rRNA has recently also been shown to
accumulate after excision as a covalently-closed circle and it has
been suggested that circle formation may in fact be an integral part

of the splicing mechanism (Cech et al., 1981). However, in yeast mitochondria, circle formation appears to be limited to the introns mentioned, so that different mechanisms of splicing must exist alongside each other. Quite what distinguishes introns that give circles from those that do not is not at all clear. Two of them (aI1 and aI2; subunit I gene) are dependent on mitochondrial protein synthesis for their excision and translation of the reading frames located within them may be a pre-requisite for splicing (see below), but the remaining two are untranslatable over the greater part of their lengths and they can be excised in the absence of mitochondrial translation. All four introns are highly homologous in a region close to their 3'-ends (Fig. 3) and they display a short homology at their 5' exon-intron borders which is not seen in other introns. Both areas could define sequences critical for excision and/or circularization and this idea has been lent credence by recent studies with splicing-deficient mutants (Schmelzer et al., 1982).

c) <u>Some introns contain information required for their own excision</u>. A puzzling observation, made during early studies on yeast mutants disturbed in their ability to synthesize cytochrome <u>b</u>, was that those with intron mutations were deficient in the splicing of precursors to the messenger RNA (mRNA) for this protein (Church et al., 1979; Van Ommen et al., 1980). At first, it was thought that the mutants, which were complementable in <u>trans</u>, defined RNA sequences within the introns that guide splicing by virtue of their primary or secondary structures. Subsequently, it became clear that these introns in fact encode proteins required for RNA processing. The case for this unusual situation is detailed in the following section.

INTRON-ENCODED RNA PROCESSING PROTEINS

a) <u>The box3 maturase</u>. Fig. 4 outlines the way in which the <u>box</u>3 (or bI2) maturase, a protein encoded in part by the second intron in the cytochrome <u>b</u> gene, is thought to act in the processing of precursor mRNAs (Lazowska et al., 1980). An early step in processing of the primary transcript of the gene is the removal of the (untranslatable) first intron by a nuclear-coded splicing enzyme. This fuses the first two exons for cytochrome <u>b</u> to a reading frame in intron 2, thus generating an mRNA for an exon-intron fusion protein. This protein, either alone or in conjunction with a nuclear-coded protein, has RNA processing activity. It mediates the splicing of intron 2 sequences, thereby destroying the mRNA that directed its synthesis. Its level within the cell is thus self-regulated, a phenomenon which has been called 'splicing homeostasis'.

Although indirect evidence for this scheme has been adduced by several groups (see Dujon, 1981 for review), direct confirmation came from DNA sequence analysis of intron 2 in both wild-type and splicing-deficient mutants (Lazowska et al., 1980). These mutants accumulate novel mitochondrial translation products, immunologically related to cytochrome <u>b</u>, whose size is dependent on the nature and position of the mutation (Kreike et al., 1979). Moreover, some are phenotypically suppressible either by mitoribosomal mutations affecting translational fidelity, or by growth in paromomycin, a

284

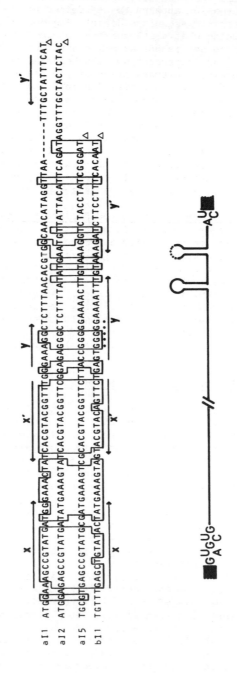

Fig. 3. Homologous sequences and secondary structures in introns that give rise to circular RNAs. The upper part shows a region of high homology at the 3'-end of those introns that accumulate after excision as single-stranded circular RNAs. Regions identical in all four sequences are boxed and splice points are indicated by ▲. X, X' and Y, Y' are sequences potentially capable of forming hairpin structures (lower panel). The dotted region in bI1 identifies a sequence altered in a splicing-deficient mutant. Adapted from Schmelzer et al. (1982).

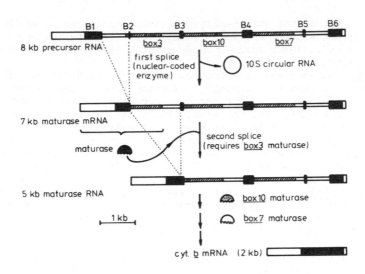

Fig. 4. A maturase model for RNA splicing in yeast
mitochondria. The figure shows the initial steps in the
processing of transcripts of the long form of the gene for
cytochrome b in yeast mtDNA and the involvement of
intron-encoded proteins in RNA splicing. Exons are shown
as solid blocks: reading frames within introns are cross-
hatched. Open bars are untranslated regions in the various
transcripts. See text for further explanation. Adapted
from Borst and Grivell (1981b).

compound thought to induce misreading by mitochondrial ribosomes.
DNA sequence analysis reveals that splicing deficiency can be
correlated with alterations in the bI2 reading frame, with both
missense and nonsense mutations being effective in blocking splicing.
The novel translation products are, therefore, likely to represent
defective maturases, incapable of inactivating the mRNA that encoded
them. Phenotypic suppression of the mutations by misreading suggests
that maturases are required in only catalytic amounts. Presumably
even a low level of misreading can provide enough active product to
overcome the mutational block.

b) The box7 maturase and the box effect. The fourth intron (bI4) of
the cytochrome b gene contains a reading frame linked in phase with
the upstream exon and capable of specifying a protein of 384 amino
acids (Nobrega and Tzagoloff, 1980). Trans-recessive, splicing-
deficient mutants have been located in the carboxy-terminal part of
this reading frame (box7; Fig. 4) and these also accumulate novel
proteins, that possibly represent defective maturases (Mahler et
al., 1982; De la Salle et al., 1982; Weiss-Brummer et al., 1982).
Unlike the situation in intron 2, these proteins are not immunologic-
ally-related to cytochrome b and they are too short (circa 27 kD) to
be fusion proteins that combine sequences from exons 1-4 and the
intronic URF. It has recently been suggested that only the carboxy-

terminal part of the URF specifies active maturase and that this
arises by proteolytic cleavage of the exon-intron fusion protein
(Mahler et al., 1982; Weiss-Brummer et al., 1982). As yet, however,
direct proof of this is lacking.

Box7 mutants are pleiotropic. Besides being deficient in
cytochrome b, they lack subunit I of cytochrome c oxidase and this
derives from incomplete or incorrect processing of the pre-mRNA for
the oxidase subunit (Church et al., 1979; Van Ommen et al., 1980).
Deletion mutants of cytochrome b are similarly oxidase-deficient and
this has led to the idea that expression of the cytochrome b gene
generates a factor required for processing of transcripts of the
oxidase gene (Alexander et al., 1980). A series of elegant genetic
experiments has traced this link in the expression of the two genes
(termed the box effect) to bI4 (Dhawale et al., 1981) and several
lines of evidence now strongly suggest that the RNA maturase encoded
by bI4 is also involved in excision of aI4, an intron which is
highly homologous to bI4 itself (De la Salle et al., 1982).

c) Other intron-encoded proteins. It seems unlikely that the principl
of splicing homeostasis is limited to the two examples given above.
Other introns in yeast, Neurospora and Aspergillus also contain
reading frames (Bonitz et al., 1980; Dujon, 1980; Waring et al.,
1981; Hensgens et al., 1982b; Netzker et al., 1982; Yin et al.,
1982) and in several cases, intron excision is dependent on mitochondrial translation (Hensgens et al., 1982a). It, therefore, seems
reasonable to assign a function in RNA processing to the products of
these URFs too, even though the evidence for this view is so far
largely circumstantial. The strongest support for maturase function
comes from the finding that the URFs display varying degrees of
sequence conservation at the amino acid level (see below). This
indicates the existence of similar constraints on protein structure
and implies that the products of the various URFs have similar
functions. This finding effectively rules out so-called 'ribosome-
stretch' models for mitochondrial splicing (Borst, 1981a; Schmelzer
and Schweyen, 1982), in which splicing is viewed as being entirely
dependent on translation per se (for example, because movement of
ribosomes over the intron may be required to induce an RNA secondary
structure favourable to splicing). In such a case, specific conservation of sequence would be expected to exist at the nucleotide rather
than protein level.

MATURASES: MODELS AND MECHANISMS

As yet, not a single active RNA maturase has been isolated. Conclusions about their involvement in splicing are, therefore, indirect,
based on the DNA sequence of the URFs themselves and the nature of
conserved sequence features within various introns. What do these
sequences tell us about maturases and the mechanism of splicing?

a) URF structure and maturase function. All URFs code for proteins
rich in basic and hydrophobic residues and the overall character of
the sequences is suggestive of globular, non-membrane-associated
proteins. Bonitz et al. (1980) were the first to note that some URFs
in yeast mtDNA display sequence homology and this observation has

been extended by Hensgens et al. (1982b) to include several other introns of the cob and oxi3 genes, the single intron of the 21S rRNA gene and an URF located in the region between the (unsplit) genes for cytochrome oxidase subunits II and III. The sequence organization of this group of URFs is shown in Fig. 5. Within the central region of each, a stretch of approx. 115 amino acids shows considerable homology at both nucleotide and amino acid levels. This conserved section is bounded in each case by nonapeptide motifs which show the consensus composition:

$$(\text{apolar})_2 \; \text{gly} \; (\text{apolar})_2 \; \binom{\text{asp}^-}{\text{glu}} \; \binom{\text{gly}}{\text{ala}} \; \text{asp}^- \; \text{gly}$$

In intron aI4, the sequences concerned are LAGLIDGDG and FIGFFDADG. We have, therefore, assigned the generic description LAGLI/DADG to them and will refer to the central region as the LAGLI/DADG block. Within this region highly similar distributions of charged residues, or of residues which disrupt α-helical structure, point to the existence of similar constraints on protein secondary structure and these are clear indicators of a crucial role of this segment in maturase function.

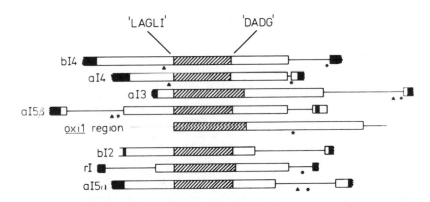

Fig. 5. Conserved sequence elements in yeast mitochondrial URFs. The figure compares URFs within genes for cytochrome b (bI2, bI4), subunit I of cytochrome c oxidase (aI3, aI4, aI5α and aI5β), the large rRNA gene (rI) and the non-intronic URF located downstream of the gene for subunit II of cytochrome c oxidase. Exons are shown by solid bars, URFs by open bars. The hatched areas represent segments of circa 300 bp, showing conserved features at the protein level. These segments are flanked by characteristic nonapeptide motifs, designated in abbreviated form as 'LAGLI' and 'DADG' (one-letter code for amino acids). ▲ and * identify sequence elements identical with, or closely resembling, the cis-dominant box9 and box2 'signal' sequences, described by De la Salle et al. (1982) and Weiss-Brummer et al. (1982). Taken from Hensgens et al. (1982b).

What the role in RNA processing may be in the case of the non-introncic URF downstream of the gene for cytochrome oxidase subunit II and of the 21S rRNA URF is not clear. The former might be involved in the processing of precursors to the mRNA for subunit II (cf. Fox and Boerner, 1980). The latter is not strictly essential for rRNA maturation, however, since cytoplasmic petites, which lack mitochondrial protein synthesis, can still excise the 21S rRNA intron precisely (Tabak et al., 1981). It is possible that the 21S rRNA URF product increases efficiency of splicing in some way.

Outside the LAGLI/DADG block, sequence comparisons bring out two additional points of interest in relation to function. First, the region immediately downstream of the block is also likely to have an important functional role. It shows signs of sequence conservation (approximately 47% in aI4 and bI4, including two insertions) and perhaps more important, numerous trans-recessive mutations affecting splicing (box3 in bI2, box7 in bI4) have been identified within it. In contrast, the region immediately preceding the LAGLI/DADG block is much less conserved, in that radical differences both in sequence and in length are found. Even between the closely related aI4 and bI4 URFs, there is only 35% identity and this requires introduction of four insertions. In the aI5β URF, this region contains a GC cluster, similar to those found at other sites in yeast mtDNA and this codes for a stretch of extra 14 amino acids (Hensgens et al., 1982b). These observations all argue against this region having an important role in the functional maturase, a view which is reinforced by an observation concerning the cis-acting box9 sequence, which is located within it (Fig. 5). This sequence is believed to be some kind of splicing signal (see below). It is identically conserved in both aI4 and bI4. Nevertheless, a deletion/-insertion has occurred in one of these frames at a site immediately preceding the box9 motif, causing a frame shift, which results in completely unrelated protein sequences upstream from this stretch.

Finally, it should not be forgotten that the box3 maturase is in reality a fusion protein, whose amino-terminal 143 residues are specified by the first two exons of the cytochrome b gene (Lazowska et al., 1980). This portion of the protein seems to contribute little to catalytic activity because missense mutations in exon 1 rarely - if ever - have an effect on splicing. However, if other URFs form part of similar fusion proteins, then the amino-terminal regions of these will differ, dependent on the location of a given URF within the structural gene and it is difficult to see how maturase activity can escape influence by such drastic differences in length, sequence and folding properties. In this context, the suggestion that the active box7 maturase arises by proteolytic cleavage of such a fusion protein acquires extra significance (Mahler et al., 1982; Weiss-Brummer et al., 1982). It is clear that future work should be directed at a better characterization of the 27-kD polypeptide, which is thought to be the defective maturase accumulated by box7 mutants and which may arise by a proteolytic cleavage upstream of the LAGLI/DADG block.

b) Mutations and maturase function: the mim 2-1 mutation. Independent confirmation that sequences within the LAGLI/DADG block play a critical role in maturase function is provided by the mim 2-1

mutation (Dujardin et al., 1982). This was isolated as one of the
many extragenic suppressors of intron splicing mutations (Groudinsky
et al., 1981) and it confers two remarkable properties on the cell.
First, in its presence, the cytochrome b gene and the RNA maturase
encoded by bI4 are no longer required for splicing of transcripts
from oxi3. Second, expression of the cytochrome b gene in mutants
with a defective bI4 maturase is once again possible. The simplest
explanation for both changes is that the mim 2-1 mutation induces
the synthesis of a novel aI4 maturase, with an activity which is
different from that in wild-type. This maturase can presumably act
as a substitute for a defective bI4 maturase (Fig. 6).

What is the mim 2-1 mutation and why should it have such
dramatic effects on the substrate specificity of the aI4 maturase?
DNA sequencing reveals a G→A transition at a site within the
LAGLI/DADG block of the aI4 reading frame, where it produces a
glutamate to lysine substitution. Dujardin et al. (1982) have
proposed that this replacement, which changes the overall charge of
the protein by +2, may bring about a local increase in positive
charge in a region of the protein that is essential for catalytic
activity. The mim 2-1 change occurs in a predicted α-helical region
of 15-18 amino acids, that occupies a homologous position in the bI4

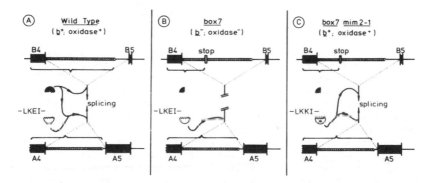

Fig. 6. A model for the box effect: an interaction between
two mitochondrial genes at the level of RNA splicing. The
figure shows portions of pre-mRNAs transcribed from the
genes for cytochrome b (cob) and subunit I of cytochrome c
oxidase (oxi3), with exons depicted by solid blocks and
intron reading frames by hatched areas. Panel A shows the
normal situation, in which the maturase encoded by cob
intron bI4 helps catalyze the excision of intron sequences
from the pre-mRNAs of both genes. In panel B a nonsense
mutation in the box7 locus has blocked synthesis of the
bI4 maturase and prevents splicing of both intron sequences
(the box effect). In panel C, synthesis of both cytochromes
has been restored by introduction of the mim 2-1 mutation
into the box7 mutant. The protein translated from aI4 has
acquired activities that allow it to replace the bI4
maturase in both its functions. These new activities
result from a glu (E) to lys (K) replacement. See text for
discussion. Adapted from Dujardin et al. (1982).

and aI4 URFs. The α-helix of wild-type bI4 is more basic than that of wild-type aI4 by 2-4 charges, dependent on strain polymorphism, so that <u>mim</u> 2-1 should make the aI4 translation product more similar to its b<u>I4</u> counterpart.

This proposal seems reasonable, but it also has its puzzling aspects. For example, one of the strain polymorphisms in wild-type bI4 takes the form of a lysine-glutamate substitution in the same α-helical stretch, only 5 amino acid residues upstream of the <u>mim</u> 2-1 change (<u>cf</u>. Nobrega and Tzagoloff, 1980), yet there are no appreciable differences in the splicing of bI4 in either strain. Moreover, <u>mim</u> 2-1 introduces a positive charge at a position occupied by glutamine in bI4, adjacent to a pre-existing positively-charged residue (lysine). Both points tend to play down the relevance of overall positive charge and they suggest that the change in activity induced by <u>mim</u> 2-1 involves more subtle interactions.

Finally, it is intriguing that of five sequence differences observed for intron bI4 between the two <u>Saccharomyces cerevisiae</u> strains D273-10B and 777-3A (De la Salle et al., 1982), four are located in the LAGLI/DADG block. The changes are without obvious effects on splicing. Perhaps a more extensive search for polymorphisms in rather more distantly-related strains will lead not only to better ideas about sequences that are important for maturase function, but also to testable hypotheses for the role of maturases in RNA splicing.

c) <u>Splice points</u>. Alignment of the sequences around splice points in eukaryotic and viral genes has led to the clear definition of a preferred primary structure at the junction (Mount, 1982). Nowhere near this amount of unanimity is found at mitochondrial splice points, however and there is little homology detectable with the eukaryotic consensus sequence (Fig. 7). Mitochondrial introns generally lack the long pyrimidine tract found at the 3'-end of introns in eukaryotic and viral genes, but some do have shorter tracts close to the 3'-exon/intron border. Certain introns, like the aI1-aI2 or aI4-bI4 pairs, show high homologies at their borders, but this may be part of the generally high level of sequence similarity exhibited by these introns, rather than a specific constraint at the splice points. This degree of diversity is perhaps to be expected, if mitochondrial splicing follows a key-in-lock type of mechanism, in which individial maturases confer intron specificity on a splicing complex. Nevertheless, in view of the underlying similarities between the various URFs, it would be somewhat surprising if common mechanisms were not to be involved in at least some steps of the splicing reaction. As the next section will show, this may well be the case.

d) <u>Cis-acting elements in RNA splicing</u>. Recent study of splicing-deficient mutants in bI4 has led to the characterization of clusters of mutations that can be distinguished from <u>box</u>7 mutations in being <u>cis</u>-dominant in complementation tests (De la Salle et al., 1982). Such mutants are, moreover, defective only in the processing of transcripts of the cytochrome <u>b</u> gene, so that they most probably define RNA sequences essential for splicing of bI4 itself rather than sequences involved in the production or specification of an

Cytochrome b

```
bI1   TATGGACAGA   gtgagacaa .... ctttcacaat   TGTCAC
bI2   ACATTGAGGT   aatataaat .... gagttacaag   GCACTA
bI3   GAGGTGGGTT   taatataga .... acattaaaag   CTCAGT
bI4   ACTTTAGGTC   aaaatatgg .... ttataaaagc   ATCCTG
bI5   GCATCTATTG   atattaaaa .... ataaattgtg   TACCTG
```

Cytochrome c oxidase sub. I

```
aI1    TTATTTAATG   gtgcgcctc .... gctatttcat   TTTTAG
aI2    ATGATTTTCT   gtgcgccgt .... gctactctac   TCTTAG
aI3    GGTTTTGGTA   accaaaaaa .... aataaaatga   ACTATT
aI4    TTCTTTGGTC   aaacagtgg .... tataacaagc   ACCCTG
aI5α   TCTCATGATT   aataaatcc .... taaataaaag   AGCTCT
aI5β   TTCCACGATA   ttaatttaa .... tttttatga    CTTACT
aI5γ   GGACATTTTC   gagcggtct .... ctatcgggat   ACTATG
```

Large rRNA (21S)

```
       CGCTAGGGAT   aatttaccc .... aaaaaatttg   AACAGG
```

Eukaryotic consensus seq.

```
               AG   gtRagt       .... Y-YYY-Yag   G
```

Fig. 7. Splice points in yeast mitochondrial genes. In
this compilation, exon sequences are shown in upper-case,
intron sequences in lower-case. Where repeated sequences
at the exon-intron boundaries prevent an unambiguous
indication of the exact splice point, the sequences have
been aligned to maximise homology and the alternative
splice points are indicated by short vertical lines on
either side of the main boundaries. Of the splice points
listed, only those for bI4, bI5 (Bonitz et al., 1982) and
aI5γ (Bonen, L., unpublished) were determined directly by
sequence analysis of mRNA. Others were deduced from
indirect evidence in one of two ways: a) sequence comparisons of 'short' and 'long' strains: bI1 and bI2 (Lazowska
et al., 1980); bI3 (Lazowska et al., 1981); rI (Dujon,
1980; Bos et al., 1980); aI5α, aI5β (Hensgens et al.,
1982b). b) Optimal sequence alignments with the unsplit
genes in human and Neurospora mitochondria (Anderson et
al., 1981; De Jonge and De Vries, 1982): aI1, aI2, aI3 and
aI4. In these cases, the splice points shown differ from
those given in the original published sequence (Bonitz et
al., 1981). At the position indicated by * in exon B4, a
G → A transition is likely to lead to a block in splicing
(De la Salle et al., 1982).

active maturase. Initially thought to represent exon-intron junction changes, these cis-dominant mutations have turned out to be situated well within the intron (Fig. 5). Two of them, designated originally as box9 and box2, are situated about 300 bp and 25-115 bp from the 5'- and 3'-splice points, respectively. Box9 thus lies within the URF, just upstream of the LAGLI/DADG block, while box2 is in the untranslatable region immediately downstream of the URF. Sequences identical to or closely resembling both elements are present in many other introns too, including those that lack URFs (Hensgens et al., 1982b). In aI4, mutations within the box9 analogue have been correlated with a block in the synthesis of subunit I of cytochrome oxidase, resulting presumably from impaired splicing (Netter et al., 1982).

The box9 sequence shows complementarity to part of the small rRNA and mutations affecting splicing alter the extent of this. This fact, together with the situation of box9 within the bI4 URF, was originally taken as support for the idea that this element could have a regulatory effect on splicing via an interaction with the mitochondrial ribosome (Jacq et al., 1982; De la Salle et al., 1982). This proposal is, however, inconsistent both with the occurence of box9-like sequences in untranslatable introns and with its location at different sites in those introns containing URFs (Fig. 5). Clearly, any function ascribed to either box9 or box2 should be independent of their position within the intron.

Perhaps one clue to what this function may be is given by the finding that a number of conserved sequence elements within introns contribute to elaborate, yet strikingly similar potential RNA secondary structures (Michel et al., 1982). Box9 and box2 can be brought into close proximity by internal folds (Hensgens et al., 1982b) and, perhaps more important still, such folds make it possible to align the splice junctions (Davies et al., 1982). It is, therefore, possible that the cis-acting function is provided by the combined sequences of both elements and that these form not a recognition site for splicing enzymes, but a structure that promotes splicing by aligning the sequences to be cut and joined. In other words, we may have come full circle back to the old idea of a guide RNA or shoe-lace model for splicing (cf. Bos et al., 1980)!

INVOLVEMENT OF NUCLEAR GENES IN MITOCHONDRIAL SPLICING

The novelty of the concept of splicing homeostasis has tended to direct attention away from the fact that the nucleus is also likely to play an important role in the control of mitochondrial splicing events. This has been clearly brought out by work with yeast. First, certain introns can be excised precisely in cytoplasmic petite mutants, which lack mitochondrial protein synthesis, showing that splicing is being carried out by nuclear-coded, imported enzymes (Church and Gilbert, 1980; Grivell et al., 1980; Halbreich et al., 1980; Merten et al., 1980; Tabak et al., 1981). Second, nuclear mutants, disturbed in mitochondrial splicing reactions, have been isolated and partially characterized (Dieckmann et al., 1982). These mutants display a pet (respiratory-deficient) phenotype and - rather surprisingly - they fall into a large number of distinct complementation groups. For example, splicing of the pre-mRNA transcribed from

the short form of the cytochrome b gene, which contains only two
introns, is dependent on functions defined by at least five complementation groups (Dieckmann, C.L. and Tzagoloff, A., personal
communication). The defects in splicing are highly specific, often
involving an inability to cut or ligate the ends of only a single
intron in the genes for either cytochrome b or subunit I of cytochrome c oxidase. Different complementation groups can affect the
same intron, which may itself - as in the case of the fourth intron
in the long cytochrome b gene - contribute its own maturase to the
splicing reaction. With such a degree of complexity, it would almost
seem as if the possibilities for control of mitochondrial splicing
have reached bureaucratic proportions! It should not be forgotten,
however, that complementation groups are not necessarily synonymous
with genes. If the enzymes which catalyze mitochondrial splicing are
multi-subunit complexes, then intragenic complementation becomes a
distinct possibility and this would simplify interpretation of the
genetic data somewhat. Information on this point should shortly be
available, since cloned gene sequences, capable of restoring the
wild-type state to pet mutants by transformation, have been isolated
and in some cases sequenced (Dieckmann et al., 1982). We can also
expect that the availability of such clones will facilitate large-
scale production of the proteins they specify, so that their action
can be studied in vitro.

WHY INTRONS IN MITOCHONDRIAL DNA?

 Three explanations have been offered for the existence of introns
in fungal mtDNAs:
        1. They could merely be a form of 'selfish' DNA, which offers
no particular advantage, but which cannot be eliminated for want of
an effective mechanism (cf. Doolittle and Sapienza, 1980; Orgel and
Crick, 1980).
        2. They could offer a selective advantage in that the increased
length of intron-containing genes may allow more effective recombination resulting in the rapid creation of new alleles, or even new
genes by exon shuffling (cf. Gilbert, 1978).
        3. They could offer advantages in terms of extra metabolic
flexibility. As noted in the previous section, the introduction of
(additional) splicing steps in the processing of precursor RNAs
transcribed from key genes could allow a more complete control over
mitochondrial gene expression. This might be of use to yeast, which
selectively suppresses the synthesis of mitochondrial components
during growth in the absence of oxygen or in the presence of ferment-
able sugars (Borst and Grivell, 1981b).
        Of these, both (1) and (2) fail completely to account either
for the highly non-random distribution of introns in yeast mitochondrial genes (Table I) or for the existence of the URFs. Furthermore,
passive retention, or retention to facilitate recombination have
additional unattractive aspects. The first, on the grounds that the
cost to the cell of maintaining and expressing the many genes
involved in splicing of the various introns is appreciable and such
a complex system must be extremely vulnerable to disruption by
mutation. The second, because its main prediction is not borne out
by facts. There is no indication that new genes have arisen in

fungal mtDNAs as a result of exon shuffling, nor is there any
evidence that fungal mitochondrially-coded proteins are evolving any
faster than their opposite numbers in mammalian mitochondria, whose
DNAs lack introns. Moreover, while exon shuffling is an excellent
means for spreading mutations conferring advantageous characteristics
such as resistance to (naturally occurring) inhibitors of mitochondrial function, there is no evidence so far that such mutations are
fixed more efficiently in split genes than in continuous ones.

These arguments strongly imply that introns and their URFs in
fungal mtDNAs are retained because they confer a selective advantage
now and in this light, proposal (3) is clearly attractive, even
though it may not apply either to all introns or to introns in all
fungal mtDNAs. For example, the gene for the large rRNA in yeast
mitochondria contains an intron plus URF, yet accurate splicing is
possible (albeit somewhat inefficiently) in the absence of the URF
product and rRNA synthesis does not seem to be selectively repress-
ible. Further, Aspergillus mtDNA has introns plus URFs in the genes
for cytochrome b and subunit I of cytochrome c oxidase, yet does not
suppress synthesis of these components even when this could be
advantageous (e.g. during the operation of shunts in the electron
transport chain). Clearly, there is still room for further speculation about possible functions of both introns and their URFs.

INTRONS AND EVOLUTION

Besides yielding information in maturase function, comparisons of
the yeast URFs have revealed features that may throw light on the
evolution of introns in mtDNA and the consequences introns may have
had for genome organization.

a) A common origin. As mentioned above, the extensive sequence
similarities exhibited by the various URFs imply a common origin,
with subsequent dispersal around the genome (Hensgens et al.,
1982b). For the URFs shown in Fig. 5, the specific relationships
derived from pairwise comparisons are given schematically in the
dendrogram in Fig. 8. They suggest the existence of two families,
which share only a low level of overall homology (16%), but which
are nevertheless likely to be related because of their conservation
of specific characteristic elements. A third, unrelated family is
formed by the introns aI1, aI2, aI5 and bI1 (see also Michel et al.,
1982).

Given a common origin, how long have the ancestral sequences
existed and when did dispersal first occur? One clue is given by the
finding that the single intron in the cytochrome b gene of A.
nidulans mtDNA is highly homologous to bI3 in S. cerevisiae and
occurs at an identical position in the gene (Lazowska et al., 1981;
Waring et al., 1981). This homology could conceivably have arisen
from recent interspecies gene transfer, but it seems more reasonable
to suppose that this intron was present in the common ancestor of
the two fungi, putting its age at 400 million years or more. Other
introns are probably more recent acquisitions, however: the introns
in the large rRNA genes of Aspergillus and S. cerevisiae occur at
identical sites, but their URFs are not obviously related (Dujon,
1980; Netzker et al., 1982).

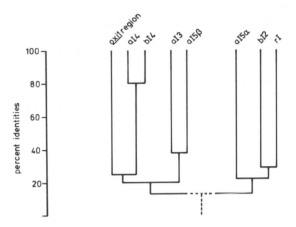

Fig. 8. URF families in yeast mtDNA. The dendrogram is a
schematic representation of relationships between the URFs
shown in Fig. 5, based on pair-wise comparisons. Taken
from Hensgens et al. (1982b).

As yet, few conclusions can be drawn about the time scale of
URF dispersal. The absence of a fossil record for the fungi precludes
estimates of the absolute rate of mutational change in mtDNA and
homologies between the URFs are not necessarily directly correlatable
with an order of appearance, since each is likely to have acquired
specialised functions and is thus subject to different selective
pressures. Further, as in other gene families, concerted evolution
mediated by gene conversion events may also have scrambled evolutionary relationships (Jeffreys , 1981).

How did dispersal occur? Various recombination or transposition
events can be envisaged, but the observation that many of the
conserved features of both intronic and free URFs occur in what
appears to be a circular permuted order has prompted the suggestion
(Bonen et al., 1982; Hensgens et al., 1982b) that transposition has
involved a circular intermediate according to a mechanism similar to
that proposed for mobile eukaryotic nuclear (pseudo)-genes (Hollis
et al., 1982). Yeast mitochondria contain stable circular RNAs
derived from some introns. It is tempting to speculate that circular
RNA intermediates with subsequent complementary DNA (cDNA) synthesis
and integration have played a role in URF dispersal.

b) Unusual codon usage in URFs: An indication of a separate genetic
origin? Most organisms examined to date show a marked bias in codon
usage (Grantham et al., 1980) and it should, therefore, be no
surprise that this also applies to yeast mitochondria. The average
GC content of this mtDNA is low and synonymous codons ending in U or
A are generally preferred (Table II). In addition, certain codons
are not used in known genes, even though the cognate tRNAs are
present. More surprising, however, is the observation (Table II)
that codon utilization in the URFs and the var1 gene, which codes
for a ribosome-associated protein, differs from that in known genes.

TABLE II

CODON USAGE IN YEAST MITOCHONDRIAL GENES AND URFs

Codon usage is given in terms of times of occurrence in genes (1743 excluding stop codons), URFs (4137 codons) and <u>var</u>1 (396 codons).

Amino acid	Codon	Genes[a]	URFs[b]	var1
Ala	GCA	54	27	0
	U	67	64	0
	C	5	11	1
	G	2	7	2
Arg	AGA	37	133	3
	G	0	7	0
	CGA	0	0	0
	U	0	15	1
	C	0	1	0
	G	0	1	1
Asn	AAU	64	422	124
	C	9	26	2
Asp	GAU	42	156	10
	C	2	9	0
Cys	UGU	12	58	1
	C	1	1	0
Gln	CAA	30	101	4
	G	4	6	0
Glu	GAA	38	125	2
	G	3	16	0
Gly	GGA	27	54	0
	U	90	132	8
	C	0	6	1
	G	6	15	1
His	CAU	47	68	4
	C	3	3	1
Ile[c]	AUU	150	307	40
	C	27	23	2
Leu	UUA	230	397	41
	G	2	15	0
Lys	AAA	31	418	35
	G	1	30	2
Met[c]	AUA	7	170	31
	G	65	41	2
Phe	UUU	72	162	5
	C	61	17	0
Pro	CCA	36	27	0
	U	39	74	4

TABLE II - continued.

Pro	C	2	7	2
	G	0	10	1
Ser	UCA	72	71	3
	U	34	79	4
	C	1	10	0
	G	0	3	0
	AGU	15	70	9
	C	0	4	0
Thr	ACA	51	83	1
	U	34	79	7
	C	1	12	0
	G	0	17	0
	CUA	15	29	2
	U	2	27	3
	C	0	1	0
	G	2	0	0
Trp	UGA	36	48	3
	G	0	6	0
Tyr	UAU	75	257	28
	C	13	19	0
Val	GUA	77	95	1
	U	43	69	4
	C	4	5	0
	G	6	7	0

a) Data are for the genes for cytochrome c oxidase subunits I-III, cytochrome b, ATPase subunits 6 and 9 (Coruzzi and Tzagoloff, 1979; Hensgens et al., 1979; Bonitz et al., 1980; Macino and Tzagoloff, 1980; Nobrega and Tzagoloff, 1980; Thalenfeld and Tzagoloff, 1980).

b) Data are for 9 intron-encoded reading frames in oxi3 (Bonitz et al., 1980; Hensgens et al., 1982b), cob (Lazowska et al., 1980; Nobrega and Tzagoloff, 1980) and the 21S rRNA gene (Dujon, 1980), together with the free URF downstream of oxi1 (Coruzzi et al., 1981). Note that the boundaries between some exons and URFs have not yet been precisely defined, so that absolute values for codon utilization may not be exact. The uncertainties do not significantly alter the overall pattern of codon usage, however.

c) AUA is assumed to encode methionine (Hudspeth et al., 1982).

For example, AGG (arg), the CGN family (arg), GGC (gly), CCG (pro), UCG and AGC (ser), ACG and CUC (thr) and UGG (trp) are found in URFs, but not in known genes: URFs use predominantly AUA for met, whereas known genes prefer AUG; URFs are strongly biassed towards the use of UUU for phe, while in known genes, usage is about equally divided between UUU and UUC.

Initial suggestions (Nobrega and Tzagoloff, 1980) that this unusual codon usage argues against a coding function for the URFs can now be discounted, as can the idea that usage is linked to tRNA abundance (cf. Bennetzen and Hall, 1982). Most telling of the many arguments against such a proposal is that yeast mitochondria employ only 25 tRNAs to read all codons, so that recognition of a rarely used codon is, more often than not, dependent on the same tRNA as used for the prevalent codon in a particular family.

A detailed analysis of the occurrence of 'unusual' codons in URFs has been carried out by Bonen et al. (1982). Dependent on location, unusual codon choice seems to be related to the following factors:

1. A lack of constraint on translational fidelity - Fox and Weiss-Brummer (1980) have shown that arrays of 5 or 6 U's in mRNA tend to create frame shifts during translation, presumably because sloppy decoding of such runs is an inherent feature of the tRNA^{phe} (Martin et al., 1978; Dujon, 1981). Low translational fidelity could apply to other codon-anticodon interactions. If so, URFs may be better able to tolerate mutations which create such codons, either because only trace amounts of their products are required for RNA processing or because only part of the reading frame is crucial for maturase structure and function. This second alternative may apply to the poorly conserved regions upstream of the LAGLI/DADG block, since these seem to be very tolerant of change, including short frame shifts (see above under MATURASES: MODELS AND MECHANISMS).

2. GC cluster insertions - Yeast mtDNA contains numerous GC-rich clusters, 20-50 bp in length, which show signs of mobility (Prunell and Bernardi, 1977; Dujon, 1980; Sor and Fukuhara, 1982). Such clusters are not found in 'regular' protein-coding genes, but they do occur in two URFs and in the var1 gene (Hensgens et al., 1982b; Hudspeth et al., 1982). Indeed, the presence or absence of one such cluster in the latter gene is responsible in part for the strain-dependent variation in size displayed by the var1 protein (Hudspeth et al., 1982). Such GC clusters introduce GC-rich codons, rarely if ever used by other genes.

3. Constraints on RNA secondary structure - Viral RNAs, which must undergo packaging in a virion, show codon preferences which are related to constraints on the maintenance of RNA secondary structure (Fiers et al., 1976). Certain of the unusual codons in URFs fall into this category (Fig. 9, Blocks 1-3). They invariably involve GC-rich codons clustered in regions displaying a high level of sequence conservation at the nucleotide level and located in areas of predicted high secondary structure. These unusual codons were probably present in the common ancestor of the various URFs, since otherwise their acquisition at so many independent sites requires an improbable degree of convergent evolution.

A number of unusual codons cannot be fitted into any of these categories. For example, the LAGLI/DADG regions of the aI4 and bI4 reading frames contain a number of rarely used third-position G or C codons. These are not clustered and they occur at different positions in the two sequences (Fig. 9). These codons could possibly represent recent acquisitions, resulting from high mutational 'noise' (cf. rates of mutation fixation in animal mtDNAs (Anderson et al., 1982;

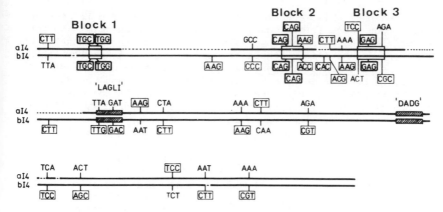

Fig. 9. Unusual codons in the intron reading frames of
yeast mtDNA. The figure shows portions of the aI4 and bI4
reading frames, aligned to display maximal homology with a
minimum of insertions or deletions (indicated by dotted
regions). Codons rarely, if ever, used in exon sequences
are boxed and those occurring at homologous sites in both
introns have been accentuated by thicker lines. The latter
codons are invariably located within regions of predicted
high secondary structure (blocks 1-3). See text for
discussion. Adapted from Bonen et al. (1982).

Brown and Simpson, 1982)), but this seems unlikely in view of the
obviously relentless drive to maximize the AT content of yeast mtDNA
(Borst and Grivell, 1978). Although the constraints that have
permitted the retention of such codons over long periods of URF
evolution are not known, Bonen et al. (1982) have suggested that
both those codons and those present in areas important for secondary
structure are the last traces of a codon strategy specific to the
URFs. This implies that the URFs once employed a codon repertoire
that was not only richer in GC than that used at present, but also
distinct from that used by 'regular' genes. As also suggested by
Hudspeth et al. (1982), we can speculate that URF sequences have a
genetic origin which is distinct from that of genes participating in
oxidative phosphorylation.

SUMMARY AND PROSPECTS

To sum up, some genes in fungal mtDNAs contain (optional) introns,
many of which encode proteins involved in RNA splicing. This matter-
of-fact statement conceals, however, some of the most intriguing
questions concerning the biological significance of a very unusual
form of gene organization. Why do such simple DNAs contain introns
at all and why is the distribution of these introns so character-
istically limited to the three genes discussed above? Why, if
optional, were the introns and the proteins they specify not eliminated long ago? Why do mitochondria make life so difficult by having
individual processing enzymes for every intron sequence spliced,
when nuclear genes can apparently be spliced much more simply?

Certain of these questions have been touched upon in previous sections, but there is obviously room both for speculation and for good experiments. More complete answers should be forthcoming in the near future, perhaps as a result of work along the following lines:

1. Introns and evolution - A systematic study of possible correlations between introns, their reading frames and the ways of life of naturally-occurring yeasts may give clues to the advantages that introns confer. The fact that introns are apparently optional in the laboratory strains looked at so far may merely mean that the differences are too subtle to have been detected by the test systems used.

2. Requirements for splicing - Cis-acting mutations have indicated which intron sequences are involved in splicing and there are hints that this involvement may be in the maintenance of a secondary structure which permits alignment of the splice junctions. A study of second-site mutations should permit verification of RNA secondary structure models. If a primary mutation affects secondary structure, it should be compensatable either by complementary base changes in duplex segments or by second-site changes which allow restoration of the original configuration.

3. Maturase action - No matter how daunting a prospect it may seem, purification of an active RNA maturase and in vitro reconstitution of the reaction it catalyzes will be essential to an understanding of the mechanism of mitochondrial splicing. This difficult task may be simplified somewhat by the use of cis-dominant splicing-deficient mutants to overproduce maturase and by the availability of a recently devised splicing assay, which measures the ability of a correctly-spliced RNA to protect a labelled, synthetic oligodeoxy-nucleotide from nuclease digestion (Tabak et al., 1981).

Additional information on maturase action should also come from an analysis of second-site mutations. In the so-called nam mutants (nuclear-accommodation of mitochondria), maturase defects have been phenotypically suppressed by second-site nuclear mutations (Groudinsky et al., 1981). Some of these are only informational suppressors (e.g. changes in key proteins of the mitochondrial ribosome), but others are likely to alter the nuclear-coded proteins which cooperate with maturase to catalyze splicing.

4. Involvement of nuclear gene products in mitochondrial splicing - With such a multiplicity of mitochondrially-coded RNA processing proteins it remains surprising that so many nuclear gene products are required to splice individual introns. Fortunately, however, it should not be long before this paradox is resolved by the in vitro study of these proteins and their mode of action. The technique of transformation-complementation, applied to nuclear splicing-deficient mutants, has already permitted isolation of a number of the nuclear genes coding for proteins required for mito-chondrial splicing (Dieckmann et al., 1982). The availability of the cloned sequences on high copy-number plasmids which can be main-tained and expressed in yeast will permit over-production of these proteins and hence facilitate their purification, so that the reactions they catalyse can be reconstructed in vitro.

REFERENCES

Alexander, N.J., Perlman, P.S., Hanson, D.K., and Mahler, H.R.,
   1980. Mosaic organization of a mitochondrial gene: evidence from
   double mutants in the cytochrome b region of Saccharomyces
   cerevisiae. Cell 20, 199-206.
Anderson, S., Bankier, A.T., Barrell, B.G., De Bruijn, M.H.L.,
   Coulson, A.R., Drouin, J., Eperon, I.C., Nierlich, D.P., Roe,
   B.A., Sanger, F., Schreier, P.H., Smith, A.J.H., Staden, R., and
   Young, I.G., 1981. Sequence and organization of the human mito-
   chondrial genome. Nature 290, 457-465.
Anderson, S., De Bruijn, M.H.L., Coulson, A.R., Eperon, I.C.,
   Sanger, F., and Young, I.G., 1982. Complete sequence of bovine
   mitochondrial DNA. Conserved features of the mammalian mitochon-
   drial genome. J.Mol.Biol. 156, 683-717.
Arnberg, A.C., Van Ommen, G.J.B., Grivell, L.A., Van Bruggen,
   E.F.J., and Borst, P., 1980. Some yeast mitochondrial RNAs are
   circular. Cell 19, 313-319.
Bennetzen, J.L., and Hall, B.D., 1982. Codon selection in yeast.
   J.Biol.Chem. 257, 3026-3031.
Bonen, L., Boer, P.H., Hensgens, L.A.M., Borst, P., and Grivell,
   L.A., 1982. URFs in yeast mitochondrial split genes. Nature,
   submitted.
Bonitz, S.G., Coruzzi, G., Thalenfeld, B.E., Tzagoloff, A., and
   Macino, G., 1980. Assembly of the mitochondrial membrane system.
   Structure and nucleotide sequence of the gene coding for subunit
   1 of yeast cytochrome oxidase. J.Biol.Chem. 255, 11927-11941.
Bonitz, S.G., Homison, G., Thalenfeld, B.E., Tzagoloff, A., and
   Nobrega, F.G., 1982. Assembly of the mitochondrial membrane
   system. Processing of the apocytochrome b precursor RNAs in
   Saccharomyces cerevisiae. J.Biol.Chem. 257, 6268 6274.
Borst, P., 1980. The optional introns in yeast mitochondrial DNA.
   In: 31. Colloquium-Mosbach 1980: Biological Chemistry of Organelle
   Formation (Eds Bücher, Th., Sebald, W., and Weiss, H.), Springer,
   Berlin, pp. 27-41.
Borst, P., 1981a. The biogenesis of mitochondria in yeast and other
   primitive eukaryotes. In: International Cell Biology 1980-1981
   (Ed. Schweiger, H.G.), Springer, Berlin, pp. 239-249.
Borst, P., 1981b. Control of mitochondrial biosynthesis. In: Cellular
   Controls in Differentiation (Eds Lloyd, C.W., and Rees, D.A.),
   Academic Press, London, pp. 231-252.
Borst, P., and Grivell, L.A., 1978. The mitochondrial genome of
   yeast. Cell 15, 705-723.
Borst, P., and Grivell, L.A., 1981a. Small is beautiful - the
   portrait of a mitochondrial genome. Nature 290, 443-444.
Borst, P., and Grivell, L.A., 1981b. One gene's intron is another
   gene's exon. Nature 289, 439-440.
Borst, P., Tabak, H.F., and Grivell, L.A., 1982. Extranuclear genes.
   In: Eukaryotic Genes: Their Structure, Activity and Regulation
   (Eds Maclean, N., Gregory, S.P., and Flavell, R.A.), Butterworth,
   London, in press.
Bos, J.L., Osinga, K.A., Van der Horst, G., Hecht, N.B., Tabak,
   H.F., Van Ommen, G.J.B., and Borst, P., 1980. Splice point
   sequence and transcripts of the intervening sequence in the

mitochondrial 21S ribosomal RNA gene of yeast. Cell 20, 207-214.

Brown, G.G., and Simpson, M.V., 1982. Novel features of animal mtDNA evolution as shown by sequences of two rat cytochrome oxidase subunit II genes. Proc.Natl.Acad.Sci.U.S. 79, 3246-3250.

Cech, T.R., Zaug, A.J., and Grabowski, P.J., 1981. In vitro splicing of the ribosomal RNA precursor of Tetrahymena: Involvement of a guanosine nucleoside in the excision of the intervening sequence. Cell 27, 487-496.

Church, G.M., Slonimski, P.P., and Gilbert, W., 1979. Pleiotropic mutations within two yeast mitochondrial cytochrome genes block mRNA processing. Cell 18, 1209-1215.

Church, G.M., and Gilbert, W., 1980. Yeast mitochondrial intron products required in trans for RNA splicing. In: Mobilization and Reassembly of Genetic Information (Eds Joseph, D.R., Schultz, J., Scott, W.A., and Werner, R.), Academic Press, New York, pp. 379-396.

Coruzzi, G., and Tzagoloff, A., 1979. Assembly of the mitochondrial membrane system. DNA sequence of subunit 2 of yeast cytochrome oxidase. J.Biol.Chem. 254, 9324-9330.

Coruzzi, G., Bonitz, S.G., Thalenfeld, B.E., and Tzagoloff, A., 1981. Assembly of the mitochondrial membrane system: analysis of the nucleotide sequence and transcripts in the oxi1 region of yeast mitochondrial DNA. J.Biol.Chem. 256, 12780-12787.

Davies, R.W., Waring, R.B., Ray, J.A., Brown, G.A., and Scazzocchio, C., 1982. Making ends meet: a model for RNA splicing in fungal mitochondria. Nature, in press.

De Jonge, J.C., and De Vries, H., 1982. The structure of the gene for subunit I of cytochrome c oxidase in Neurospora crassa mitochondria. Curr.Genet., in press.

De la Salle, H., Jacq, C., and Slonimski, P.P., 1982. Critical sequence within mitochondrial introns: Pleiotropic mRNA maturase and cis-dominant signals of the box intron controlling reductase and oxidase. Cell 28, 721-732.

Dhawale, S., Hanson, D.K., Alexander, N.J., Perlman, P.S., and Mahler, H.R., 1981. Regulatory interactions between mitochondrial genes: interactions between two mosaic genes. Proc.Natl.Acad.Sci. U.S. 78, 1778-1782.

Dieckmann, C.L., Pape, L.K., and Tzagoloff, A., 1982. Identification and cloning of a yeast nuclear gene (CBP1) involved in expression of mitochondrial cytochrome b. Proc.Natl.Acad.Sci.U.S 79, 1805-1809.

Doolittle, W.F., and Sapienza, C., 1980. Selfish genes, the phenotype paradigm and genome evolution. Nature 284, 601-603.

Dujardin, G., Jacq, C., and Slonimski, P.P., 1982. Single base substitution in an intron of oxidase gene compensates splicing defects of the cytochrome b gene. Nature 298, 628-632.

Dujon, B., 1980. Sequence of the intron and flanking exons of the mitochondrial 21S rRNA gene of yeast strains having different alleles at the omega and rib-1 loci. Cell 20, 185-197.

Dujon, B., 1981. Mitochondrial genetics and functions. In: Molecular Biology of the Yeast Saccharomyces: Life Cycle and Inheritance (Eds Strathern, J.N., Jones, E.W., and Broach, J.R.), Cold Spring Harbor Laboratory, Cold Spring Harbor, pp. 505-635.

Earl, A.J., Turner, G., Croft, J.H., Dales, R.B.G., Lazarus, C.M.,

Lünsdorf, H., and Küntzel, H., 1981. High frequency transfer of species specific mitochondrial DNA sequences between members of the Aspergillaceae. Curr.Genet. 3, 221-228.

Fiers, W., Contreras, R., Duerinck, F., Haegeman, G., Iserentant, D., Merregaert, J., Min Jou, W., Molemans, F., Raeymakers, A., Van den Berghe, A., Volckaert, G., and Ysebaert, M., 1976. Complete nucleotide sequence of bacteriophage MS2 RNA: primary and secondary structure of the replicase gene. Nature 260, 500-507.

Fox, T.D., and Boerner, P., 1980. Transcripts of the OXI-1 locus are asymmetric and may be spliced. In: The Organization and Expression of the Mitochondrial Genome (Eds Kroon, A.M., and Saccone, C.), North-Holland, Amsterdam, pp. 191-194.

Fox, T.D., and Weiss-Brummer, B., 1980. Leaky +1 and -1 frameshift mutations at the same site in a yeast mitochondrial gene. Nature 288, 60-63.

Fox, T.D., and Leaver, C.J., 1981. The Zea mays mitochondrial gene coding cytochrome oxidase subunit II has an intervening sequence and does not contain TGA codons. Cell 26, 315-323.

Gilbert, W., 1978. Why genes in pieces? Nature 271, 501.

Grantham, R., Gautier, C., and Gouy, M., 1980. Codon frequencies in 119 individual genes confirm consistent choices of degenerate bases according to genome type. Nucl.Acids Res. 8, 1893-1912.

Grivell, L.A., 1982a. Mitochondrial DNA. Sci.Am., in press.

Grivell, L.A., 1982b. Restriction and genetic maps of yeast mitochondrial DNA. In: Genetic Maps: A Compilation of Linkage and Restriction Maps of Genetically Studied Organisms (Ed. O'Brien, S.J), Vol. 2, Lab. of Viral Carcinogenesis, Natl. Cancer Institute, Natl.Inst. of Health, Frederick, N.D., pp. 221-235.

Grivell, L.A., Arnberg, A.C., Hensgens, L.A.M., Roosendaal, E., Van Ommen, G.J.B., and Van Bruggen, E.F.J., 1980. Split genes on yeast mitochondrial DNA: organisation and expression. In: The Organization and Expression of the Mitochondrial Genome (Eds Kroon, A.M., and Saccone, C.), North-Holland, Amsterdam, pp. 37-49.

Groudinsky, O., Dujardin, G., and Slonimski, P.P., 1981. Long range control circuits within mitochondria and between nucleus and mitochondria. II. Genetic and biochemical analyses of suppressors which selectively alleviate the mitochondrial intron mutations. Mol.Gen.Genet. 184, 493-503.

Hahn, U., Lazarus, C.M., Lünsdorf, H., and Küntzel, H., 1979. Split gene for mitochondrial 24S ribosomal RNA of Neurospora crassa. Cell 17, 191-200.

Halbreich, A., Pajot, P., Foucher, M., Grandchamp, C., and Slonimski, P., 1980. A pathway of cytochrome b mRNA processing in yeast mitochondria: specific splicing steps and an intron-derived circular RNA. Cell 19, 321-329.

Hensgens, L.A.M., Grivell, L.A., Borst, P., and Bos, J.L., 1979. Nucleotide sequence of the mitochondrial structural gene for subunit 9 of yeast ATPase complex. Proc.Natl.Acad.Sci.U.S. 76, 1663-1667.

Hensgens, L.A.M., Arnberg, A.C., Roosendaal, E., Van der Horst, G., Van der Veen, R., Van Ommen. G.J.B., and Grivell, L.A., 1982a. Variation, transcription and circular RNAs of the mitochondrial gene for subunit I of cytochrome c oxidase. J.Mol.Biol., in press.

304

Hensgens, L.A.M., Bonen, L., De Haan, M., Van der Horst, G., and
  Grivell, L.A., 1982b. The sequence of two optional introns in the
  mitochondrial gene for subunit I of cytochrome c oxidase reveals
  homologies among URF-containing introns and strain-dependent
  variation in flanking exons. Cell, in press.
Hollis, G.F., Hieter, P.A., McBride, O.W., Swan, D., and Leder, P.,
  1982. Processed genes: a dispersed human immunoglobulin gene
  bearing evidence of RNA-type processing. Nature 296, 321-325.
Hudspeth, M.E.S., Ainley, W.M., Shumard, D.S., Butow, R.A., and
  Grossman, L.I., 1982. Location and structure of the var1 gene on
  yeast mitochondrial DNA: Nucleotide sequence of the 40.0 allele.
  Cell 30, 617-626.
Jacq, C., Pajot, P., Lazowska, J., Dujardin, G., Claisse, M.,
  Groudinsky, O., De la Salle, H., Grandchamp, C., Labouesse, M.,
  Gargouri, A., Guiard, B., Spyridakis, A., Dreyfus, M., and
  Slonimski, P.P., 1982. Role of introns in the yeast cytochrome b
  gene: cis- and trans-acting signals, intron manipulation,
  expression and intergenic communications. In: Mitochondrial Genes
  (Eds Slonimski, P., Borst, P., and Attardi, G.), Cold Spring
  Harbor Laboratory, Cold Spring Harbor, pp. 155-183.
Jeffreys, A.J., 1981. Recent studies of gene evolution using recombi-
  nant DNA. In: Genetic Engineering (Ed. Williamson, R.), Vol. 2,
  Academic Press, London, pp. 1-48.
Kreike, J., Bechmann, H., Van Hemert, F.J., Schweyen, R.J., Boer,
  P.H., Kaudewitz, F., and Groot, G.S.P., 1979. The identification
  of cytochrome b as a mitochondrial gene product and immunological
  evidence for altered apocytochrome b in yeast strains having
  mutations in the cob region of mitochondrial DNA. Europ.J.Biochem.
  101, 607-617.
Lazowska, J., Jacq, C., and Slonimski, P.P., 1980. Sequence of
  introns and flanking exons in wild type and box3 mutants of
  cytochrome b reveals an interlaced splicing protein coded by an
  intron. Cell 22, 333-348.
Lazowska, J., Jacq, C., and Slonimski, P.P., 1981. Splice points of
  the third intron in the yeast mitochondrial cytochrome b gene.
  Cell 27, 12-14.
Leaver, C.J., and Gray, M.W., 1982. Mitochondrial genome organization and expression in higher plants. Ann.Rev.Plant Physiol. 33,
  373-402.
Macino, G., and Tzagoloff, A., 1980. Assembly of the mitochondrial
  membrane system. Sequence analysis of a yeast mitochondrial ATPase
  gene containing the oli2 and oli4 loci. Cell 20, 507-517.
Mahler, H.R., Hanson, D.K., Lamb, M.R., Perlman, P.S., Anziano,
  P.Q., Glaus, K.R., and Haldi, M.L., 1982. Regulatory interactions
  between mitochondrial genes: Expressed introns - their function
  and regulation. In: Mitochondrial Genes (Eds Slonimski, P.,
  Borst, P., and Attardi, G.), Cold Spring Harbor Laboratory, Cold
  Spring Harbor, pp. 185-199.
Martin, R.P., Sibler, A.P., Schneller, J.M., Keith, G., Stahl,
  A.J.C., and Dirheimer, G., 1978. Primary structure of yeast
  mitochondrial DNA coded phenylalanine tRNA. Nucl.Acids Res. 5,
  4579-4592.
Merten, S., Synenki, R.M., Locker, J., Christianson, T., and
  Rabinowitz, M., 1980. Processing of precursors of 21S ribosomal

RNA from yeast mitochondria. Proc.Natl.Acad.Sci.U.S. 77, 1417-1421.

Michel, F., Jacquier, A., and Dujon, B., 1982. Comparison of fungal mitochondrial introns reveals extensive homologies in RNA secondary structure. Biochimie, in press.

Mount, S.M., 1982. A catalogue of splice junction sequences. Nucl. Acids Res. 10, 459-472.

Netter, P., Jacq, C., Carignani, G., and Slonimski, P.P., 1982. Critical sequences within mitochondrial introns: cis-dominant mutations of the 'cytochrome-b-like' intron of the oxidase gene. Cell 28, 733-738.

Netzker, R., Köchel, H.G., Basak, N., and Küntzel, H., 1982. Nucleotide sequence of Aspergillus nidulans mitochondrial genes coding for ATPase subunit 6, cytochrome oxidase subunit 3, seven unidentified proteins, four tRNAs and L-rRNA. Nucl.Acids Res. 10, 4783-4794.

Nobrega, F.G., and Tzagoloff, A., 1980. Assembly of the mitochondrial membrane system. DNA sequence and organization of the cytochrome b gene in Saccharomyces cerevisiae D273-10B. J.Biol.Chem. 255, 9828-9837.

Orgel, L.E., and Crick, F.H.C., 1980. Selfish DNA: the ultimate parasite. Nature 284, 604-607.

Prunell, A., and Bernardi, G., 1977. The mitochondrial genome of wild type yeast cells. VI. Genome organization. J.Mol.Biol. 110, 53-74.

Sanders, J.P.M., Heyting, C., Verbeet, M.Ph., Meijlink, F.C.P.W., and Borst, P., 1977. The organization of genes in yeast mitochondrial DNA. III. Comparison of the physical maps of the mitochondrial DNAs from three wild-type Saccharomyces strains. Mol.Gen.Genet. 157, 239-261.

Schmelzer, C., and Schweyen, R.J., 1982. Evidence for ribosomes involved in splicing of yeast mitochondrial transcripts. Nucl.Acids Res. 10, 513-524.

Schmelzer, C., Schmidt, C., and Schweyen, R.J., 1982. Identification of splicing signals in introns of mitochondrial split genes. Mutational alterations in intron bI1 and secondary structure of related introns. Nucl.Acids Res. 10, in press.

Slonimski, P., Borst, P., and Attardi, G., 1982. Mitochondrial Genes, Cold Spring Harbor Laboratory, Cold Spring Harbor.

Sor, F., and Fukuhara, H., 1982. Nature of an inserted sequence in the mitochondrial gene coding for the 15S ribosomal RNA of yeast. Nucl.Acids Res. 10, 1625-1633.

Tabak, H.F., Van der Laan, J., Osinga, K.A., Schouten, J.P., Van Boom, J.H., and Veeneman, G.H., 1981. Use of a synthetic DNA oligonucleotide to probe the precision of RNA splicing in a yeast mitochondrial petite mutant. Nucl.Acids Res. 9, 4475-4483.

Tabak, H.F., Grivell, L.A., and Borst, P., 1982. Transcription of mitochondrial DNA. In: CRC Critical Reviews in Biochemistry (Ed. Fasman, F.G.), CRC Press, Boca Ratton, in press.

Thalenfeld, B.E., and Tzagoloff, A., 1980. Assembly of the mito-chondrial membrane system. Sequence of the oxi2 gene of yeast mitochondrial DNA. J.Biol.Chem. 255, 6173-6180.

Turner, G., Earl, A.J., and Greaves, D.R., 1982. Interspecies variation and recombination of mitochondrial DNA in the

<u>Aspergillus nidulans</u> species group and the selection of species-specific sequences by nuclear background. In: <u>Mitochondrial Genes</u> (Eds Slonimski, P., Borst, P., and Attardi, G.), Cold Spring Harbor Laboratory, Cold Spring Harbor, pp. 411-414.

Van Ommen, G.J.B., Boer, P.H., Groot, G.S.P., De Haan, M., Roosendaal, E., Grivell, L.A., Haid, A., and Schweyen, R.J., 1980. Mutations affecting splicing and the interaction of gene expression of the yeast mitochondrial loci <u>cob</u> and <u>oxi</u>3. <u>Cell</u> 20, 173-183.

Ward, B.L., Anderson, R.S., and Bendich, A.J., 1981. The mitochondrial genome is large and variable in a family of plants (<u>Cucurbitaceae</u>). <u>Cell</u> 25, 793-803.

Waring, R.B., Davies, R.W., Lee, S., Grisi, E., McPhail Berks, M., and Scazzocchio, C., 1981. The mosaic organization of the apocytochrome <u>b</u> gene of Aspergillus nidulans revealed by DNA sequencing. <u>Cell</u> 27, 4-11.

Weiss-Brummer, B., Rödel, G., Schweyen, R.J., and Kaudewitz, F., 1982. Expression of the split gene <u>cob</u> in yeast: Evidence for a precursor of a 'maturase' protein translated from intron 4 and preceding exons. <u>Cell</u> 29, 527-536.

Yin, S., Burke, J., Chang, D.D., Browning, K.S., Heckman, J.E., Alzner-DeWeerd, B., Potter, M.J., and RajBhandary, U.L., 1982. <u>Neurospora crassa</u> mitochondrial tRNAs and rRNAs: Structure, gene organization and DNA sequences. In: <u>Mitochondrial Genes</u> (Eds Slonimski, P., Borst, P., and Attardi, G.), Cold Spring Harbor Laboratory, Cold Spring Harbor, pp. 361-373.

Genes: *Structure and Expression*
Edited by A. M. Kroon
© 1983 John Wiley & Sons Ltd.

ASSEMBLY OF MITOCHONDRIAL PROTEINS

Bernd Hennig and Walter Neupert

INTRODUCTION

The eucaryotic cell is organized by a variety of membranes: the plasma membrane which forms the border of the cell and various intracellular membranes which delimit the various organelles. In most cells intracellular membranes greatly exceed the plasma membrane with respect to surface area. The diverse membranes of the cell have properties in common which are the basis for compartmentation: a) membranes possess identity, i.e. with few exceptions each particular membrane of a cell has a unique protein composition; b) membranes have continuity, not only in space but also in time, i.e. membranes are formed by insertion of newly made components into preexistent membranes; c) membranes display selective permeability, i.e. they prevent the free diffusion of most molecules between the two separated spaces.

A membrane gives the compartment which it encloses identity: It determines the protein composition of the luminal or cisternal space ("matrix") of that organelle. This is because the proteins of the various organelles and their membranes - with only very few exceptions - are not synthesized in the same compartments where their functional sites are located. Rather, they are synthesized on cytoplasmic ribosomes and have to be selectively translocated across the diverse membranes. Organelle membranes therefore must not only have devices to specifically recognize proteins which are destined for the compartment they enclose, but also have mechanisms to translocate these proteins across the lipid bilayer. This is quite remarkable in light of the fact that many of these proteins are not only large but also very hydrophilic. Therefore, translocation of newly formed proteins across membranes and their intracellular sorting are quite puzzling phenomena.

I. A GENERAL VIEW ON MEMBRANE AND ORGANELLE ASSEMBLY

According to our present knowledge, two different mechanisms are involved in the translocation of proteins across membranes and their insertion into membranes (1). Cotranslational mechanisms are employed with proteins which traverse the membrane of the endoplasmic reticulum (ER). In contrast, the import of proteins into organelles such as mitochondria, chloroplasts, and probably peroxisomes and glyoxysomes is apparently the result of posttranslational processes. Some proteins of the ER and the plasma membrane appear to utilize this latter mode of transport as well.

Cotranslational Transport: The Translocation of Proteins Is Coupled to the Elongation of the Nascent Polypeptides.

Many proteins are translated on polysomes tightly associated with the ER. They are cotranslationally inserted into or transferred across the membrane, i.e. this process is concomitant with polypeptide chain elongation. Glycosylation of nascent polypeptide chains has been observed, and it is generally accepted that glycosylation occurs only in the luminal space of the ER (2). Completed polypeptides are never seen on the cytosolic side of the membrane which binds the ribosomes. Cotranslationally transported proteins can be synthesized _in_ _vitro_ in cell-free translation systems as complete precursors but they never cross the membrane of vesicles derived from the ER unless those vesicles are present during translation.

A detailed mechanism by which cotranslationally transported proteins reach their proper positions was first proposed in the "signal hypothesis" and has been extensively modified in response to new data (3). A large body of evidence now supports a mechanism which entails the following steps: Synthesis of the polypeptide begins on free ribosomes. The nascent polypeptide carries an amino-terminal presequence of 15 - 30 amino acid residues representing the "signal sequence". The signal sequence protrudes out of the ribosome when a polypeptide of about 60 - 70 amino acid residues in length has been synthesized. An oligomeric "signal recognition protein" (SRP) binds to the freely accessible signal sequence and arrests further elongation of the polypeptide. The SRP - polysome complex then interacts with an SRP receptor or "docking protein" at the ER membrane (4). This releases the arrest of elongation and results in the nascent polypeptide chain penetrating the membrane.

Recognition in this cotranslational transfer thus seems to involve two specific interactions: the signal sequence interacts with the SRP, and the SRP interacts with its receptor on the membrane. Further proteins such as the "ribophorins" may stabilize the ribosome - membrane interaction (5). The signal sequence is cleaved off the polypeptide before peptide elongation is completed. This is accomplished by the "signal peptidase", an integral membrane protein which is probably located at the luminal side of the ER membrane (6). According to information inherent in the structure of the poly-peptide chain, the protein will be transferred either into the lumen of the ER or integrated into the membrane.

Proteins obeying this mechanism are not only destined for integration into the membrane or for the luminal compartment of the ER itself. For a large number of proteins, traversing the ER mem-brane is only the first step in a complicated process by which they

are allocated to destinations such as the Golgi complex, lysosomes
and plasma membrane, or outside of the cell in the case of secreted
proteins. They reach these final locations in a transport process
involving a flow of membranes, shuttling between several organelles
and the plasma membrane (7). The cell must have quite a few mechanisms to correctly sort out these proteins after the initial, common
step of traversing the ER membrane. However, our knowledge about the
molecular rules organizing this traffic is scarce as yet.

Posttranslational Transport: The Translocation of Completed Precursor Proteins.

In contrast to the cotranslational transport described above,
posttranslationally transferred proteins are synthesized essentially
on free polysomes (1). They are released from the ribosomes into the
cytosol as completed polypeptide chains. The primary translation
products differ, as far as we know, in structure and properties from
their mature counterparts. In particular, precursors of most proteins carry aminoterminal presequences. The newly synthesized polypeptides enter extraorganellar pools of free precursors from which
they are rapidly cleared by uptake into the appropriate organelles.
The precursors interact directly with the target membranes, apparently via specific receptor sites. During translocation across the
membrane, the precursors are processed to the mature proteins. A
specific endopeptidase removes the presequences of larger precursors. Synthesis of the polypeptides on the one hand and translocation of these polypeptides across the membranes and processing to
the mature forms on the other hand are clearly separate events.
This kind of precursor - product relationship is the most characteristic feature of posttranslational transport.

Assembly of Mitochondria Depends on Two Separate Genetic Systems.

The general features of posttranslational protein transport across membranes described above pertain also to the membranes of mitochondria and chloroplasts. However, these organelles are peculiar in an important respect: Their biosynthesis and maintenance depends on the expression of two different genetic systems in the cell. One of these is the DNA located in the cell nucleus; the other is the small DNA enclosed within mitochondria and chloroplasts themselves (8).

It is estimated that mitochondria contain several hundred different proteins. Some 5 - 7 proteins which are coded by mitochondrial DNA (mtDNA) in diverse species have been identified (cf. section VI). MtDNA apparently codes for a few additional proteins whose functions are not known. For example, human cells may contain up to eight additional proteins coded by mtDNA since complete sequence analysis of human mtDNA has revealed a total of 13 open reading frames (9). This number varies with different species but it is obvious that the majority of mitochondrial proteins are coded and synthesized by the nucleocytoplasmic system. Considering the total protein mass of mitochondria, about 95 % is of cytoplasmic origin and becomes imported into mitochondria.

The processes which interconnect the two genetic systems and coordinate the expression of genes for mitochondrial proteins are largely unknown. An important role may be played by the mitochondrial RNA-polymerase which transcribes the mitochondrial genes and which itself is coded by a nuclear gene. It is formed according to the cellular demands for mitochondrial gene expression (10).

Mitochondrial Proteins are Located in Four Different Submitochondrial Compartments.

It is another important aspect of mitochondria that these

organelles are composed by two strongly differing membranes whereas only single membranes surround most other organelles like peroxisomes, glyoxysomes, and lysosomes. The two mitochondrial membranes allow the division of mitochondria into four different compartments: the outer membrane, the inner membrane, the intermembrane space which lies between the outer and inner membrane, and the matrix which is enclosed by the inner membrane.

The function of the outer membrane is not really understood. In contrast to almost all other cellular membranes, it is permeable in an unspecific manner to molecules with molecular weight up to 2,000 – 6,000 daltons. This unusual permeability is due to large channels formed by the porin, the most abundant protein in the mitochondrial outer membrane (11). Relatively few other proteins have been identified in this membrane. One of these is the mitochondrial cytochrome b_5 which participates in the rotenone-insensitive NADH-oxidase pertinent in the outer membrane. It is closely related to the microsomal cytochrome b_5 present in the membrane of the ER.

The inner membrane, on the other hand, is quite rich in proteins compared to other cellular membranes. It contains a number of carrier systems such as the ADP/ATP carrier as it is a typical biological membrane impermeable to charged and/or polar molecules of low molecular weight. The various carrier systems allow the passage of the multitude of low-molecular-weight metabolites required by enzymes of the mitochondrial matrix. The inner membrane is usually extensively invaginated, thereby providing a large area containing the enzymes and components of cellular respiration (e.g. cytochrome oxidase and the cytochrome bc_1-complex) and oxidative phosphorylation (i.e. the oligomycin-sensitive ATPase) which consist of numerous hydrophobic subunits. The surface area of the inner membrane, i.e. the number and size of invaginations or "cristae", varies among mitochondria of different cells and is affected by the metabolic state of a cell. As an extreme example, mitochondria of glucose-repressed yeast and promitochondria of anaerobic yeast are practically devoid of cristae (12).

The space between the two mitochondrial membranes, the inter-membrane space, contains soluble enzymes, e.g. adenylate kinase, sulfite oxidase (rat liver) or cytochrome b_2 (yeast). The latter two proteins interact with cytochrome $\underline{c}$, a protein contained in mito-chondria of all species. Cytochrome $\underline{c}$ is confined to the intermem-brane space too, but is loosely attached to the surface of inner membrane where it mediates electron transport between the cytochrome bc_1-complex and cytochrome oxidase.

The innermost space of mitochondria, the matrix, contains a large number of soluble and hydrophilic enzymes, in particular those of the citric acid cycle and the urea cycle. The various proteins engaged in replication and transcription of mtDNA (e.g. mtRNA-poly-merase) and intramitochondrial protein biosynthesis (i.e. mitochon-drial tRNA-synthetases, subunits of mitochondrial ribosomes, etc.) also reside in this compartment.

Biogenesis of mitochondria occurs by insertion of individual components into preexistent mitochondria, i.e. mitochondria multiply by growth and division (13). The ultrastructure of mitochondria and the protein composition of the different submitochondrial com-partments are subject to change depending on the developmental fate of a cell. The transfer of the several hundred mitochondrial pro-teins into the various submitochondrial compartments is a complex process. Several major questions concerning the mechanism(s) of this assembly may be asked. We will focus on those for which answers have begun to emerge. These include: How do precursors of mitochondrial proteins differ from their mature forms ? What is the evidence that these precursors are posttranslationally imported into mitochon-dria ? How are the precursors recognized by mitochondria so that transport is specific ? What are the mechanisms by which these proteins are translocated into the different mitochondrial compartments ?

II. SYNTHESIS AND POSTTRANSLATIONAL TRANSPORT OF MITO-CHONDRIAL PRECURSOR PROTEINS

Two different approaches have been used to study synthesis of mitochondrial proteins and their translocation into mitochondria. The first relies on experiments performed in vitro employing cell-free translation and transfer systems. The second is based on experiments performed in vivo, i.e. employing whole cells. The following conclusions have been derived from both kinds of approaches: a) mitochondrial proteins are synthesized as extramitochondrial precursors, b) the precursors of most mitochondrial proteins are larger than the mature forms, and c) the completed precursors are transported into mitochondria.

Most Proteins are Synthesized as Larger Precursors in Cell-free Systems.

Cell-free translation systems are frequently employed to analyse the properties of precursors as they are released from ribosomes. In such systems the mRNA can be from a different organism than the ribosomes and cofactors and mitochondria are absent during protein synthesis. This minimizes the danger that the primary translation products undergo some posttranslational modification which is part of their intracellular biogenetic pathway. Preparations of osmotically lysed reticulocytes or extracts of wheat germs are used for most studies because these systems are standardized and can be efficiently programmed with mRNA from any eucaryotic organism. Efficient translation of mitochondrial proteins, however, was also achieved using homologous cell-free systems dependent on endogenous mRNA. Mitochondrial proteins are synthesized in the presence of a radioactive amino acid, usually methionine or leucine. The radioactively labelled proteins are collected from the translation mixtures by immunoprecipitation or immunoadsorption onto Staphylococcus protein A using specific antibodies against defined proteins isolated from mitochondria. The immunoprecipitates are analysed by polyacrylamide gel electrophoresis in the presence of sodium dodecylsulfate (SDS), which dissociates the polypeptide - antibody complexes. This separates the proteins according to the molecular weights of their subunits. Autoradiography (or fluorography) of the dried gels serves to locate the labelled proteins. By this procedure, the protein synthesized in vitro can be compared with the corresponding mature form present in mitochondria.

It has turned out that the majority of mitochondrial proteins, whether membrane bound or soluble polypeptides, are formed as precursors which are larger than the mature proteins by some 500 to 10,000 daltons (Table). Although mobility differences on SDS gels must be interpreted with care since a number of factors do influence this parameter, it was demonstrated in several cases that these larger apparent molecular weights are actually due to additional amino acid sequences. As far as studied, these sequences are located at the aminoterminus. During import into mitochondria they are proteolytically removed (cf. section V).

Several large protein complexes inserted in the mitochondrial inner membrane consist of different subunits most of which are coded for by nuclear genes (e.g. 4 subunits of cytochrome oxidase, some 7 subunits of the cytochrome bc_1-complex, and at least 7 subunits of ATPase). It was investigated whether they are synthesized as polyprotein precursors similarly to certain viral capsid proteins. Evidence from several studies demonstrates that the different subunits are synthesized as individual precursors (14,15). The presequence of one precursor, that of the subunit 9 of <u>Neurospora</u> ATPase, was deduced from the corresponding DNA sequence (16). This sequence of 66 residues is quite hydrophilic and positively charged, whereas the mature protein (81 amino acid residues) is very hydrophobic and deeply imbedded in the membrane.

Some Precursors are Just as Large as their Mature Forms.

In a few cases, comparison of precursor and mature form does not reveal a molecular weight difference. Two such proteins, namely the ADP/ATP carrier and cytochrome <u>c</u>, were thoroughly investigated and it was shown that the failure to detect a difference in molecular weights between these precursors and their mature proteins was not an artifact of the cell-free systems or of gel electrophoresis. In these experiments radioactively labelled N-formyl-methionyl-tRNA served to selectively label the initiator methionine in reticulocyte lysates. Aminoacylation and N-formylation of Met-tRNA from calf was performed by employing enzymes from E. coli. N-formyl-methionine is not removed by the peptidase which normally cleaves off the unformy-lated initiator methionine specified by the AUG codon in eucaryotic systems.

The ADP/ATP carrier, a major integral protein of the inner membrane was transported <u>in vitro</u> into mitochondria without losing this radioactive label (17). This shows that import of this particular precursor occurs without removal of any amino acid from its amino terminus. In contrast, the radioactive N-formylmethionine of the larger precursor of ornithine carbamyoltransferase disappears when this protein is processed during import into mitochondria (18).

Cytochrome <u>c</u> is also synthesized without a presequence. Experiments with the N-formylated precursor of cytochrome <u>c</u> led to similar results as described for the ADP/ATP carrier (19). In the case of

TABLE

Cytoplasmic Precursors of Mitochondrial Proteins

Protein	Species	Mature Protein (subunit size) $MW_{app.}$ (kD)	Precursor (extension) $MW_{app.}$ (kD)
OUTER MEMBRANE:			
Porin	Neurospora	32	–
INTERMEMBRANE SPACE:			
Cytochrome c	Neurospora, rat	11.7[*]	–
Cytochrome b_2	yeast	58	10 (7)[**]
Sulfite oxidase	rat	55	4
INNER MEMBRANE:			
ADP/ATP carrier	Neurospora	32	–
Cyt. c oxidase,			
subunit IV	yeast, rat	16.5	3
subunit VI	yeast	12.5	7.5
Cyt. bc_1-complex,			
Cytochrome c_1	yeast, Neurospora	31[*]	7 (4)[**]
subunit 7	Neurospora	12	0.5
ATPase,			
subunit 2	yeast	54	2
subunit 9	Neurospora	8	6
MATRIX:			
Ornithine carbamoyl-transferase	rat	39	4
Carbamoylphosphate synthetase	rat	160	5
Citrate synthase	Neurospora	45	2
mtRNA-polymerase	yeast	45	2

[*] Polypeptide without covalently linked heme.

[**] Data in brackets refer to intermediates during processing.

This table describes only proteins discussed in the text.

cytochrome c from yeast or rat, a comparison of the amino acid sequence of the mature protein with that derived from the nucleotide sequence of the DNA is possible. This comparison confirms that no additional amino acids are present at the aminoterminal or the carboxyterminal end of the primary translation product.

It would appear that while proteolytic cleavage often occurs, it is not an obligatory step in protein transfer across mitochondrial membranes. Additionally, the conclusion can be drawn that the absence of a presequence in precursors is not related to a particular submitochondrial site to which these precursors are transported: porin is translocated into the mitochondrial outer membrane, cytochrome c into the intermembrane space, and the ADP/ATP carrier into the mitochondrial inner membrane.

Properties of Precursors Suggest the Existence of Conformational Differences from their Mature Forms.

Precursors are found in the cytosolic fraction after being released from the ribosomes. Apparently even the most insoluble membrane proteins can exist in some kind of soluble form when they are on their way into the mitochondria. Precursors to insoluble membrane proteins perhaps form protein micelles: Analysis of the precursors of inner membrane proteins, such as the ADP/ATP carrier and the subunit 9 of ATPase, indicates that they occur not as monomers but in an aggregated state in the postribosomal supernatant (20). The precursors of soluble matrix proteins, e.g. carbamoylphosphate synthetase and ornithine carbamoyltransferase from rat liver, were reported to form large aggregates as well (21). This formation of soluble aggregates must have a basis in the peculiar conformations of precursors.

There is further evidence to suggest that precursor proteins differ in conformation from the mature forms. The ADP/ATP carrier,

for example, does not bind to hydroxylapatite after being solubi-
lized from mitochondria with detergents. In contrast, its precursor
is firmly bound by hydroxylapatite. With other precursors, differences in their affinities to antibodies against the mature forms are
observed. The strongest difference in this respect is exhibited by
cytochrome c. This protein is synthesized as apocytochrome c, i.e.
the polypeptide without the covalently attached heme group. Antibodies against the mature form, holocytochrome c, were obtained
which do not precipitate apocytochrome c synthesized in cell-free
systems. On the other hand, antibodies against apocytochrome c which
was prepared from holocytochrome c by removal of the heme group
recognize the biosynthetic precursor, but do not precipitate holocytochrome c (26). Strong conformational differences between holocytochrome c and chemically prepared apocytochrome c are revealed by
physicochemical measurements. Obviously it is the heme group which
governs the different folding of the polypeptide chain in apocytochrome c and holocytochrome c.

Free Ribosomes are Engaged in the Synthesis of Mitochondrial Proteins.

The question whether free or membrane-bound ribosomes synthesize cytoplasmic precursors of mitochondrial proteins is of importance, since synthesis on free ribosomes points to a posttranslational mode of transport. Two types of experiments were carried
out to answer this question: First, free and membrane-bound polysomes were isolated and synthesis of polypeptide chains was
completed in homologous or heterologous postribosomal supernatants
in presence of radioactive amino acids. Second, mRNA was isolated
from free and membrane-bound polysomes and translated in reticulocyte lysates. With both approaches free polysomes were found to be
the major site of synthesis of mitochondrial proteins in several

independently performed studies.

There are, however, observations which cannot be easily reconciled with free ribosomes being the exclusive sites of synthesis of mitochondrial proteins: Yeast mitochondria were found to be associated with cytoplasmic ribosomes. Electron micrographs of yeast spheroplasts treated with cycloheximide to arrest protein synthesis, and of isolated mitochondria derived from such cells revealed ribosomes lined up on the mitochondrial outer membrane (22). They were observed especially at sites where outer membrane and inner membrane were in close apposition. Evidence discussed here make it very unlikely that these ribosomes are engaged in cotranslational transport of mitochondrial proteins. It is, however, possible that the nascent chains of these ribosomes interact with specific receptors on the mitochondrial surface (cf. section III). This could also explain observations that certain mitochondrial proteins (e.g. subunit IV of cytochrome oxidase) are synthesized to a significant proportion on membrane-bound polysomes in yeast (23) or rat liver (24), though this is not obligatorily required for translocation of these proteins into mitochondria.

Precursors are Posttranslationally Transferred into Isolated Mitochondria.

Several independent investigations have demonstrated that precursors of various different proteins are transferred into mitochondria when the postribosomal supernatants of cell-free translation systems after mRNA-directed protein synthesis are incubated together with intact isolated mitochondria. A number of observations suggest that translocation of precursors across the mitochondrial membrane(s) actually takes place in such reconstituted systems: 1) Precursors are processed to the sizes of the mature proteins. For some proteins it is proven that proteolytic cleavage does occur at the correct sites. 2) In contrast to the precursors present in the supernatant, the transferred proteins are resistant to added proteases. Apparently they have crossed at least the mitochondrial outer membrane. 3) The transferred proteins acquire properties of the mature assembled counterparts. The ATPase subunit 9, for example, which is a subunit of the F_0 part of the F_1F_0 complex, is transferred in vitro into mitochondria and can then be precipitated with antibodies against the F_1-part of the F_1F_0 complex (20). This indicates that transfer in such a reconstituted system can lead to the formation of mature proteins which have the ability to interact with other subunits of the whole, functional complex.

Studies with Whole Cells Confirm the Existence of Precursors and their Posttranslational Import into Mitochondria.

Important information on the transfer mechanism is obtained by studying the kinetics of assembly of mitochondrial proteins in whole cells. For this purpose the following techniques can be employed: Cells are pulse-labelled with a radioactive amino acid and mitochondrial proteins are analysed during a chase period, i.e. after adding excess unlabelled amino acid. This is done by immunoprecipitating defined proteins either from homogenates of whole cells or from subcellular fractions. All results obtained by this experimental approach are only compatible with a posttranslational mechanism of precursor transport into mitochondria: 1) Immunoprecipitation from whole cell extracts leads to the detection of the same larger precursor proteins which are also detected after synthesis in cell-free systems. 2) In most cases these precursor molecules are converted to the mature forms with a rather short half life (1 - 3 min). 3) The appearance of pulse-labelled proteins in the mitochondria shows a characteristic lag phase as compared to the kinetics of labelling of cytosolic proteins. This indicates that the newly synthesized proteins first have to pass through an extramitochondrial precursor pool before they enter the mitochondria. These lag phases differ among various mitochondrial proteins, suggesting different sizes of the extramitochondrial pools. 4) A sudden block of cytoplasmic protein synthesis by cycloheximide leads to an immediate block in the labelling of cytosolic proteins, but labelled proteins continue to appear in mitochondria for a certain period (25). Apparently precursor proteins are imported into mitochondria in the absence of protein synthesis.

The observation that complete precursor molecules occur in vivo is certainly one of the strongest arguments that precursor synthesis and transfer into mitochondria are separate events. These observations convincingly rule out a cotranslational transport of mitochondrial proteins. They fully agree with the conclusions derived from experiments with cell-free systems.

III. INTERACTION OF PRECURSORS WITH MITOCHONDRIAL RECEPTOR SITES

One can expect that the transfer of proteins into mitochondria is a process which involves several distinct steps. In order to understand this process on a molecular basis, experimental systems must be available which allow a resolution of these steps. The establishment of cell-free systems employing precursors synthesized in vitro and their transport into isolated mitochondria facilitates such investigations. The first step in the transfer process is the recognition of mitochondrial precursors by the mitochondrial surface.

Translocation of Precursors into Mitochondria Can be Arrested at their Specific Binding to the Mitochondrial Surface.

To study the interaction between precursors and mitochondria in detail one must be able to inhibit the translocation. Several procedures can be utilized for that purpose. One which is generally applicable to precursors relies on the observation that the binding of precursors to the mitochondria appears to be less dependent on temperature than the translocation through or insertion into the membrane. Another approach takes advantage of the observation that the import of most proteins into mitochondria requires an electrical potential across the inner membrane as will be described in detail below. When the membrane potential is dissipated, transfer is halted at the level of precursors being bound to mitochondria.

However, a major drawback in analysing the recognition step is the availability of precursors in only minute amounts. Almost all precursors can only be obtained by synthesis in cell-free systems and the amounts produced are not sufficiently large to study binding to mitochondrial recognition sites in desirable detail. Fortunately there is one precursor which can be obtained in practically unlimited amounts. This is apocytochrome c. It can be prepared by removal of the covalently bound heme group from holocytochrome c by chemical means. Apocytochrome c prepared in this way can be radioactively labelled by reductive methylation or by iodination and is bound to mitochondria, translocated across the outer membrane, converted to holocytochrome c, and competes in this respect with apocytochrome c synthesized by cell-free translation. The translocation of apocytochrome c into mitochondria is inhibited by the heme analogue deuterohemin (26). Apocytochrome c is tightly bound to the mitochondrial surface under this condition. Thus, interaction of apocytochrome c with the mitochondrial surface can be studied in experiments similar to those carried out to elucidate the interaction of peptide hormones with their receptors on cell surfaces.

Precursors are Bound to Mitochondria via Specific Receptor Sites.

The following characteristics of precursor binding to the surface of intact mitochondria have emerged:

Binding is rapid and tight, and the number of binding sites on
mitochondria is limited.   Precursors bind to mitochondria at a rate
which is sufficiently large to account for the rate of transport in
vivo. Once bound, precursors are not removed when the mitochondria
are washed in the medium in which binding has been performed.
Binding sites for apocytochrome c have been titrated. The Scatchard
plot revealed that apocytochrome c binds with high affinity (K_A= 2.2
x 10^7 M^{-1} in the case of Neurospora) to mitochondria (27). The
binding sites are located on the mitochondrial surface, probably at
the cytoplasmic face of the outer membrane, since bound precursors
are sensitive to added proteases in contrast to precursors transferred into mitochondria. There are about 100 pmol high-affinity
binding sites for apocytochrome c per mg mitochondrial protein.

Binding is specific and functionally related to transfer of
precursors into mitochondria. Apocytochromes c from various species
(e.g. Neurospora, yeast, horse, parsnip) have different affinities
to apocytochrome c binding sites on mitochondria (27). Non-mitochondrial proteins, such as bacterial apocytochrome c (Paracoccus deni-
trificans) or the precursor to glyoxysomal isocitrate lyase do not
bind to mitochondria. Also, mature proteins do not interfere with
the binding of the precursor forms. Apocytochrome c bound to mitochondria in the presence of deuterohemin, which inhibits translocation of this particular precursor, is taken up into mitochondria
and converted to holocytochrome c when the inhibition due to this
heme analogue is relieved by addition of excess protohemin (26). The
precursor of the ADP/ATP carrier bound to de-energized mitochondria

becomes internalized when the membrane potential is restored (34). The precursor to ATPase subunit 9 is internalized under these conditions and processed to the size of the mature form .

More than One Kind of Receptor is Present on the Mitochondrial Surface.

The results of the binding studies imply the existence of receptors on the mitochondrial surface. However, no such receptor has been isolated yet and many questions remain open: What is the chemical nature of these receptors ? How many different kinds exist ? How is the binding to a receptor related to the translocation across the membrane(s) ? Although no definite answers can be given, some interesting data are available concerning these points.

When mitochondria are treated with protease (trypsin or proteinase K) before they are employed in in vitro transfer experiments, they lose the ability to bind precursors and to import them. This may mean that the receptors are proteins. This finding and the evidence accumulated in the binding experiments suggest that receptors are exposed at the cytoplasmic face of the outer membrane.

There is information that mitochondria have more than one kind of receptor: Apocytochrome c employed in large amounts saturates its own binding sites but does not interfere with the binding of various other precursors, e.g. porin, ADP/ATP carrier, ATPase subunit 9, and cytochrome c_1 (28). On the other hand, it seems extremely unlikely that each of the several hundred mitochondrial proteins has its own receptor. Two observations suggest the existence of a very limited number of different and evolutionary conserved recognition mechanisms. First, transfer in vitro of most proteins studied so far does not exhibit species specificity, i.e. precursors of mitochondrial proteins from Neurospora can be transferred into mitochondria isolated from yeast, rat liver or guinea pig heart. Second, a precursor

protein from one type of cell, e.g. the ornithine carbamoyltransferase from rat liver, can be imported into the mitochondria isolated from another type of cell, e.g. rat kidney cells, which do not
contain this protein in vivo (29).

Which Part of a Precursor is Recognized by the Mitochondrial Receptors ?

Mature proteins do not compete with their precursors for
transfer and precursors processed in vitro are not taken up by
mitochondria, perhaps with the controversial exception of the mature
form of aspartate aminotransferase (30). However, this failure by no
means allows the conclusion that addressing of precursors to mitochondria occurs via the presequences of precursors. Rather, the
function of these sequences may be of a quite different nature: They
could serve to alter the conformation of a mitochondrial protein in
such a way that a part of its structure is exposed which can
interact with a receptor. The presequences could also have the
function of altering the proteins in such a way that the precursors
outside the mitochondria cannot become functionally active, for
instance because the catalytic site is changed or because a cofac-
tor cannot be bound by the polypeptide. This would correspond to the
well established situation with zymogens whose prosequences shield
the catalytic sites and removal of the prosequences accompanied by a
conformational change leads to the activation of the enzymes (e.g.
the conversion of trypsinogen to trypsin).

Mutual recognition between precursors and receptors must also
occur with precursor proteins which do not possess additional
sequences, in this case obviously by some part of the sequence
present in the mature protein. Cytochrome c provides important
information in this respect. The mature form, holocytochrome c, is
not bound by the receptor. Apparently the corresponding binding

domain is accessible at the surface of the molecule in apocytochrome c but not in holocytochrome c. Thus, the strong positive
charge of cytochrome c cannot be solely responsible for binding to
the receptor since this is practically the same in precursor and
mature form.

In contrast to holocytochrome c, the three-dimensional structure of apocytochrome c is not known. However, some indications as
to the structural part involved in recognition with its mitochondrial receptor may be obtained from a comparison of the amino acid
sequences of the same 90 different cytochromes c so far determined.
The N-terminal part preceding the heme-binding region (which is at
positions 14 - 17 according to standardized nomenclature) is apparently not involved in receptor binding. Several yeast mutants
contain holocytochrome c with amino acid exchanges in this particular region, and even a complete deletion of the first 11 sequence
positions was found (31). Hence, the absence of this region in
apocytochrome c does not interfere with the import of apocytochrome
c into mitochondria and its conversion to holocytochrome c. Recent
data indicate that the receptor-binding structure is located in the
C-terminal half of the sequence where an extremely conservative and
hydrophobic sequence of some 10 amino acids is present (32).
However, one has to take into consideration that the binding domain
is a three-dimensional structure and different distant parts of the
molecule could contribute to it.

IV.   TRANSLOCATION OF PRECURSORS INTO THE DIVERSE
      SUBMITOCHONDRIAL COMPARTMENTS

The step which follows selective binding of precursors to
mitochondria, i.e. the translocation into one of the four submitochondrial compartments, is presumably of considerable complexity.

The mechanisms involved in a) the insertion of proteins into the outer membrane, b) their translocation across the outer membrane into the intermembrane space, c) the insertion into the inner membrane, and d) the translocation across both membranes, i.e. into the matrix, are poorly understood. It is, for instance, not clear whether a protein destined for the matrix crosses outer membrane and inner membrane in one, two, or even more distinct steps. It has been repeatedly speculated that the two mitochondrial membranes can come into close contact or may even fuse. "Contact sites" or "fusion sites" were observed upon electron microscopy of sectioned mitochondria (22). It is an open question, however, whether these structures are really related to protein transport. An understanding of the precise transfer mechanism would require exact knowledge of the conformation of a precursor and its changes during interaction with the membrane(s).

A Membrane Potential is Required for Translocation of Most Mitochondrial Proteins.

It has been observed that posttranslational transfer of precursors and their proteolytic processing are blocked when whole cells are exposed to an uncoupler of oxidative phosphorylation such as CCCP (carbonyl cyanide m-chlorophenylhydrazone). Subsequent experiments with isolated mitochondria confirmed that the import of precursors is blocked when the mitochondria are de-energized (e.g. refs. 15,28,33). As already described above, the uncouplers do not inhibit binding of the precursors to the mitochondria but do inhibit their transfer across the outer membrane. However, these observations do not allow one to discriminate whether the electrical membrane potential, the proton motive force, or ATP is the primary source of energy involved in protein translocation: Uncoupling of mitochondria not only dissipates the membrane potential but also results in the induction of intramitochondrial ATPase activity, i.e. the reversed action of ATP synthase, thus lowering the level of ATP in the matrix.

The following experiments employing transfer in vitro (34) identified the membrane potential as the primary source of energy for translocation of extramitochondrial precursors across the mitochondrial membranes:

Conditions were created under which the mitochondrial membrane potential was low, but the level of ATP in the mitochondria was high. The membrane potential was dissipated with protonophores (CCCP or dinitrophenol) or with an ionophore (valinomycin plus K^+), and oligomycin was added to inhibit ATP degradation by the oligomycin-sensitive ATPase. Furthermore, ATP was added in high concentration (5 mM) to the mixture. It is known that ATP is readily imported into the matrix via the ADP/ATP carrier when mitochondria are uncoupled. This does not occur in mitochondria which have a normal membrane potential (150 - 200 mV) since the ADP/ATP carrier is electrogenic and a potential positive outside favours the export of ATP from mitochondria in exchange against ADP. Therefore, in the presence of uncoupler and oligomycin, higher ATP levels than in respiring coupled mitochondria can be obtained. With all the precursors tested, import into mitochondria was blocked under such conditions and the precursors remained at the mitochondrial surface.

On the other hand, conditions were created under which mitochondria maintain a membrane potential but the matrix ATP level is far below normal. Since the direct determination of mitochondrial matrix ATP is difficult in the complex in vitro system, the ATP level was measured indirectly by following a reaction requiring ATP in the matrix, namely mitochondrial protein synthesis. ATP within the mitochondria is derived from two sources: phosphorylation of ADP by the ATP synthase and import from the cytosol via the ADP/ATP carrier. There are specific inhibitors for both processes, i.e. oligomycin and carboxyatractyloside. Simultaneous addition of the two inhibitors to mitochondria had no effect on protein import into mitochondria, it did however inhibit intramitochondrial protein synthesis.

Nigericin, an ionophore which exchanges H^+ versus K^+ and therefore does not affect the membrane potential but dissipates the proton gradient, did not interfere with the transfer of precursors into mitochondria. Thus it is apparently the electrical membrane potential that is required for protein import. A membrane potential can be generated in mitochondria in two ways: by electron transport and by the reversed action of ATP synthase, i.e. ATP hydrolysis by the oligomycin-sensitive ATPase. Hence, it is easily explained that inhibitors of respiration alone do not (or only weakly) inhibit protein transfer and are effective only in combination with oligomycin.

Which Role Does the Membrane Potential Play in the Translocation of Precursors ?

The translocation of many mitochondrial proteins into mitochondria was found to depend on the energization of the inner membrane. Does this apply to all precursors ? At least the translocation of two proteins is independent from a membrane potential. One example is the porin of the outer membrane (35). This protein is inserted into mitochondria without passing through a membrane. In this context it is interesting that the microsomal cytochrome b_5, which is closely related to the mitochondrial cytochrome b_5 present in the outer membrane, is also synthesized on free ribosomes without a presequence and is posttranslationally inserted into the membrane of the ER (36). Insertion of these proteins into their respective membranes may occur by self-assembly with subunits of these proteins preexistent in the target membrane, and it might be expected that all proteins destined for insertion into the mitochondrial outer membrane obey this particular mechanism of posttranslational transport. If this suggestion would prove to be correct it would be needless to postulate a special receptor for the assembly of newly formed receptors in the mitochondrial outer membrane.

The second protein whose import is independent from energization of mitochondria is cytochrome c (28). In this case, one could argue that there is no reason for such a dependence since cytochrome c must only be translocated across the outer membrane, whereas the membrane potential is confined to the inner membrane. However, other intermembrane proteins such as cytochrome b_2 in yeast and sulfite oxidase in rat liver are imported only into energized mitochondria (37, 38). Substantial evidence has been presented in favour of an import pathway of cytochrome b_2 which involves a 'detour' of the precursor into the inner membrane. This apparently rather complex pathway involves proteolytic processing of the precursor in two separate steps (cf. section V).

What is the role of the membrane potential in the assembly of those proteins which are inserted into the inner membrane either transiently or permanently ? One possibility is that the membrane potential provides the energy for translocation. The precursor - receptor complex or a complex between precursor and a hypothetical "translocator" protein could respond to the membrane potential in such a way that the precursor is transferred across the membrane(s). Another possibility is that the energy for transmembrane transfer is provided primarily by refolding of the polypeptide chain of the precursor and that the membrane potential serves to trigger such refolding events (39). In this context one should remember that a membrane potential is also required in the export of periplasmic proteins across the plasma membrane of gram-negative bacteria (40).

On the other hand, transport of precursor proteins into chloroplasts was reported to depend on ATP in the chloroplast stroma space (41). It remains to be determined whether this observed difference reflects a genuine difference in the assembly mechanism of mitochondrial and chloroplast proteins.

Translocation of Cytochrome c is Coupled to Heme Attachment to the Precursor.

In contrast to most other mitochondrial proteins, neither a membrane potential nor proteolytic processing is required for import of cytochrome c into mitochondria. How is apocytochrome c translocated across the outer membrane ?

The following view concerning assembly of cytochrome c can be proposed from experimental results (Fig.1): Apocytochrome c is bound to its receptor at the outer surface of mitochondria in such a way that the thiol groups of its heme-binding cysteine residues become exposed at the intermembrane face of the outer membrane. The heme group becomes linked to these cysteines via thioether bonds aided by an enzyme contained in the intermembrane space. The covalent attachment of the heme group forces the polypeptide chain to refold and by this refolding the polypeptide is pulled through the outer membrane. The properly folded holocytochrome c is trapped in the intermembrane space. It associates with its functional binding sites on the outer face of the inner membrane, where it mediates electron transport as a component of the mitochondrial respiratory chain.

The evidence for such a pathway can be summarized as follows: The covalent attachment of heme to apocytochrome c is apparently mediated by an enzyme. Protohemin, but not protoporphyrin IX, is linked to apocytochrome c in a stereospecific reaction (26). This reaction is inhibited by certain heme analogues (e.g. deuterohemin, mesohemin) but not by others (e.g. hematohemin). The converting enzyme, cytochrome c heme lyase, is presumably contained in the intermembrane space since neither the cytosol nor the isolated mitochondrial outer membrane or inner membrane appears to contain an activity converting apocytochrome c to holocytochrome c. Inhibition of heme attachment causes inhibition of the translocation of apocytochrome c across the mitochondrial outer membrane and leads to accumulation of apocytochrome c at the mitochondrial surface (cf. section III). This inhibition can be releaved by excess protohemin. The covalently linked heme group strongly affects the conformation of the polypeptide (42). Denatured holocytochrome c rapidly resumes the native conformation when the denaturing conditions have been abandoned. Finally, holocytochrome c cannot penetrate the mitochondrial outer membrane (43). Extraction of holocytochrome c from mitochondria and insertion of exogenous holocytochrome c into the mitochondria is possible only after rupture of the mitochondrial outer membrane.

As already mentioned and as will be dicussed in the next section, the assembly of other mitochondrial heme proteins such as cytochrome b_2 or cytochrome c_1, the latter of which also contains a covalently attached heme, follow a different and much more complicated mechanism than the assembly of cytochrome c.

V. PROTEOLYTIC PROCESSING OF PRECURSORS

We have discussed above that most mitochondrial precursor proteins are formed as larger precursors (cf. Table). Hence, they are assembled in mitochondria with concomitant proteolytic removal of their presequences. Where in the mitochondria does this processing occur, and which particular protease is involved ? What is the role of this cleavage in the translocation ?

FIGURE 1:   Assembly of Cytochrome c
             (Proposed Mechanism)

STEP 1: Cytochrome c is synthesized as a precursor, apocytochrome c, on free ribosomes and released into the
cytosol.

STEP 2: Apocytochrome c is bound and properly arranged at
the surface of the mitochondrial outer membrane (OM) by a
specific receptor which either itself might form a pore in
the outer membrane or is associated with such a pore. The
receptor-bound apocytochrome c exposes the heme-binding
cysteine residues through this pore in the intermembrane
space.

STEPS 3 and 4: Cytochrome c heme lyase, an intermembrane
enzyme, accepts protoheme provided by the ferrochelatase
and attaches it in a stereospecific reaction to the
apocytochrome c.

STEP 5: The covalently linked heme forces the polypeptide
chain of cytochrome c to wrap around the heme, thereby
pulling the polypeptide completely through the membrane.

STEP 6: The mature protein, holocytochrome c, is entrapped
in the intermembrane space and binds at the surface of the
inner membrane (IM), associating with the pertinent
components of the respiratory chain.

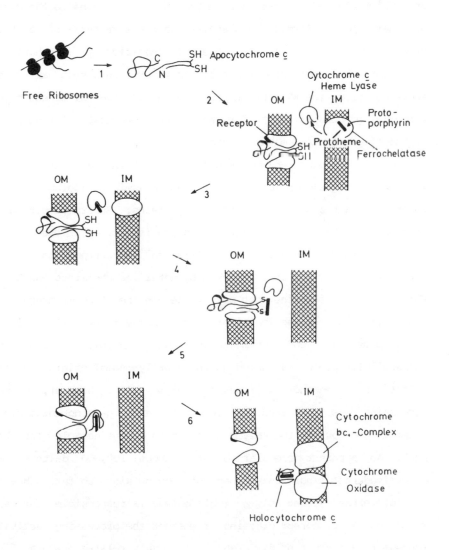

Free Ribosomes

Apocytochrome c

Cytochrome c Heme Lyase

Receptor

Proto-porphyrin

Protoheme

Ferrochelatase

Cytochrome bc₁-Complex

Cytochrome Oxidase

Holocytochrome c

An Enzyme Specifically Processing Mitochondrial Precursor Proteins Resides in the Mitochondrial Matrix.

In a number of studies a protease was extracted from mitochondria which cleaved precursors of mitochondrial proteins to the sizes of their mature forms. This enzyme meets the criteria of a true processing protease: a) It processes specifically mitochondrial precursors but not precursors of other proteins such as those secreted across the ER. b) It processes a number of different mitochondrial precursors. c) It does not degrade precursors further than to the sizes of their mature forms.

The processing enzyme is a soluble protein and thus differs from the "signal peptidases" of the endoplasmic reticulum and of bacteria which are integral membrane proteins. Subfractionation of mitochondria has revealed that the enzyme is located in the matrix (44). In order to be active it requires Zn^{++} or certain other divalent metal ions and it is blocked by metal ion chelators such as EDTA or o-phenanthroline, but not by the non-chelating m-phenanthroline. Various protease inhibitors, e.g. phenylmethylsulfonyl fluoride, pepstatin, and chymostatin, which inhibit a variety of intracellular proteases including those of lysosomal origin, do not affect the mitochondrial processing enzyme (45). However, normal processing is inhibited by leupeptin and p-aminobenzamidine. Attempts to purify the enzyme have led to a considerable enrichement. An apparent molecular weight of about 108,000 daltons was determined but a pure enzyme has not been obtained so far. Thus it is not clear whether only a single enzyme is responsible for all mitochondrial precursor proteins or whether the processing activity represents a mixture of different, yet closely related enzymes.

Some Precursors Undergo a Two-Step Processing.

The occurence of the processing enzyme in the matrix has important implications. Apparently any precursor requiring proteolytic processing must be transferred into the matrix or at least inserted into the inner membrane in such a way that the presequence is exposed to the matrix side before proteolytic processing can occur. This is 'en route' for precursors destined for the matrix itself such as ornithine carbamoyltransferase (29) or citrate synthase (46). However, a complex picture emerged for cytochrome b_2 and cytochrome c_1. The former protein is a soluble intermembrane enzyme, the latter one an integral membrane protein which faces the intermembrane space. These two cytochromes are apparently processed by two successive proteolytic events, since an intermediate form between the original precursor and the mature form is transiently generated (15, 37).

It is quite likely that they represent true intermediates in the assembly pathway since they are detected both in vivo and in vitro. The first step which leads to the intermediate forms requires energization of the inner membrane. In the case of cytochrome c_1, the first proteolytic processing step precedes the covalent attachment of the heme group. Heme deficiency leads to accumulation of the intermediate form. The submitochondrial location of the second processing step remains to be determined. The protease which is involved in this second step is apparently different from the matrix protease utilized in the first step. The intermediates, but not the mature proteins, are formed when the precursors are incubated solely with the processing protease prepared from mitochondrial matrix.

A hypothetical mechanism for the transfer of precursors into mitochondria by the various discussed pathways is presented in Fig. 2.

FIGURE 2:   Mechanisms Involved in Transfer of
Various Precursors into Mitochondria
(Hypothetical Sequence of Events)

STEP 1: Extramitochondrial precursors are recognized by specific receptors at the mitochondrial surface.

STEPS 2 and 3: The outer membrane (OM) and the inner membrane (IM) of mitochondria come into contact at certain sites and form "fusion sites". The precursor - receptor complex reorients in this area, perhaps aided by an hypothetical "translocator" protein. Either the formation of the "fusion sites" or the reorienting of the protein complexes in the fused membrane areas (or both events) depend on the electrical potential across the inner membrane.

STEPS 4 A-D: Precursors are (transiently or permanently) inserted in the inner membrane, processed, and allocated to their final destinations according to their particular properties. A: The precursor refolds and is inserted in the inner membrane without proteolytic processing (e.g. the ADP/ATP carrier). B: The precursor is attacked by a processing enzyme contained in the matrix. After removal of the presequence the protein is relocated into the intermembrane space, a step which may entail a second processing event (e.g. cytochrome b_2). C: The proteolytically processed precursor occupies its final topological position in the inner membrane (e.g. subunit 9 of the ATPase). D: The proteolytically processed precursor is discharged into the matrix (e.g. citrate synthase).

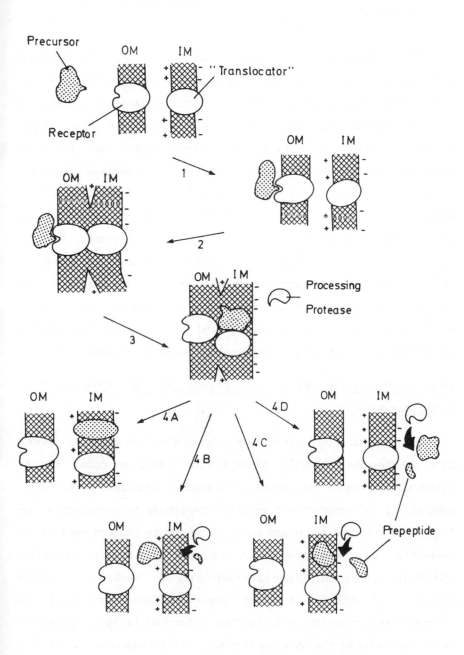

VI.   ASSEMBLY OF PROTEINS   SYNTHESIZED   WITHIN THE   MITO-
CHONDRIA

The   intramitochondrial   genetic system uses a   codon   language
slightly   different   from that used by the nucleocytoplasmic   system
(47).   Therefore,   no simple exchange of translatable information is
possible   between   the two systems,   neither in vivo nor   in   vitro.
Structural   genes   on   mtDNA code for a few proteins   of   the   inner
membrane, i.e. for three subunits of cytochrome c oxidase   (subunits
I,   II,   and III),   one or two subunits of the   oligomycin-sensitive
ATPase (subunit 6;   in yeast also subunit 9), and one subunit of the
bc_1- complex (subunit 3,   i.e.   cytochrome b). Furthermore, one protein   of   the small subunit of mitochondrial ribosomes is   coded   on
mtDNA at least in yeast and Neurospora. Since the amino acid sequences   can   be deduced from the known nucleotide sequences of   several
mtDNAs (i.e.   man,   bovine,   mouse,   and in part yeast) the complete
structures of the primary translation products are known.

Intramitochondrially   synthesized   proteins may be formed as   larger
precursors as well.

How   are   the intramitochondrially synthesized proteins   assembled   ?   The   three subunits of cytochrome c oxidase   coded   for   by
mitochondrial   genes   are formed as separate   translation   products.
Subunit   II   of cytochrome c oxidase from bovine is   apparently   not
formed   as a larger precursor because the mature protein retains the
N-terminal formylmethionine,   i.e. the amino acid by which the mitochondrial   system   initiates   translation.   In   contrast,   a   larger
precursor   of   subunit   II of cytochrome c oxidase   from   yeast   is
observed after translation in isolated mitochondria (48).   According
to the sequence of the structural gene,   the presequence (about   1.5
kD)   of   the precursor protein does not display   exceptionally   high

apolarity nor does it display other unusual features. A similarly confusing result is obtained comparing subunit I of cytochrome c oxidase from beef heart and Neurospora (49): A larger precursor is apparently formed in the mold but not in beef heart. No simple explanation is available for this heterogenous picture. It is also not known wich protease is involved in the processing of the larger precursors.

Cytochrome b is present as the apoprotein in mitochondria from heme-deficient yeast cells. The accumulated apoprotein has the same apparent size as the mature protein (50), but it is not known whether a proteolytic processing precedes its accumulation. Although the nucleotide sequence of the mitochondrial gene specifying the amino acid sequence of cytochrome b from yeast has been determined, the presence of a presequence in the primary translation product remains an open question since the aminoterminal sequence of the mature protein is not known yet.

Attachment of mitochondrial ribosomes at the surface of the inner membrane was observed and on the basis of genetic data it has been suggested that this interaction is functionally important (51). However, it is not known whether the intramitochondrially synthesized proteins are integrated into the inner membrane by cotranslational or by posttranslational events, or whether both modes of transport coexist.

How is the Assembly of Mitochondrially and Cytoplasmatically Synthesized Proteins Interconnected ?

Many mitochondrial proteins are assembled not as separate entities but as components of large protein complexes. The two genetic systems of the cell contribute proteins to complexes of the respiratory chain and the ATPase, which are assembled in the inner membrane, and to the ribosomes, which are assembled in the matrix.

Assembling the various subunits destined for a particular complex must be a cooperative process since the subunits pertinent in a complex are present in stoichiometric amounts.

One example is the ribosomal protein which is coded for by a mitochondrial gene (var1) in yeast and which is part of the small subunit (37 S) of mitochondrial ribosomes. This protein is apparently indispensable for the correct assembly of the other ribosomal proteins, all of which are imported from the cytoplasm. When this mitochondrially synthesized protein is defective or absent this leads to an arrest in the final assembly of the small ribosomal subunit at the stage of a 30 S ribonucleoprotein particle (52) which not only lacks the var1 protein but also certain cytoplasmically synthesized proteins.

An especially interesting example of the interrelationship between the two genetic systems in the assembly of mitochondria is the biogenesis of subunit 9 of the oligomycin-sensitive ATPase. This protein is coded on nuclear DNA in all species studied so far, with the exception of yeast where it is coded on mtDNA. In Neurospora a nuclear as well as a mitochondrial gene are present but the mitochondrial one appears to be silent (53). The nuclear gene product, which is translated on cytoplasmic ribosomes and posttranslationally imported into mitochondria, is formed as a precursor carrying a transient presequence which is roughly as large as the mature protein itself (14). In yeast, where this protein is synthesized in the mitochondrial matrix, the primary translation product has no presequence (54). Yet, in both cases the protein is transported to the same destination, namely inserted into the mitochondrial inner membrane. Why is it formed as a larger precursor here, but in the same size as the mature form there ? We do not know the answer. This is all the more remarkable since the larger extramitochondrial precursor of ATPase subunit 9 from Neurospora, which has been

synthesized in a cell-free translation system, can be translocated
into isolated mitochondria from yeast and correctly processed to the
mature size (55).

VII. CONCLUSIONS

The biogenesis of some fourty cytoplasmically synthesized mitochondrial proteins has been investigated so far. These studies have
established that these proteins are formed as extramitochondrial
precursors and that they are posttranslationally imported into
mitochondria. Their uptake into mitochondria requires specific
interaction with the mitochondrial surface. This apparently involves receptors. The mechanisms by which precursors are posttranslationally translocated across the membranes are not clearly understood yet. Obviously there is no uniform pathway for the translocation process. Rather, the details of translocation vary considerably
with different proteins: Most precursors contain presequences of
various length which are proteolytically processed in either one or
two steps, but a few precursors lack a presequence. Import of most
precursors into mitochondria depends on the electrical potential of
the inner membrane, but a few precursors which are destined for the
outer membrane or the intermembrane space do not require a membrane
potential for assembly. Far less is known about the events in the
assembly of intramitochondrially synthesized proteins.

As far as one can judge presently, no correlation exists between transfer of precursors into a particular submitochondrial compartment and any special sequence of events during translocation. As
a common theme, however, in all cases irreversible steps such as
proteolytic processing, covalent modification, or substantial refolding occur during translocation in order to trap the proteins in
their proper submitochondrial locations.

ACKNOWLEDGMENTS

We are grateful to Martin Teintze for help in preparing the manuscript. We would also like to appreciate support of the authors' research by grants from the Deutsche Forschungsgemeinschaft.

REFERENCES

1. Sabatini, D. D., Kreibich, G., Morimoto, T., and Adesnik, M., J. Cell Biol. 92, 1-22 (1982)

2. Hanover, J. A. and Lennarz, W. J., Arch. Biochem. Biophys. 211, 1-19 (1981)

3. Walter, P. and Blobel., G., J. Cell Biol. 91, 557-561 (1981)

4. Meyer, D.I., Krause, E., and Dobberstein, B., Nature 297, 647-650 (1982)

5. Kreibich, G., Czako-Graham, M., Grebenau, R., Mok, W., Rodriguez-Boulan, E., and Sabatini, D. D., J. Supramol. Structure 8, 279-302 (1978)

6. Walter, P., Jackson, R. C., Marcus, M. M., Lingappa, V. R., and Blobel, G., Proc. Natl. Acad. Sci. U.S. 76, 1795-1799 (1979)

7. Rothman, J. E., Science 213, 1212-1219 (1981)

8. Wallace, D. C., Microbiol. Rev. 46, 208-240 (1982)

9. Anderson, S., Bankier, A. T., Barrell, B. G., de Bruijn, M. H. L., Coulson, A. R., Drouin, J., Eperon, I. C., Nierlich, D. P., Roe, B. A., Sanger, F., Schreier, P. H., Smith, A. J. H. , Staden, R., and Young, I. G., Nature 290, 457-465 (1981)

10. Lustig, A., Levens, D., and Rabinowitz, M., J. Biol. Chem. 257, 5800-5808 (1982)

11. Zelman, L. S., Nikaido, H., and Kagawa, Y., J. Biol. Chem. 255, 1771-1774 (1980)

12. Damsky, C. H., J. Cell. Biol. 71, 123-135 (1978)

13. Luck, D. J. L., Proc. Natl. Acad. Sci. U.S. 49, 233-240 (1963)

14. Lewin, A. S., Gregor, J., Mason, T. L., and Schatz, G., Proc. Natl. Acad. Sci. U.S. 77, 3998-4002 (1980)

15. Teintze, M., Slaughter, M., Weiss, H., and Neupert, W., J. Biol. Chem. 257, 10364-10371 (1982)

16. Viebrock, A., Perz, A., and Sebald, W., The EMBO J. 1, 565-571 (1982)

17. Zimmermann, R., Paluch, U., Sprinzl, M., and Neupert. W., Eur. J. Biochem. 99, 247–252 (1979)

18. Kraus, J. P., Conboy, J. G., and Rosenberg, L. E., J. Biol. Chem. 256, 10739–10742 (1981)

19. Zimmermann, R., Paluch, U., and Neupert, W., FEBS Letters 108, 141–146 (1979)

20. Schmidt, B., Hennig, B., Zimmermann, R., and Neupert, W., J. Cell Biol., in press

21. Miura, S., Mori, M., Amaya, Y., Tatibana, M., and Cohen, P., Biochem. Internat. 2, 305–312 (1981)

22. Kellems, R. E., Allison, V.F., and Butow, R. A., J. Biol. Chem. ?49, 3297–3303 (1974)

23. Ades, I. Z. and Butow, R. A., J. Diol. Chem. 255, 9918–9924 (1980)

24. Northemann, W., Schmelzer, E., and Heinrich, P. C., Eur. J. Biochem. 119, 203–208 (1981)

25. Hallermayer, G., Zimmermann, R., and Neupert, W., Eur. J. Biochem. 81, 523–532 (1977)

26. Hennig, B. and Neupert,W., Eur. J. Biochem. 121, 203–212 (1981)

27. Hennig, B., Koehler, H., and Neupert, W., paper submitted for publication

28. Zimmermann, R., Hennig, B., and Neupert, W., Eur. J. Biochem. 116, 455–460 (1981)

29. Morita, T., Miura, S., Mori, M., and Tatibana, M., Eur. J. Biochem. 122, 501–509 (1982)

30. Marra, E., Doonan, S., Saccone, C., and Quagliariello, E., Eur. J. Biochem. 83, 427–435 (1978)

31. Sherman, F., and Stewart, J. W., in: The Biochemistry of Gene Expression in Higher Organisms (Pollak, J. K. and Lee, J. W., eds.) p. 56–86 (1973) Australia and New Zealand Book Comp., Sydney

32. Matsuura, S., Arpin, M., Hannum, C., Margoliash, E., Sabatini, D. D., and Morimoto, T., Proc. Natl. Acad. Sci. U.S. 78, 4368–4372 (1981)

33. Mori, M., Morita, T., Miura, S., and Tatibana, M., J. Biol. Chem. 256, 8263–8266 (1981)

34. Schleyer, M., Schmidt, B., and Neupert, W., Eur. J. Biochem. 125, 109–116 (1982)

35. Freitag, H., Janes, M., and Neupert, W., Eur. J. Biochem. 126, 197–202 (1982)

36. Rachubinski, R.A., Verma, D. P. S., and Bergeron, J. J. M., J. Cell. Biol. 84, 705–716 (1980)

37. Gasser, S. M., Ohashi, A., Daum, G., Böhni, P. C., Gibson, J., Reid, G. A., Yonetani, T., and Schatz, G., Proc. Natl. Acad. Sci. U.S. 79, 267-271 (1982)

38. Ono, H., and Ito, A., Biochem. Biophys. Res. Commun. 107, 258-264 (1982)

39. Wickner, W., Science 210, 861-868 (1980)

40. Daniels, Ch. J., Bole, D. G., Quay, S. C., and Oxender, D. L., Proc. Natl. Acad. Sci. U.S. 78, 5396-5400 (1981)

41. Grossmann, A., Bartlett, S., and N.-H. Chua, Nature 285, 625-628 (1980)

42. Timkovich, R., in: The Porphyrins, vol.7, part B (Dolphin, D., ed.), p. 241-294 (1979) Academic Press, New York

43. Wojtczak, L. and Sottocasa, G. L., J. Membrane Biol. 7, 313-324 (1969)

44. Boehni, P., Gasser, S., Leaver, Ch., and Schatz, G., in: The Organization and Expression of the Mitochondrial Genome (Kroon, A. M. and Saccone, C., eds.) 423-433 (1980) Elsevier/ North-Holland Biomed. Press Amsterdam, New York

45. Miura, S., Mori, M., Amaya, Y., and Tatibana, M., Eur. J. Biochem. 122, 641-647 (1982)

46. Harmey, M. A. and Neupert, W., FEBS Letters 108, 385-389 (1979)

47. Gray, M. W., Can. J. Biochem. 60, 157-171 (1982)

48. Sevarino, K. A. and Poyton, R. O., Proc. Natl. Acad. Sci. U.S. 77, 142-146 (1980)

49. van't Sant, P., Mak, J. F. C., and Kroon, A. M., Eur. J. Biochem. 121, 21-26 (1981)

50. Clejan, L., Beattie, D. S., Gollub, E. G., Liu, K.-P., and Sprinson, D. B., J. Biol. Chem. 255, 1312-1316 (1980)

51. Spithill, T. W., Trembath, M. K., Lukins, H. B., and Linnane, A. W., Molec. Gen. Genetics 164, 155-162 (1978)

52. Maheshwari, K. K., Marzuki, S., and Linnane, A. W., Biochem. Internat. 4, 109-115 (1982)

53. van den Boogaart, P., Samallo, J., and de Agsteribbe, E. , Nature 298, 187-189 (1982)

54. Macino, G. and Tzagoloff, A., J. Biol. Chem. 254, 4617-4623 (1979)

55. Schmidt, B., Hennig, B., and Neupert, W., unpublished results

Genes: *Structure and Expression*
Edited by A. M. Kroon
© 1983 John Wiley & Sons Ltd.

THE NON-UNIVERSALITY OF THE GENETIC CODE

Albert M. Kroon

Laboratory of Physiological Chemistry, State University
at Groningen, Bloemsingel 10, 9712 KZ GRONINGEN,
The Netherlands

Cecilia Saccone

Istituto di Chimica Biologica, Università di Bari,
Via Amendola 165/A, 70126 BARI, Italy

INTRODUCTION

The genetic code is the cryptography contained in the sequence of
the bases in DNA and its messenger RNA-transcripts that specifies
the sequence of amino acids in proteins. It is a triplet code, *i.e.*
a group of three bases is actually coding for one amino acid. The
fascinating experiments and pioneering work for breaking the code
have been elegantly described by Khorana and by Nirenberg in their
1968 Nobel lectures. The genetic code was deciphered for the larger
part by using synthetic RNA as template in a reconstituted system
for protein synthesis. Homopolymers and random polymers as well as
copolymers with defined repeating sequences, which were obtained by
either biochemical or organic chemical synthesis, served as messenger RNAs in these studies. The observation that trinucleotides are
able to promote the binding of aminoacyl-tRNA to ribosomes in the
absence of protein synthesis added a further tool in the search for
the genetic code. By using trinucleotides of a defined sequence
specific binding of a single labeled amino acid esterified with its
transfer RNA could be obtained for about fifty of the sixty-four
trinucleotides. To make a long story short the combined results are
given in Table 1.
Any textbook of biochemistry, genetics or cell biology will present
the data of Table 1. The features of the genetic code are well known
and obvious from the table. The code is degenerate: most of the
amino acids are coded by more than one triplet. In the cases of
arginine, leucine and serine even six codons are available for each
of these amino acids. In the case of serine the degeneracy concerns
the bases of the first, of the second as well as of the third position. In most instances, however, the degeneracy is restricted to
the base at the 3' end and is either complete (four codons for one
amino acid) or restricted to the pyrimidines (U and C) or purines
(A and G). Isoleucine is the only amino acid with an U/C/A degeneracy in the third position. The biological significance of the degeneracy of the genetic code may be that it permits a certain degree of

TABLE 1. The genetic code

first base (5' end)	Second base				third base (3' end)
	U	C	A	G	
U	phe	ser	tyr	cys	U
	phe	ser	tyr	cys	C
	leu	ser	stop	stop	A
	leu	ser	stop	trp	G
C	leu	pro	his	arg	U
	leu	pro	his	arg	C
	leu	pro	gln	arg	A
	leu	pro	gln	arg	G
A	ile	thr	asn	ser	U
	ile	thr	asn	ser	C
	ile	thr	lys	arg	A
	met*	thr	lys	arg	G
G	val	ala	asp	gly	U
	val	ala	asp	gly	C
	val	ala	glu	gly	A
	val	ala	glu	gly	G

*Initiation codon as well

mutational variation of the base sequence of DNA without deleterious changes in the amino acid sequences of the proteins encoded in the DNA. Further features of the genetic code are that decoding starts at a fixed point, namely at the AUG triplet for methionine, and that three out of the sixty-four triplets serve as termination signals.

The most intriguing feature of the genetic code is its universality. Higher organisms, bacteria, viruses and phages all employ the same code. Why the genetic code has remained invariable through evolution is an interesting question. For viruses and phages it is obvious that they need the same code as the hosts of which they are using the decoding apparatus. For the success of recombinant-DNA technology aimed at *e.g.* the synthesis of human-type proteins by bacteria this universality is also useful if not indispensable.

THE MITOCHONDRIAL GENETIC CODE

The availability of sensitive and fast methods for analysing the base
sequence of nucleic acids has induced an avalanche of information on
the structure of genes. For mammalian mitochondria the complete
genomes are available for various species, including man. Also for
various yeasts and moulds large parts of the mitochondrial genome
have been sequenced. One of the surprising findings of these analyses
is that the genetic code used by mitochondria differs from the universal code of all organisms. Even more surprising is, perhaps, the
finding that the codes used by mitochondria is not identical for different organisms (Barell et al. 1980; Bonitz et al. 1980). The information available to date is summarized in Table 2.

TABLE 2. Deviations of the mitochondrial genetic code
from the universal code in various organisms

codon	Universal code	Mammalian mtDNA	Yeast mtDNA	N.crassa mtDNA	Paramecium mtDNA	Plant mtDNA
UGA	stop	trp	trp	trp	trp	stop
CUN*	leu	leu	thr	leu	n.d.	leu
CGG	arg	arg	arg	arg	n.d.	trp
AUA	ile	met	met	ile	n.d.	ile
AGR	arg	stop	arg	arg	n.d.	arg

*N = any of the four bases; R = purine; n.d. = no specific
data available.

The stopcodon UGA is used to code tryptophane in the mitochondrial
translation products of most organisms; not, however, in plant mitochondria (Leaver and Gray, 1982). Plant mitochondria follow the universal code quite closely, the only exception being the use of the
CGG codon for tryptophane. For Paramecium mtDNA only little information is available as yet (Seilhamer and Cummings, 1982). The AUA
codon specifies methionine in mammalian mtDNA and is also thought to
have an initiator function in this system (Barrell et al. 1980). Recently Butow and his coworkers have clearly shown that AUA codes for
methionine also in yeast (Hudspeth et al. 1982) in contrast to previous assumptions (Bonitz et al. 1980). In the mammalian mtDNA the
AGA and AGG do not code for arginine but are supposed to serve a
termination function (Barrell et al. 1980). The use of the CUN codon
family for threonine has been found for the yeast *Saccharomyces cerevisiae*. The corresponding mitochondrial threonine transfer RNA has
an extended anticodon loop of eight instead of seven bases and contains the UAG leucine-type anticodon (Li and Tzagoloff, 1979). It is
not yet clear if it concerns an incidental insertion mutation in
Saccharomyces only or that also other yeasts follow the same decoding
policy for CUN.

Mitochondrial transfer RNAs. The differences between the universal
and mitochondrial genetic codes mentioned above all concern the code

itself. As became obvious already from what was said about the diverging use of the leucine codons for threonine in *Saccharomyces cerevisiae*, the code as such offers only one side of the medal. The other – certainly as important – is the decoding system, the set of transfer RNAs available for the actual translation of the few(!) mitochondrial messages into polypeptides. It is known for a long time already that the number of tRNA genes encoded in mtDNA is limited and lower than 32, the minimal necessay number for decoding the complete universal code according to the rules of the wobble hypothesis (Crick, 1966). It is now well established that animal mtDNA and yeast mtDNA contain 22 and 24 transfer RNA genes respectively (Attardi *et al*. 1980; Bonitz *et al*. 1980). It is, perhaps with the exception of *Tetrahymena* (*cf*. De Vries and Van 't Sant, p. 255), very unlikely that tRNA species are imported from the cytosol into the mitochondria. None the less it is clear that the tRNAs together are able to decode 60 codons in the case of animal mtDNA and 62 codons of yeast mtDNA. As already mentioned the universal codons AGA and AGG for arginine are stopcodons in animal mtDNA, that contains only one arginine tRNA gene with the sequence TCG at the anticodon position. This accounts for one of the two tRNA genes difference between animals and yeast. There is, furthermore, only one gene for a methionine tRNA on the animal mtDNA versus two such genes on yeast mtDNA. In animal mitochondria the one methionine tRNA plays a role in chain initiation as well as in chain elongation.

The differences of the universal and mitochondrial genetic codes and the ability of mitochondria to recognize all codons with a limited number of tRNAs are the consequence of the unusual structural properties of mitochondrial tRNAs, particularly in the anticodon-sequence. It was suggested by Barrell *et al*. (1980) that anticodons with a U in the wobble position can recognize the four codons of a genetic box (*cf*. Table 1) either by a U:N wobble or by what they called a "two out of three" base interaction. In all mitochondrial tRNA genes sequenced so far, those decoding a box of four codons indeed have a T at the 5' end (Barrell *et al*. 1980; Cantatore *et al*. 1982; Heckman *et al*. 1980). For a number of *Neurospora crassa* mt-tRNAs RNA-sequence data reveal that the U at the wobble position is unmodified in these cases. On the other hand, in at least two other tRNAs that decode a set of two codons with a purine at the 3' end, the U present in the wobble position of the anticodon is modified (Heckman *et al*. 1980). The sequence presented for the anticodon of the tRNA for the CGN arginine genetic box in yeast does not match this rule (Bonitz *et al*. 1980). Furthermore, mitochondria from *Aspergillus nidulans* have retained two glycine tRNAs to decode the GGN genetic box, one with U and one with A at the wobble position (Köchel *et al*. 1981).

So far only the anticodon part of the mitochondrial tRNAs was discussed. Apart from variations in the overall G+C content, the structure of mitochondrial tRNAs of lower eukaryotes reasonably fits the structure of other tRNAs in nature. The stem and loop structures of the clover leaf are in general of the normal size, although any given mitochondrial tRNA may show one or more odd characteristics especially in the dihydro-U loop and -stem. Animal mitochondrial tRNAs are much more divergent. One of the serine tRNAs is even entirely lacking its dihydro U-loop (Anderson *et al*. 1981).

THE EVOLUTION OF THE GENETIC CODE

Being faced with the differences between the universal and the mitochondrial codes and codon-recognition systems the question arises as
to whether the mitochondrial coding-decoding system has to be considered more highly evolved or just regarded as a remnant of an
ancestral type of code of a more primitive type. Assuming that the
endosymbiont hypothesis for the origin of mitochondria and plastids
is correct, none of the two alternatives is very attractive (for
recent review, see Gray and Doolittle, 1982). For, in that case we
have to assume that at a certain moment after the symbiotic event
genes of the endosymbiont have moved to the host genome (Harrington
and Thornley, 1982). At the time of transfer the coding and decoding
policy of endosymbiont and host had to be highly comparable if not
equal. Only after the final settlement of genes either code could
evolve autonomously.

An interesting discussion about the evolution of organelle DNAs is
given by Wallace (1982). According to his model organelle DNAs of
unicellular organisms remained free to evolve, whereas those of
multicellular organisms retained the "primitive traits present at
the onset of multicellularity in the late Precambrian". In this con-
ception the mammalian mitochondrial genome represents a less evolved
set of genes than any other genome, if we exclude viral genomes from
our comparison. It is difficult to judge these considerations for
their true value, because it is not at all clear how the chloroplast
and plant mitochondrial genomes have escaped the retention of the
socalled primitive traits. In this context it is interesting to put
Wallace's model to the test with respect to the phenomenon of mitochondrial introns as described by Grivell *et al.* elsewhere in this
volume (pp. 279-306).

Apart from the question about the evolution of the genomes as a
whole one may wonder if the reasoning of Wallace holds for the
coding properties as such. It was suggested previously that the universal genetic code might have evolved from an archetypal code of 16
quartets using the recognition procedure presently used by the mitochondrion for decoding a genetic box containing four codons for a
single amino acid. By assuming duplications and various mutations the
present assignments of the universal code were obtained stepwise. One
of the intermediary steps would have led to a combination of only
quartets and doublets, a situation which is actually present in mammalian and yeast mitochondria (Jukes, 1982). Comparisons of the properties of various aminoacyl-tRNA synthetases (Lapointe, 1982) and of
mitochondrial tRNA genes (Cedergren, 1982) are consistent with duplications in relation to this archetypal code. It is, furthermore,
interesting to note that this primitive code was postulated to specify only 15 different amino acids. For serine there were postulated
two quartets: UCN and AGN (N = U, C, A or G). It is noteworthy, that
statistical analysis of mammalian mitochondrial genomes reveal, that
the mitochondrial UCN and AGY codons are not interchangeable (Henaut,
1982).

Because of the above arguments it is tempting to consider the present-
day mitochondrial genetic codes as ancesters of the universal code.

However, the question arises immediately why the plastid genomes and also the plant mitochondrial genome have evolved further. The same holds for the two glycine tRNA genes in *Aspergillus nidulans* mtDNA already mentioned (Köchel *et al.* 1981). An alternative way of approaching the data is assuming that the mitochondria represent an evolutionary simplification in which a minimum number of tRNAs have been conserved (Bonitz *et al.* 1980). This simplification holds a fortiori for the mammalian mtDNA. The limited role of mitochondrial protein synthesis and the seeming lack of regulatory functions of mitochondrial tRNAs may have enabled this reduction of the number of tRNA genes.

CODON USAGE

It is well known that the relative amounts of the various isoacceptor tRNA species vary within cells. This holds for prokaryotic as well as for eukaryotic cells. As a consequence one may expect that the expression of the genetic information contained in a messenger RNA is influenced by the codons used as compaired to the cellular concentration of the isoacceptor tRNAs. The base sequences of many genes are now available and have been compiled (Grantham, 1978; Grantham *et al.* 1980 and 1981). The majority, if not all, genes coding for polypeptides show a significant bias in the choice from the degenerate triplets available to code a particular amino acid.

In a recent study Gouy and Gautier (1982) have analysed the genetic code usage in more than eighty genes of *E. coli*. They could confirm a strong relationship between the codon composition and the expressivity of a gene. Even for genes of one operon such a correlation could be established. An interesting example offers the unc-operon, that codes for the eight polypeptides of the membrane-bound ATP-synthase complex. The stoichiometry of the subunits in this complex, designated α, β, γ, δ, ϵ, a, b and c, is most likely 3:3:1:1:1:1-2: 1-2:6. The expressivity of the various genes of this operon based on the codon usage in relation to abundancy, respectively rarity of the cognate tRNAs correlates well with the copy number of each of the subunits in the final complex. It appears then that the expression of any gene in a cell with a particular pattern of available isoacceptor tRNA is directly depending on its codon usage. It is interesting to note in this context that the DNA has clear perferences for asymmetric usage of its four bases and this dinucleotide asymmetry does not stem from and is not guided by the messenger-functions in a particular DNA sequence (Nussinov, 1981).

For the yeast *Saccharomyces cerevisiae* a strong codon bias has been reported (Bennetzen and Hall, 1982). Also in this case it appears that genes that are strongly expressed, such as those for the glyceraldehyde-3-phosphate dehydrogenase and the alcohol dehydrogenase isozyme I, have a very strong preference for the codons complementary to the anticodons of the most abundant isoacceptor tRNAs. Genes with a lower level of expression are much less biased. It is clear therefore that the use of codons is nonrandom. The given examples of the relation between codon usage and gene expressivity in unicellular organisms already illustrate this. However, also in multicellular organisms codon usage is highly selective. Within the scope of this chapter the phenomenon can not be treated thoroughly. Suffice it to quote Strehler and North (1982): "The most parsimonious interpretation of the data is that codon usage is among the most highly conserved of genetic qualities and that this conservatism reflects a cell-specific codon-language that is maintained over billions of generations because any 'silent' mutations which result in the presence of an untranslatable word in a cell will be lethal in the homozygous condition. Such mutations are not unlike non-sense mutations in prokaryotes, except that the set of non-sense words is cell-type specific".

It has been observed that codons which can mutate to a termination signal in one step are used less frequent. On this basis it has been postulated that the use of such pretermination codons is avoided in evolution. However, this is certainly not the only selective pressure (Golding and Strobeck, 1982). It was already mentioned that certain codons may represent non-sense words for a certain cell type. One would expect than that also the codons that can change into such non-sense words by a single mutation should be likewise avoided as pseudo-preterminators. Also selection against codons for amino acids which can deteriorate polypeptide structure and function such as *e.g.* proline may be operative. It is clear that the various constraints are not necessarily compatible or may be even opposite. For instance, the leucine pretermination codons may be used preferentially if mutations to proline are deleterious.

Besides the selective pressures discussed codon usage is strongly influenced by the mutation rate. But one should realize that these two processes are interweaved. This can be illustrated by recalling the differences between the universal and mitochondrial decoding systems. In the latter case there are no isoacceptor tRNAs. As a consequence 'silent' mutations of mtDNA do not demand for another tRNA during gene expression. It is well known that the mutation rate for mitochondrial DNA is very high. We have compared sequence data for various mitochondrial genes of rat obtained in our own groups with those of mouse and ox from the literature. For the three large subunits of cytochrome *c* oxidase, which are mitochondrial gene products and which together represent 1003 amino acids, 489 nucleotide differences between rat and mouse and 624 differences between rat and ox were recorded. By far the most of these differences concerned 'silent' mutations; 454 in the rat/mouse and 502 in the rat/ox comparison. Within both the category of 'silent' and the category of replacement mutations transitions prevailed against transversions. It is tempting to assume that the relative excess of 'silent' mutations is owing to the fact that the decoding system is unequivocal in the sense that only one tRNA is present for a particular amino acid in each genetic box.

As outlined in the chapter by Grivell *et al.* (pp. 279-306) the codon usage in the assigned and the unassigned reading frames (ARFs and URFs) of yeast mtDNA is not identical. The consequence for expressivity of ARFs and URFs is not clear. The choice of the degenerate base may influence the rate of transcription or regulate transcription either through the secondary structure of the messenger or due to the socalled "codon context effect", indicating that the efficiency of reading a particular codon can depend on the neighbouring codons.

For the ARFs and URFs of mammalian mtDNA codon usage is not significantly biased.

CONCLUDING REMARKS

The recent discovery of the deviating mitochondrial genetic code was used as a motive to discuss the universality of the classical genetic code. This universality is relative in a way also for the organisms using this code, because of the limitations of decoding in certain cells and organisms. This understanding has been borne out in the past decade mainly by careful analysis of the many data on nucleotide sequences of genes from various organisms. The knowledge that the coding and decoding rules are not completely watertight stems of course from much earlier, because the phenomenon of suppression has been described quite long ago. Recent studies show that changes in the tRNAs may combine the gain of suppressor activity with a loss of ability for wobbling, suggesting that changes in overall tRNA conformation or base modification may influence codon recognition (Murgola, 1981). Also for the yeast *Saccharomyces cerevisiae* interesting new information on suppression has been obtained at the level of the nuclear (Ono *et al*. 1981) and the mitochondrial genome (Fox and Staempfli, 1982). The recognition of four-base codons by the frameshift suppressor *suf J* offers a further example of deviating, albeit incidental recognition (Bossi and Roth, 1981). It is furthermore quite clear that the translation efficiency of transfer RNAs is also influences by the base sequence of the complete anticodon loop and -stem (Yarus, 1982). It is interesting to note that the process of suppression is even used in an attempt to manipulate tRNA genes in an approach to gene therapy for β-thalassaemia (Temple *et al*. 1982).

REFERENCES

Anderson, S., Bankier, A.T., Barrell, B.G., De Bruijn, M.H.L., Coulson, A.R., Drouin, J., Eperon, I.C., Nierlich, D.P., Roe, B.A., Sanger, F., Schreier, P.H., Smith, A.J.H., Staden, R. and Young, I.G., 1981. Sequence and organization of the human mitochondrial genome. Nature, 290, 457-465.

Attardi, G., Cantatore, P., Ching, E., Crews, S., Gelfand, R., Merkel, C., Montoya, J. and Ojala, D., 1980. The remarkable features of gene organization and expression of human mitochondrial DNA, in: The organization and expression of the mitochondrial genome (Eds., Kroon, A.M. and Saccone, C.) Elsevier/North-Holland, Amsterdam, 103-119.

Barrell, B.G., Anderson, S., Bankier, A.T., De Bruijn, M.H.L., Chen, E., Coulson, A.R., Drouin, J., Eperon, J.C., Nierlich, D.P., Roe, B.A., Sanger, F., Schreier, P.H., Smith, A.J.H., Staden, R. and Young, I.G., 1980. Different pattern of codon recognition by mammalian mitochondrial tRNAs. Proc. Natl. Acad. Sci. USA, 77, 3164-3166.

Bennetzen, J.L. and Hall, B.D., 1982. Codon selection in yeast. J. Biol. Chem., 257, 3026-3031.

Bonitz, S.G., Berlani, R., Coruzzi, G., Li, M., Macino, G., Nobrega, F.G., Nobrega, M.P., Thalenfeld, B.E. and Tzagoloff, A., 1980. Codon recognition rules in yeast mitochondria. Proc. Natl. Acad. Sci. USA, 77, 3167-3170.

Bossi, L. and Roth, J.R., 1981. Four-base codons ACCA, ACCU and ACCC are recognized by frameshift suppressor sufJ. Cell, 25, 489-496.

Cantatore, P., DeBenedetto, C., Gadeleta, G., Gallerani, R., Kroon, A.M., Holtrop, M., Lanave, C., Pepe, G., Quagliariello, C., Saccone, C. and Sbisa, E., 1982. The nucleotide sequence of several tRNA genes from rat mitochondria: common features and relatedness to homologous species. Nucl. Acids Res. 10, 3279-3289.

Cedergren, R.J., 1982. An evaluation of mitochondrial tRNA gene evolution and its relation to the genetic code. Can. J. Biochem., 60, 475-479.

Crick, F.H.C., 1966. Codon-anticodon pairing: the wobble hypothesis. J. Mol. Biol., 19, 548-555.

Fox, T.D. and Staempfli, S., 1982. Suppressor of yeast mitochondrial ochre mutations that maps in or near the 15S ribosomal RNA gene of mtDNA. Proc. Natl. Acad. Sci. USA, 79, 1583-1587.

Golding, G.B. and Strobeck, C., 1982. Expected frequencies of codon use as a function of mutation rates and codon fitnesses. J. Mol. Evol., 18, 379-386.

Gouy, M. and Gautier, C., 1982. Codon usage in bacteria: correlation with gene expressivity. Nucl. Acids Res., 10, 7055-7074.

Grantham, R., 1978. Viral, prokaryote and eukaryote genes contrasted by mRNA sequence indexes. FEBS Lett., 95, 1-11.

Grantham, R., Gautier, C. and Gouy, M., 1980. Codon frequencies in 119 individual genes confirm consistent choises of degenerate bases according to genome type. Nucl. Acids Res., 8, 1893-1912.

Grantham, R., Gautier, C., Gouy, M., Jacobzone, M. and Mercier, R., 1981. Codon catalog usage is a genome strategy for gene expressivity. Nucl. Acids Res., 9, r43-r74.

Grantham, R., Gautier, C., Gouy, M., Mercier, R. and Pave, A., 1980. Codon catalog usage and the genome hypothesis. Nucl. Acids Res., 8, r49-r62.

Gray, M.W. and Doolittle, W.F., 1982. Has the endosymbiont hypothesis been proven? Microbiol. Rev., 46, 1-42.

Harrington, A. and Thornley, A.L., 1982. Biochemical and genetic consequences of gene transfer from endosymbiont to host genome. J. Mol. Evol., 18, 287-292.

Heckman, J.E., Sarnoff, J., Alzner-de Weerd, B., Yin, S. and Rajbhandary, U.L., 1980. Novel features in the genetic code and codon reading patterns in Neurospora crassa mitochondria based on sequences of six mitochondrial tRNAs. Proc. Natl. Acad. Sci. USA, 77, 3159-3163.

Henaut, A., 1982. The serine codons UCX and AGY are not interchangeable in the mammalian mitochondrial genomes. Bioscience Reports, 2, 515-519.

Hudspeth, M.E.S., Ainley, W.M., Shumard, D.S., Butow, R.A. and Grossman, L.I., 1982. Location and structure of the var-1 gene on yeast mitochondrial DNA: nucleotide sequence of the 40.0 allele. Cell, 30, 617-626.

Jukes, T.H , 1982. Possible evolutionary steps in the genetic code. Biochem. Biophys. Res. Commun., 107, 225-228.

Khorana, H.G., 1972. Nucleic acid synthesis in the study of the genetic code, in: Nobel Lectures Physiology or Medicine 1963-1970, Elsevier Publishing Company, Amsterdam, 341-369.

Köchel, H.G., Lazarus, C.M., Basak, N. and Küntzel, H., 1981. Mito-chondrial tRNA gene clusters in Aspergillus nidulans: organization and nucleotide sequence. Cell, 23, 625-633.

Lapointe, J., 1982. Study of the evolution of the genetic code by comparing the structural and catalytic properties of the amino-acyl-tRNA synthetases. Can. J. Biochem., 60, 471-474.

Leaver, C.J. and Gray, M.W., 1982. Mitochondrial genome organization and expression in higher plants. Ann. Rev. Plant Physiol., 33, 373-402.

Li, M. and Tzagoloff, A., 1979. Assembly of the mitochondrial mem-brane system: sequences of yeast mitochondrial valine and an unusual threonine tRNA gene. Cell, 18- 47-53.

Murgola, E.J., 1981. Restricted wobble in UGA codon recognition by glycine tRNA suppressors of UGG. J. Mol. Biol., 149, 1-13.

Nirenberg, M., 1972. The genetic code, in: Nobel Lectures Physiology or Medicine 1963-1970, Elsevier Publishing Company, Amsterdam, 372-395.

Ono, B.I., Wills, N., Stewart, J.W., Gesteland, R.F. and Sherman, F., 1981. Serine-inserting UAA suppression mediated by yeast tRNA[Ser]. J. Mol. Biol., 150, 361-373.

Seilhamer, J.J. and Cummings, D.J., 1982. Altered genetic code in Paramecium mitochondria: possible evolutionary trends. Mol. Gen. Genet., 187, 236-239.

Strehler, B. and North, D., 1982. Cell-type specific codon usage and differentiation. Mech. Ageing Dev., 18, 285-313.

Temple, G.F., Dozy, A.M., Roy, K.L. and Kan, Y.W., 1982. Construction of a functional human suppressor tRNA gene: an approach to gene therapy for β-thalassaemia. Nature, 296, 537-540.

Yarus, M., 1982. Translational efficiency of transfer tRNAs: uses of an extended anticodon. Science, 218, 646-652.

Wallace, D.C., 1982. Structure and evolution of organelle genomes. Microbiol. Rev., 46, 208-240.

Index

ADP/ATP carrier: 316
alternate oxidase: 260
cAMP, see cyclic AMP
amphipotential proteins: 160
animal viruses: 59-61; 176; 186
anticodon: 128; 350
antiproliferative agents: 260
antirepressor of bacteriophage p22: 216
Aspergillus: 265; 281; 294; 352
assembly of mitochondrial proteins: 307-346
ATPase comple: 253; 317

Bacillus stearothermophilus: 105
Botryodiplodia: 266
box effect: 283; 289
box3 maturase: 283
box7 maturase: 285

carbamoylphosphate synthetase: 317
β-casein gene: 191
cDNA library: 207
cell convection: 178
chloramphenicol: 87; 260
chloramphenicol acetyltransferase: 177
chloroplast ribosomes: 72-73
chromatin: 1-42; 43-70; 174
chromomeres: 30
circular RNA: 282
citrate synthase: 317
cloacin: 88
cloned genes: 172
cloning vectors: 174
codon context effect: 353
codon usage: 296; 349; 352
colicin: 88
conformational differences: 318
cotransformation: 176
cotranslational secretion: 155
cotranslational transport: 308
cyclic AMP: 196